VALUE-ADDED PARTNERING AND INNOVATION IN A CHANGING WORLD

International Series on Technology Policy and Innovation

General Editors:

Manuel V. Heitor, Secretary of State for Science, Technology, and Higher Education, Portugal

David V. Gibson, IC^2 Institute, The University of Texas at Austin, Texas

Pedro Conceição, Deputy Director, Office of Development Studies, United Nations Development Program

VALUE-ADDED PARTNERING AND INNOVATION IN A CHANGING WORLD

Edited by
Marina van Geenhuizen, Dariusz M. Trzmielak,
David V. Gibson, Maciej Urbaniak

Purdue University Press
West Lafayette, Indiana

Printed in the United States of America.

ISBN 978-1-55753-513-9

Library of Congress Cataloging-in-Publication Data

Value-added partnering and innovation in a changing world / edited by Marina van Geenhuizen ... [et al.].
p. cm.
ISBN 978-1-55753-513-9
1. Technological innovations--Government policy. 2. Technology and state. 3. Academic-
industrial collaboration. 4. Strategic alliances (Business) I. Geenhuizen, M. S. van
HC79.T4V355 2008
338.8'7--dc22

2008012997

CONTENTS

Foreword

Transfer of knowledge encourages both private and public sector research organizations to apply their inventions to commercial projects, which are crucial to the development of new ideas, technologies, products and services. The transformation intensity in Poland, one of European Union member countries, accelerates development of new technologies and attracts many organizations to cooperate within the innovation field. Poland needs acceleration in terms of employment growth and added value for the small and medium sized enterprises sector. Continuous rapid development and technology transfer improvement have crucial impact on the network building by many innovative organizations. In the real world nothing is possible without the cooperation which provides added value for the partners. This cooperation is a vital resource for improvement of performance and competition on the market. Any new organization, business, governmental or academic must learn how to operate on the market and put together all the information that is needed if they are expected to be internationally oriented. Without this knowledge and skills, organizations are never going to find value added partnering. The connection between organization's innovations, its culture and development policy are vital for networking. Factors involved in network building include: scientific and industrial potential, supply of innovations, relationship between science and industry, know-how, the quantity and quality of academic institutes capable of being involved in the networking process, economic structure, and the level of management of particular partners, the quality of human resources, the infrastructure, as well as innovation environment and its sources of financing and financial position, sociological factors (ethics, skills to cooperate, notice opportunities etc.), legal regulatory environment, local economic and social policies concerning innovation, and the need for a proper strategy for international development. Innovation strategies are not universal however, and the particularities of various regions need to be taken into account in formulating a proper strategy for the transfer of knowledge from the scientific community to business. This latter aspect plays a significant role in either accelerating or thwarting the networking of any organization in a country like Poland.

Many innovations are dependent on factors that generate the ideas and their capacity to be absorbed. These in turns are dependent on economic policy directed towards the science and high-tech sectors. Within the context of an overall technology and innovation policy, both academia and business need to develop a network policy. This aspect of economic policy needs to focus on stimulation of high-tech enterprises by the creation and support of programs which encourage scientists and entrepreneurs to cooperate and prepare them for competition on the European and international markets. Poland scientific and high-tech sectors dealing with many technology entrepreneurship developments are quite young, but the network building projects are being widely accepted. With the trend towards market liberalization and globalization, the government policy of early support for new technologies and inventions is being carefully adjusted to insure the impact of a significant growth in technologies commercialization. One of the main goals of any support program is to force the scientists and entrepreneurs to cooperate. Support for science and business alliances is an integral part of a many government's strategies, and it will be implemented to meet high-tech sector needs for development, allow fast transfer of knowledge and, as many cases show, allow companies to achieve sustainable competitive advantage.

The strategic goal of this publication is to create a network between organizations and authors involved in publishing processes. "Value added partnering" has become fundamental motto shaping the thoughts and discussions presented in the chapters. A second initiative for writing this book was the meeting and deliberations on technology policy and innovation in Poland, Łódź in 2005.

The national strategy created by Polish government in 2003, aimed at creating key organizational structures which would increase prosperity, focus on firms creating new technologies and innovations. These structures included the Innovation Center and the High-Tech Accelerator at University of Łódź. One of the Innovation Center and High-Tech Accelerator's primary goals was the transfer and commercialization of the technologies developed by national organizations for the purpose of creating jobs and accelerating the economy growth. The fundamental aim was to create an infrastructure which would exploit research potential and increase economic innovation, optimizing the potential of R & D units, increasing innovation level in enterprises, and triggering the adjustment of the existing level of scientific research to the needs of the market. It was the source of inspiration for enterprise incubators, technology accelerators, R & D parks, technology parks, economic zones, and industrial parks. It created a climate for initiating a whole new high-technology business. In addition, it brought attention of international network developed by University of Texas at Austin, which organized a meeting on technology policy and innovation two years later. This international meeting in Łódź was co-organized by the University of Łódź and City of Łódź, which gathered representatives from the spheres of public administration, business, science and academia from a over eighteen countries.

This publication presents views and research results, and initiation of wide-range discussion concerning: the national science, innovation systems and policies to support networking and innovation, regional innovation systems, clusters and technology applications, research and development, value-added partnering, knowledge Management and Entrepreneurial Innovation strategies. Academia differs from business in this respect that the goal of innovation is to teach entrepreneurship in the best possible way, but the innovation is often the centerpiece of strained relations between universities and entrepreneurs.

The research done by the authors showed us that various aspects of innovation and technology policy are typically diffused across the organizations, countries and regions. Consequently, researchers' intent is to present a set of science and innovations ideas useful to international readers. There have been many contributors to this book, who shared theirs problems, attempted to provide solutions and judgments of success.

The chapters presented in this publication are based on the studies carried out by representatives of ten countries. This book would not have been possible without the support of two organizations: IC2 Institute at the University of Texas at Austin and Innovation Center at the University of Łódź. Working with various local, state, and federal government organizations and international networks to assist in providing incentives and resources for knowledge and the best practices development, helped better attract international researchers and authors to take part in the publication. The authors and chief editors have established international network that has made impact on their international cooperation on technology policy and innovation.

In my capacity as Vice Rector of the University of Łódź I would like to take this opportunity to thank the editors of this volume, the members of the University of Texas at Austin, Technical University at Delft and the University of Łódź, whose work and assistance with gathering presented chapters make this publication possible. The high quality of chapters in this publication constitutes proofs the excellent cooperation between the University of Texas at Austin, Technical University at Delft and University of Łódź. I would like to thank the authors of the enclosed chapters for sharing their insights and country perspective on the main themes. The representatives from spheres of public administration, business, science, and academia from many countries: Denmark, Germany, The Netherlands, Norway, Poland, Portugal, Spain, The United Kingdom and The United States took part in the discussion on the role of science and technology in socio-economic development. I am convinced that this cooperation will bear fruitful in the near future in the form of a number of specific international research projects. The conclusion that one can draw from this publication is that cooperation between different organizations and researchers would facilitate a dialogue on innovation and pool of knowledge together for everyone's benefits. The European, Japanese and American innovation dialogue will be a focused platform for better connection of different innovation players at different levels, from the regions, industry and policy makers.

Prof. dr. Wojciech Katner
Vice-rector for International Affairs, University of Łódź

Dr. Dariusz Trzmielak
Innovation Center Director
American-Polish Offset Program
University of Texas –
University of Łódź

1

Introduction

Value-Added Partnering and Innovation in a Changing World

Marina van Geenhuizen
Dariusz M.Trzmielak
Maciej Urbaniak

PARTNERING AND INNOVATION IN A CHANGING WORLD

In view of the current socio-economic changes, in which innovation is a key driving force behind the knowledge-based economy, this volume addresses the question which challenges are faced by value-added networking and innovation in a way that contributes to the commercialization of new knowledge and the adoption of firm strategies leading to a stronger and sustainable performance. This broad question has motivated the work behind the present volume that is centered in the relationships between *technology, policy and firms' management* in a context increasingly influenced by dynamics of interacting networks. underlying processes of knowledge accumulation and utilization in learning economies and societies.

In the current uncertain competitive environment driven by emerging technologies, the majority of technology firms are dependent on the people, capital funds and co-operation with other organizations. Thanks to networking, teams of innovators are able to start their own business and existing technology firms are able to increase innovation intensity and remain competitive in global markets (e.g., Teece et al., 1997). Networking may provide various important benefits, like access to funds, markets, human resources and knowledge, and innovative culture, which enable innovators to elaborate their ideas and even build a prototype. The normal, logical step in creating a new technology, new firm, or new product is to seek outside expertise, management and financing in the form of access to a capital fund established for such a purpose. Such funds may be either public or private, but access thereto depends to a large extent on the knowledge of both the market and conditions for investment (Hisrich and Petrs, 2002).

Networking, however, does not happen of itself. Teams of innovators or small firms need to find the type of partners and modes of networking (equity-based, contractual, informal, etc.) that fit their specific need for resources and also provide a balanced distribution of benefits between the partners (e.g. Alvarez and Barney, 2001). In addition, networks may differ in terms of support or benefits dependent upon specific social characteristics of the relationships, like tightness and strength, and heterogeneity of the partners (e.g., Burt, 1992; Uzzi, 1996; Granovetter, 1973). From this perspective, attention is called for a potential decrease of benefits from network relationships if these relationships become too tight and too strong, and prohibit the absorption of new knowledge and information in the firms. In fact, the ability to form and manage network relationships by technology firms, be-it contractual or informal, is increasingly recognized as a distinctive organizational capability that needs to be developed because it provides the mechanism for assessing and re-configuring resources in response to changing circumstances.

Cooperation networks are of key significance in the globalization of technology, which includes not only its global distribution but the distribution of scientific research as well. Firms locate their activities in places where they can create and hook into chain reactions of events and where there is a good climate for innovative activity. The geographical entity to foster such places in public policy and public-private initiatives seems to be the region and this has led to the concept of regional innovation systems (RIS) in the early 1990s (Asheim and Gertler, 2005; Cooke, 2001). Accordingly, regional innovation systems as systems of networks of various actors surrounded by supportive knowledge organizations, are seen as important bases of economic coordination and governance at the meso-level between the national and the local (cluster) level.

Networking is also required for the rapid transfer of knowledge and innovation in building up a knowledge economy. Knowledge, which is one of the fundamental resources conditioning the development of new technology, arises in the minds of both inventors and entrepreneurs. Taking into account the ever shorter life cycle of products and inventions, it is critical that ways are found to quickly put new solutions into practice. Within firms, therefore an increased attention has grown for "soft assets" like the absorptive capacity (Cohen and Levinthal, 1990); knowledge management, including fostering a learning culture (Salleh and Goh, 2002); and quality systems, particular those technical ones that are coupled with support and reward elements serving the human and cultural dimension.

NATIONAL INSTITUTIONS AND POLICIES TO SUPPORT NETWORKING AND INNOVATION

Innovation critically depends on the availability of new knowledge but also on a broad range of learning and interactive networking. These phenomena cannot be seen in isolation but must be seen in the context of the national innovation system concerned. Actors like firms, governments, universities, intermediaries and banks, etc., the interactive relationships between them and the institutions underpinning these relationships, together form the innovation system (Carlsson et al., 2002). 'Innovation system' is a broad concept, but there is a set of characteristics upon which most researchers agree. Lundvall summarizes them as follows (Lundvall, 1992). The central focus is on technological innovation, but organizational and institutional change is considered to be important as well. The viewpoint is holistic, meaning that many determinants and their relationships are included in the analysis, and the perspective is historical meaning that innovation is seen as an evolutionary and path dependent process. It is possible that at one moment in time one system is better suited for stimulating certain technological developments, whereas later on another system performs better. The importance of learning, and especially of interactive learning, is stressed in the innovation system approach.

Subsystems of national innovation systems may be distinguished on the basis of a specific technology and, as above indicated, on the basis of regions. Technology subsystems may differ between countries on the basis of approaches used in the adoption process (top-down versus bottom-up) and involvement and interactive learning between relevant actors. The dynamic interaction in innovation systems causes that considering particular system aspects or elements in isolation leads to poor results of analysis and a wrong content of policy.

In Western Europe and North America, value-added partnering increasingly occurs between universities, industries and governments. One of the clearest changes over time is a blurring of the edges between the functions of each of the three partners. Thus, universities have become more entrepreneurial, whereas large companies have started knowledge production-based education in campuses hosting training in specific academic fields; and governments are increasingly engaged in enhancing the knowledge-based economy by improving conditions for entrepreneurial innovation (Etzkovitz, 2002). Particularly the role of universities has changed in the past decade, with additional tasks in the entrepreneurial sphere, aside from the traditional ones in education and research (e.g., Shane, 2004). There seems no end in the above changes, meaning a continued integration and hybridization of functions in the coming years, although the starting point and pace of changes may differ across countries.

LEARNING IN REGIONS AND CITIES

The growing specialization and innovation in production and the increased global competition have caused an increased need for knowledge. Knowledge is not only a vital or supportive component in manufacturing and services, it also serves learning in public policy. Part of the learning in public policy is prompted by the need for governments to embrace bottom-up developments and to adopt various ways of interactive learning, like in the development of industrial clusters and in the implementation and utilization of new telecommunication infrastructure (see also, Van Geenhuizen and Nijkamp, 2006). However, what matters is the quality of learning, such as through a sufficient professionalism and critical self-reflection in a public policy context. In the context of global value chains in which small firms in clusters in developing countries operate, the quality of learning resides in a direct connection with global knowledge (not merely through local buyers).

Information and communication technology (ICT) infrastructure may influence the performance and growth of cities and communities in many ways. First, the potentials of cities significantly increase for large and quick data transmission and use by technology firms and research institutes, like in global grid cluster computing. Secondly, ICT and related technologies in informatics, multimedia etc. constitute a basis for new firm establishment and for restructuring product-markets and business models of existing firms, thus enabling the urban economy to renew its structure. Thirdly, the performing of particular tasks by city departments may increase in quality, and urban regeneration may be supported using ICT, like in improving traffic flow (fighting congestion); watching services to prevent crime; specific tasks in healthcare and education for particular groups; supply of e-governance services to the public, and in particular areas, support in participatory urban planning (Cohen et al., 2005; Conroy and Evans-Cowley, 2006).

However, a strong connectivity through hard infrastructures is not sufficient in bringing about economic growth and an increased quality of life in communities. There is no simple linear cause-effect model of the introduction of ICT in the human society and economy. Rather, the outcome is the result of a complex interplay between the technologies and various socio-economic forces in the recipient economy and policy unit (Graham and Marvin, 1996). In addition, it is increasingly realized that the design of ICT networks need to be viewed from broader human centered systems (e.g., Moutinho and Heitor, 2005). Much depends on the organizational quality and learning attitude in the recipient communities and governments concerned. Such institutional factors are also known as "soft infrastructure" (World Bank, 2006) and may also include elements of social and cultural life. Currently, soft infrastructure is seen as just as important as hard infrastructure in the development of urban regions.

KNOWLEDGE MANAGEMENT AND FIRM'S INNOVATION AND GROWTH STRATEGIES

The question which types of strategies, innovation, knowledge, networking etc. determine growth and productivity of firms has been frequently addressed in the management and business literature. This question gets the more pressing today in an increasingly competitive world, particularly in places where many firms lack behind in innovativeness.

Of course there is not one answer to this question because firms' internal processes and external environments may strongly differ between industry sectors, between firms with different size and ages, and between countries. Factors of influence range from a strong ambition of the entrepreneurs to grow to impacts of self-reinforcing mechanisms of growth following a shift in strategy. What is clear so far is that R&D matters in productivity but that the results may be different between different innovation strategies (radical and incremental innovations) and strategies focussing on volume efficiency or volume differentiation. The finding of circular processes indicates that the introduction of new strategies, i.e. concerning learning, type of innovation, etc. always needs to be evaluated in the broader context of the firm and its development.

In the context of R&D and innovation an increased attention is being paid to what is named absorptive capacity and knowledge management, both internal qualities of firms. According to Cohen and Levinthal (1990) absorptive capacity - defined as an organization's ability to recognize the value of new external knowledge and to assimilate and apply it effectively - is a critical part of an organization's potential to innovate. There are signs in empirical studies (in this volume, Chapter 17 by *Dautel*) that such a capacity indeed influences the benefits from new knowledge. A connected issue is knowledge management. In its most basic form, knowledge management can be seen as the explicit and deliberate building, renewal and application of knowledge to maximize a company's knowledge-related effectiveness and returns from its knowledge assets. Knowledge management is connected with different types of innovation, organizational adaptation and effective decision-making (e.g. Salleh and Goh, 2002). Knowledge management has an important role to play in the introduction and improvement of management systems. Empirical studies have just started to attempt to measure and investigate the causal relations in comparative studies. Measurement issues constitute a major challenge in this type of research.

INTRODUCTION TO THE VOLUME

This volume is organized in three main parts (Table 1) Part I, on national science and innovation systems, contains a focus on various relations between universities, business and government, on factors in the development of national brands based on value-added products, and on factors at work in technology subsystems in the national innovation system. Part II examines the need for learning in different contexts with an emphasis on specific types of learning, like in the building of a regional innovation system and in the adoption of new ICT applications in cities. Part III has a strong focus on internal features and factors of firms, like growth strategies and R&D, and how these may contribute to productivity and innovativeness.

The geographic scope of the volume is quite broad by including Poland as the host country of the conference, various other EU countries like the Netherlands, Denmark, Portugal, Luxembourg, and Italy, and the US, Japan and Indonesia.

Part I: National Science and Innovation Systems and Policies

Chapter 2 to Chapter 4 deal with national innovation and national science systems, by focusing on institutions concerning university-industry relations and scientific output on the level of countries and of universities. In Chapter 2, *David Mowery* surveys evidence from the US and other industrial economies on the relationship between university research and industrial innovation. The author gives particular attention to the Bayh-Dole Act in the US and its importance as a role model in various countries. The Bayh-Dole Patent and Trademark Amendments Act (1980) provided blanket permission for performers of federally funded research to file for patents on the results of such research and to grant licences for these patents to other parties. However, different from what is often thought, this act modified but did not fundamentally alter the longstanding structure of incentives and constraints supporting collaboration between university and industrial researchers. Mowery illustrates this development with evidence from *Massachussets Institute of Technology*, *Stanford University* and the *University of California-Berkeley*.

Table 1
Structure and summary of the book

Author(s)	Subject (specific nature of study)	Countries/focus
Part I		
Mowery	National innovation systems and university-industry relations	US and other OECD countries; in US focus on MIT, Stanford and Berkeley
Horta & Veloso	National science systems and scientific production in small economies	Small countries, e.g. Belgium, Netherlands, Portugal, Taiwan, Singapore
Elena	National conditions for and models of intellectual capital management at universities	Europe, with a focus on Autonomous University of Madrid
Colombo & Grilli	A national support system for high-tech start-ups (empirical-modelling)	Italy
Domanski & Herezniak	Development of a country brand and critical conditions	Poland
Kamp	National wind turbine innovation subsystems and lessons to be drawn from success	Netherlands and Denmark
Trzmielak	National institutions in knowledge transfer, particularly concerning university spin-offs	Poland, with focus on Lodz
Part II		
Markowski	Critical conditions in designing regional innovation policy, including learning	Poland

Amano	A local Japanese cluster policy based on a specific Japanese-Finnish project	Japan, with a focus on Sendai
Van Geenhuizen & Indarti	Learning by small firms in developing countries and the relation with innovation intensity	Jepara furniture cluster in Indonesia
Moutinho	Success and failure in the EU digital cities program	Various regions in Portugal, with a focus on Aveiro
Van Geenhuizen	Digital infrastructure, location behaviour of firms, and potentials of different regions	Large cities in the core area and a peripheral region in the Netherlands
Part III		
van Loon	Needs for management to include people in process-focussed approaches	Web-based survey, in frame of a knowledge society community portal (EU)
Urbaniak	Critical conditions in the role of knowledge management in managing systems' improvement	Organizational learning
Graversen & Mark	Return on R&D capital and R&D factors influencing productivity of firms (empirical-modelling)	Denmark
Dautel	R&D impact on innovativeness (empirical modelling)	Luxembourg
Graversen & Mark	Factors governing choices R&D partnering (empirical-modelling)	Denmark
Simon	Factors behind steady firm growth and different ways of growth	Case studies from Germany
Zegveld & Den Hartigh	Relations between exploration and exploitation strategies in the context stability and change (empirical-modelling)	A case study and a sample of firms listed on the Stock Exchange in the Netherlands
Den Hartigh, Langerak & Zegveld	Positive feedback effects in firms' strategies and performance (empirical modelling)	A sample of firms in the Netherlands listed on the Stock Exchange
Gwarda-Gruszczynska	A diagnosis of firms' innovation strategies and environment, and opportunities for improvement	Poland

In Chapter 3, *Hugo Horta* and *Francisco Veloso* examine the scientific output of small countries and analyze the structure of the science system to identify each country's specific influence of the national institutional system. The authors make use of data on scientific publications and citations and data on R&D investment in exploring the relation between level of R&D funding and scientific output. Their results suggest that countries with both high and low R&D funding face a wide range of scientific output. Although investment is important, it is the process of knowledge production and accumulation (in other words, learning) that matters. This situation is illustrated with countries facing different science and innovation systems and different efforts to specialize or broaden in terms of disciplines.

In Chapter 4, *Susana Elena* addresses the increased importance and rationale of managing intellectual capital (IC). It is the intense transformation process in which European universities have been involved and the increasing role in the knowledge-based economy which evidence the need to introduce new IC management systems, both to improve internal management and to provide better information to stakeholders. The author observes quite some variation between European universities in managing intellectual capital and presents an in-depth study of the *Autonomous University of Madrid (UAM)* revealing practical problems of harmonisation of the systems. IC management ranges from a radical shift within universities through new legislation to the creation of individual models on a voluntary basis. The author concludes that the development of a common framework under different institutional conditions will cause several methodological problems (measurement and data collection), preventing benchmarking analysis and comparative studies. An analysis of the approaches observed seems to be crucial in the new Higher Education scenario.

In Chapter 5, *Massimo Colombo* and *Luca Grilli* address the effectiveness of the national support system for high-tech start-ups in Italy. First, the authors give a classification of direct policy measures granted by Italian authorities, e.g., on the basis of main objective, evaluation method, and target group, and this is followed by the estimation of an econometric model to identify whether particular new technology-based firms have benefited from such policy measures, particularly public funds. The findings contribute to the view that direct horizontal subsidies have been quite inefficient in supporting such firms. To improve the situation, the authors promote the idea of vertical measures that match with specificities of high-technology start-ups, including indirect support schemes in which selection and monitoring are delegated to specialized institutions, like technology incubators. In addition, they forward the development of an efficient venture capital industry as an important policy target.

In Chapter 6, *Tomasz Domanski and Marta Hereźniak* examine strategies for emerging economies to develop a country brand. They argue that a unique country brand may significantly contribute to creating and sustaining the competitive advantage of emerging economies and facilitate the latter's transition from a supplier of commodities and unbranded goods to a brand-owner. The authors distinguish between two strategies in creating positive brands, that is, to use an overall positive image of the country and associate this with certain products and industries, and to create a few high profile corporate or product brands on which a country's brand can be established. The authors present results from the "Brand for Poland" project (2002/2003) and conclude that building a country's brand requires to build a system that strengthens, stimulates and highlights the strongest brands, and to fulfil several tasks by firms and public actors. In Chapter 7, *Linda Kamp* addresses differences in the development of national innovation subsystems with implications for the success of innovation. The subsystems are concerned with large-scale and small-scale wind turbines. The author examines why the Netherlands was far less successful in developing wind power than Denmark in the period 1970-2000. A focus on large science-based technology development (science-push) and a top-down innovation strategy, including a strong government involvement and a weak involvement of intended buyers, have prevented a successful innovation in the large-turbine subsystem. By contrast, a bottom-up strategy in the Danish small-turbine innovation subsystem enabled to leave initiatives to companies, turbine owners and researchers, etc. and allowed networks of these actors to build a strong knowledge base. The author concludes that a complete and well-functioning innovation system is a prerequisite for successful innovation, meaning that all major actors have to be actively involved from an early stage in facilitating interactive learning.

Dariusz Trzmielak, in Chapter 8, shows that the commercialization of new technology is correlated with macro and micro trends. Macro trends shape the new sectors and micro trends enrich the innovation potential. The fundamental issue in new project management in the high-tech sector is concerned with 'soft' factors that stimulate an increased effectiveness. The author argues that educational and scientific achievements are not only important to the academic community but to the business community as well. The chapter is divided into three parts: theoretical, focusing on various innovation and knowledge management issues; analytical, examining the possibility of starting a new venture on technology-based ideas; and, a case study of Polish-American programs launched in 2003 as an example of building knowledge and best practices within academia and business organizations. Polish-American programs have had a great impact on Łodz and on the Polish university community, but first of all on the support of entrepreneurship in Poland.

Part II Learning and ICT in Regional Innovation Systems and Clusters

The need for high-quality learning, both in the business world and in public policy in a region, is in the centre of Part II. *Tadeusz Markowski*, in Chapter 9, gives a reflection on the concept of Regional Innovation Systems. Despite the lack of comprehensive empirical tests, this concept is attractive in policies of managing regional development because it takes uncertainty in the environment into consideration and complies with current paradigms of knowledge-based economic growth and development through networks. The success of a Regional Innovation System policy applied to Poland depends upon the level of learning, the development of social capital and the innovativeness of local and regional authorities. Accordingly, the author recommends the creation of a civil corps of local government officers in key regional fields selected on a high degree of professionalism. In addition, the development of competitive clusters in Regional Innovation Systems is preferably based upon already existing initial formations of such clusters. In Chapter 10, *Gen Amano* examines a local industrial policy in Japan that connects with abroad to gain international competitiveness, i.e., the Sendai-Finland Wellbeing Project. In Japan, such new initiatives need to be considered in the context of a recent shift from a strong national power (as in technopolis and cluster policies) towards decentralization to local governments. Sendai, traditionally acting as a branch office economy, is nowadays involved in implementing direct overseas relationships. The plan includes a clustering policy for the health equipment industry and the establishment of a healthcare centre in Sendai. A major element is co-incubation, meaning that the cities of Sendai and Oulu in Finland mutually support each others' incubated firms. This process draws on the co-location in Sendai of research facilities and (potential) users of health equipment, ensuring a mutual learning. The project also draws on the idea of co-branding in which the Japanese cluster intends to utilize the existing brand power of a Scandinavian cluster. In Chapter 11, *Marina van Geenhuizen* and *Nurul Indarti* address ways of learning by small manufacturers in the furniture cluster of Jepara in Indonesia. The study draws on extensive fieldwork and is one of the first of its kind. It reveals a dominance of in-house learning and knowledge gaining from buyers (contractors), with a limited use of sources that connect with global knowledge. The outcomes support the idea that this learning is associated with an overall low level of newness in the market place. A small segment of more innovative firms, however, tend to be stronger knowledge-oriented and less focused on learning from buyers; they rely more on global knowledge sources like exhibitions and government agencies. The authors argue that increasing the overall level of innovativeness in furniture clusters like Jepara is necessary to respond to an increased global competition and a growing depletion of forests. Government support is needed to promote a shift in learning but also to improve other conditions enabling small firms to escape their relatively poor and low innovative position as subcontractors.

Next, *Jose Luiz Moutinho* examines success and failure in the Digital Cities Program in Portugal (Chapter 12). A discussion of strategies for the Information Society is followed by a discussion of hurdles that need to be taken. The Portuguese approach was relatively broad by including, for example, quality improvement of the life in cities (supply of local administration services), contributing to the development of peripheral areas, and fighting information exclusion. The evidence, with a focus on the Aveiro Digital City project, suggests the need to broaden the technocratic paradigm of technical change with a system design of fostering societal developments. Accordingly, the cultural and social shaping of information technologies requires the design of human-centered systems to support community-building activities. Enabling people to interact more easily and purposefully with each other and the surrounding world and to become more productive, requires a change of the mind-set of users and designers of the new technology. In Chapter 13, *Marina van Geenhuizen* addresses potential impacts from ICT infrastructure in two different regions, i.e., core metropolitan regions and more remote regions. The spatial formation (layout) of advanced digital infrastructure and location preferences of young innovative companies are explored by taking the Netherlands as an example. Despite early connections with the newest digital infrastructure, large cities may loose young innovative companies to adjacent (suburban) areas. It is only the largest city (Amsterdam) that is endowed with agglomeration advantages nowhere else available (partially linked with highest level digital nodes). The results indicate that young innovative companies leaving urban regions are not attracted by more remote regions. In the last regions, however, opportunities may be created by utilizing emptiness or remoteness as an asset in developing particular information-intensive activity, like in observatories and related research (astronomy, earth observation, air traffic, etc.). Such a development justifies a connection with the highest-bandwidth in the national grid and may help to increase growth of SMEs, provided that the requirements of institution building, good governance and mutual learning in the region are satisfied.

Part III R&D, Partnering, Knowledge Management and Firms' Strategies

This part has a specific focus on firms, their management and strategies, and their innovative and growth performance. Chapter 14 and Chapter 15 are concerned with different aspects of management, particularly issues of knowledge management in maintaining and growing 'soft assets' of technology firms. In Chapter 14 *Han van Loon* presents the results of a survey undertaken at Knowledge Board (the knowledge society community portal) to find indicators for management strategies. The results show the importance of people to the success of firms' organisation and of process assessment in determining process capability and improvements. The author concludes that the introduction of the improvements in itself does not lead to positive effects. It has to be supported and rewarded. Process assessment itself and the associated standards need to be expanded to take people and culture into consideration and develop the organisation. In Chapter 15, *Maciej Urbaniak* addresses the highly important role of knowledge management in processes of introduction, maintenance and improvement of management systems, particularly in the implementation of quality management. Knowledge management is essential to the success and sustainability of the business through identifying those products, service offerings, processes, and performance attributes that are unique to the business. Business Excellence Performance systems are challenging the status quo and enhance change by utilizing learning to create innovation and improvement opportunities. The author also addresses the importance of continuous improvement and learning as integral parts of management systems. These activities require analysis and problem solving everywhere within the firm, and encourage change and innovation as regards how work is organized and conducted. Further, customer-driven quality management emphasizes the "positive side of quality", that is enhancement, new services, and customer relationship management. It success depends heavily on creativity, more so than on steps to reduce errors and defects that rely on well-defined techniques. The essential ingredient for benefiting from knowledge management is the establishment of a learning culture of an organization. The author concludes that if firms believe that they already know the best way or that the 'best practice' is not appropriate to them, then sustainable improvement will just not happen.

In Chapter 16, Chapter 17 and Chapter 18 various aspects and implications of R&D are discussed. In Chapter 16, *Ebbe Krogh Graversen* and *Michael Mark* estimate an econometric model to identify the influence of R&D on productivity among firms in Denmark. The model uses a modified Cobb-Douglas production function. Present values of firms' accumulated R&D investments, together with interaction effects, are included directly in the production function. The estimation draws on data derived from a large national survey. The authors observe that R&D active firms have a much higher value added per employee than R&D inactive firms. In addition, there is a significant positive return on R&D capital in terms of productivity. The results lead the authors to recommend that future policy preferably supports possibilities of industrial sectors to become and remain innovative, and to increase the amount of R&D activity overall. A major way is to increase the educational level in firms by employing R&D competent employees, including upgrading the level of existing employees.

In Chapter 17, *Vincent Dautel* examines the impact of R&D aspects on the innovative performance of firms. His analysis is based on micro-data from the Community Innovation Survey (CIS3) carried out in Luxembourg and encompasses various regression models. Aside from R&D aspects, the models include firm's characteristics and the firm's environment. The results show that R&D engagement allows for added-value creation through new products as well development of absorptive capacity (a more fruitful use of external knowledge). For manufacturing firms, R&D also generates opportunities for co-operation with scientific partners (in the service sector higher educated employees generate such co-operation). However, co-operation without distinction between types of partners does not have any significant effect on product innovation. The author argues that co-operation with scientific partners is partially linked with a given level of R&D and also with a firm's knowledge resources. Overall, the results highlight the importance of R&D investment above a critical level.

In Chapter 18, *Ebbe Krogh Graversen* and *Michael Mark* go back to conditions under which R&D networks come into being among firms in Denmark. They use logistic models to estimate the influence of firm and business characteristics on R&D partnering and on the choice of particular R&D partners, i.e., from Denmark or abroad and from the private or public sector. The estimations draw on a database of R&D active firms selected from a large national survey. The factors explored include four sets, i.e., basic firm characteristics, like age, size, and geographic location; R&D profile, like location vis-à-vis research centres and R&D type; R&D competence and absorption capacity; and, business sector characteristics, like type of sector and market concentration. Basic firm characteristics tend to affect participation in R&D cooperation, whereas the R&D profile tends to affect engaging in R&D cooperation with all types of partners. In addition, firms with permanent R&D competences have a relatively high probability of engaging in R&D co-operation. Business sector characteristics, however, tend to have a minor influence on this co-operation. The authors conclude with policy suggestions, for example, to increase co-operation through aspects related to the absorption capacity of firms.

Chapters 19, Chapter 20 and Chapter 21 deal with long-term growth patterns of firms. In Chapter 19, *Herbert Simon* examines so-called Hidden Champions, i.e. firms facing a continuous growth at rather modest annual rates over a long time, and explains this growth by the unrelenting will to grow of the entrepreneurs. Using four case studies of Hidden Champions, the author argues that a continuous growth can be realized in mature markets as well as in growth markets, and that there are many ways to grow, like adding new products, entering new regional markets, and addressing new customer markets or industries. Growth may be realized by taking shares from competitors through competitive advantages or through technological leadership in a new market. However, the author sees one clear limitation to growth and that is that growth should be profitable. In Chapter 20, *Marc Zegveld* and *Erik den Hartigh* analyse growth as the outcome of a balance between exploration and exploitation. Exploration focuses on search, variation and radical changes, while exploitation focuses on alignment, refinement, or incremental changes. The authors carry out an empirical test to validate the proposed exploration-exploitation-stability framework using a sample of firms in the Netherlands for the period 1988-1997. It appears that stability on exploration is more durable than on exploitation. In both cases stability pays off; however, a shift in exploration generates a significantly higher impact on the development of cash flow.

In the next Chapter (21) *Erik den Hartigh, Fred Langerak* and *Marc Zegveld* examine firms' strategies and performance from the perspective of positive feedback mechanisms. The authors test Verdoorn's law at the firm level by examining the circular relationship between strategies of volume-efficiency and volume-differentiation, growth of output, effects of scale and learning, and productivity growth. In particular, they analyze three strategic value drivers to generate operational cash flow, i.e., volume, efficiency and differentiation. Drawing on a sample of firms in the Netherlands and a case study of one firm over the years 1983-2002, the authors observe that firms face productivity losses through negative induced and exogenous learning effects. Firms also realize productivity growth (and performance growth) through increase of scale and autonomous learning effects. These effects follow from a volume-strategy aimed at internalizing market growth. The results in this study comply with those of previous cross-industry and cross-regional research.

In the final chapter, *Edyta Gwarda-Gruszczynska* presents an analysis of the institutional and industrial environment and of innovation strategies of Polish firms to clarify the relatively low level of innovation of Polish firms today. On the basis of a SWOT study performed by the Polish government, the author discusses various lines of improvement by firms, like in long-term strategic thinking concerning innovation and in cooperation with suppliers, customers and universities and other knowledge centres. The author also briefly addresses the new policy of the Polish government (2007) to support the new innovation strategies.

FUTURE RESEARCH

Many of the studies in this volume are single-country or single technology studies, like the ones on R&D productivity (Denmark) and on innovation subsystems (wind-turbine and ICT). The applicability of their results could be investigated in the context of other countries and other technologies. In addition, the plea for networking leaves one key question open and that is concerned with the conditions under which the positive influence of networking changes into a negative influence, i.e., manifesting itself in growing barriers to connect with new knowledge and in less effective learning.

Another research path is emerging and that is concerned with the non-tangible aspects or 'soft factors' of business organisations, addressed in various chapters of the volume. Comparative analysis of different types of management systems and an estimation of the contribution of these systems to turnover and innovation are necessary to better underpin advice on best practices. However, systematic empirical research is facing many measurement and data-collection issues. This is also true for empirical research of absorptive capacity, a feature that needs to be measured more precisely such that its impact on innovation can be determined in a better way.

An international view is crucial and this volume provides a very timely reminder of the complexity and magnitude of the reform task and continuous adjustment of entrepreneurship and public policy ahead of us. Importantly though, this volume demonstrates the willingness and capacity for collaborative activity and creative learning. Indeed there is clear optimism for genuine and effective transformation and continuous adjustment.

BIBLIOGRAPHY

Alvarez, S. and J. B. Barney, (2001), "How Entrepreneurial Firms Can Benefit from Alliances with Large Partners", *Academy of Management Executive* 15 (1), pp. 139-148.

Arthur, W. B., (1996), "Increasing Returns and the New World of Business", *Harvard Business Review*, 74, July-August, pp. 100-109.

Asheim, B. T. and M. S. Gertler, (2005), "The Geography of Innovation: Regional Innovation Systems", [in:] J. Fagerberg, D. C. Mowery and R. R. Nelson (eds.), *The Oxford Handbook of Innovation*, Oxford University Press, Oxford, pp. 291-317.

Burt, R. S., (1992), *Structural Holes. The Social Structure of Competition*. Harvard University Press, Cambridge.

Carlsson, B., S. Jacobsson, M. Holmen and A. Rickne, (2002), "Innovation Systems: Analytical and Methodological Issues", *Research Policy*, Vol. 31, pp. 233-245.

Cohen, G., M. van Geenhuizen and P. Nijkamp, (2005), "ICT as a Contributing Factor to Urban Sustainable Development: Policy Making in Dutch Cities", [in:] M. van Geenhuizen, D. Gibson and M. Heitor (eds), *Regional Development and Conditions for Innovation in the Network Society,* Purdue University Press, West Lafayette, pp. 99-117.

Cohen, W. M. and D. A. Levinthal, (1990), "Absorptive Capacity: a New Perspective on Learning and Innovation", *Administrative Science Quarterly*, Vol. 35, pp. 128-152.

Cooke, P., (2001), "Regional Innovation Systems, Clusters, and the Knowledge Economy", *Industrial and Corporate Change*, Vol. 10 (4): pp. 945-974.

Conroy, M. M., and J. Evans-Cowley, (2006), "The Growth of E-government in Municipal Planning", *Journal of Urban Technology*, 13 (1): pp. 81-107.

Etzkovitz, H., (2002), "Incubation of Incubators: Innovation as a Triple Helix of University-Industry-Government Networks", *Science and Public Policy*, 29 (2): pp. 115-128.

Geenhuizen, M. and P. Nijkamp, (2006), "Learning Regions in an Evolutionary Context: Policymaking for High Technology Firms",

International Journal of Entrepreneurship and Innovation Management, Vol. 6 (3): pp. 265-282.

Graham, S. and S. Marvin (1996) *Telecommunications and the City. Electronic Spaces, Urban Places,* Routledge, London.

Granovetter, M. A., (1973), "The Strenght of Weak Ties", *American Journal of Sociology*, Vol. 78 (6): pp. 1360-1380.

Hisrich R. D., and M. P. Peters, (2002), *Entrepreneurship,* McGraw-Hill Higher Education.

Lundvall, B-A. (ed.), (1992), *National Systems of Innovation – Towards a Theory of Innovation and Interactive Learning,* Pinter Publishers, London.

Moutinho, J and M. Heitor, (2005), "Building Human Centered Systems Based on Communication Infrastructures: Evidence from Portugal", [in:] S. Marshall, W. Taylor and X. Yu (eds.), *The Encyclopedia of Developing Regional Communities with ICT*, Hershey, Idea Group.

Salleh, Y. and W. K. Goh, (2002), "Managing Human Resources Toward Achieving Knowledge Management", *Journal of Knowledge Management*, Vol. 6 (5), pp. 457-468.

Shane, S., (2004), *Academic Entrepreneurship. University Spinoffs and Wealth Creation*, Edward Elgar, Cheltenham.

Teece, D. J., G. Pisano, and A. Shuen, (1997), "Dynamic Capabilities and Strategic Management", *Strategic Management Journal*, 18 (7), pp. 509-533.

Uzzi, B., (1996), "The Sources of Consequences of Embeddedness for the Economic Performance of Organizations: the Network Effects", *American Sociological Review* 61, pp. 674-698.

World Bank (2007) Local Economic Development (LED) website, at http://web.worldbank.org/wbsite/external/topics/exturbandevelopment/extled/

PART I
NATIONAL SCIENCE AND INNOVATION SYSTEMS AND POLICIES

2

University-Industry Relationships in the Knowledge-Based Economy: How Useful is the "Bayh-Dole Model"?

David C. Mowery

INTRODUCTION

The 21st century is the century of knowledge-based competition among nations. More than ever, the prosperity of nations depends on the ability of public and private institutions, policies, managers, and workers to mobilize and exploit knowledge-based capabilities and assets. Although natural resources play an important role in economic competitiveness, the ability of even resource-rich economies to raise and sustain their citizens' living standards depends on the ability to combine natural resources and the created resources of knowledge and human capital.

Governments in the industrial and industrializing economies have faced the competitiveness challenge by strengthening policies and institutions that support innovation. One such institution is the research university, and OECD governments have launched a number of policy initiatives that seek to enhance the contribution of national university systems to industrial innovation. Many of these initiatives have been influenced by the Bayh-Dole Act of 1980, which policymakers in the United States and elsewhere argue has galvanized technology transfer and commercial application of academic research advances in U.S. industry.

There are a number of reasons for the recent policy focus on research universities. Considerable evidence (Narin et al., 1997; Hicks et al., 2001) suggests that the dependence of technological innovation on advances in science has increased in recent decades, a considerable change from the "trial and error" character of innovation in the late 19th and early 20th centuries. Universities also play a unique role in both research and training, and their ability to expose graduates to the frontiers of scientific research provides a powerful mechanism for the transfer of knowledge and technology. Governmental laboratories rarely combine research and education to a comparable degree and therefore frequently face forbidding obstacles to the transfer and application of their research results. At least some expert groups (See OECD, 2003) also credit universities with greater flexibility and capabilities to form teams in new, multidisciplinary research fields.

The decline of the large central research laboratories of many industrial firms in the industrial economies means that a greater share of basic research in many of these economies now is performed within their university systems, and policies to strengthen links with these research performers therefore assume greater importance. Declines in the importance of these privately funded research performers have been paralleled in many OECD economies with declining public budgets for R&D in government laboratories and in some fields of "mission-oriented" research, and technology transfer from universities is sometimes characterized as a means to capture greater private funding for fundamental research[1]. Finally, universities are viewed as productive sources of new firms, so-called "spinoffs", that are believed to be effective vehicles for technology commercialization and job creation.

There are also reasons for skepticism about these premises and the policies that are founded on them. First, the "science-dependence" of innovation in fact varies considerably among fields—for example, it appears to be greater in biomedical research than in information technology. The flexibility and responsiveness of universities to research opportunities also vary considerably, and may be quite sensitive to the overall structure of national policies toward funding and administration of higher education. The relationship between academic and industrial research as well as the contributions of university research to the formation of new firms often depend critically on institutions or policies external to the university, notably the operation of labor markets for scientists and engineers or the existence of venture-capital finance.

There is also reason for policymakers outside of the United States to be cautious in either attributing U.S. "success" in university-industry technology transfer to the Bayh-Dole Act, or in seeking to implement similar policies very different institutional environments. Many of these initiatives fail to take into account the interactive nature of university-industry research relationships,

---

[1] "Since mission-oriented public research no longer plays a pivotal role within ISRs [industry-science relationships], new market-friendly co-ordination must be implemented, with greater involvement of the financial sector, especially venture capital", (OECD, 2003).

which embody considerable feedback and iteration, rather than a unidirectional flow of fundamental knowledge into industry application. In addition, the existence of multiple channels of interaction and knowledge flow between academia and industry often is overlooked in reforms that seek to strengthen patent protection for university inventions without considering the implications of such steps for the operation of other channels of interaction. Finally, the importance of the overall structure of higher education systems, as well as the importance of institutions external to universities, is overlooked in many of the reforms proposed or initiated elsewhere in the OECD.

This paper surveys the evidence from the United States and other industrial economies on the relationship between university research and industrial innovation. I devote particular attention to the Bayh-Dole Act, in large part because of its importance as a "role model" for policy elsewhere in the OECD. The historical antecedents to the Act, as well as more recent shifts in policy by leading U.S. universities and growing criticism from U.S. industry of university patenting and licensing policies, provide additional issues for consideration in discussions of "reform" of university-industry relationships.

COMPARING NATIONAL SYSTEMS OF HIGHER EDUCATION

The limited data on the role of national higher education systems as R&D performers highlight other cross-national contrasts, including differences in their significance within the overall national R&D enterprise, their scale, their roles as employers of researchers, and their relationships with industry. As Figure 1 shows, the role of universities as R&D performers (measured in terms of the share of national R&D performed within higher education) is greatest in Italy, the Netherlands, and Canada, all of which are economies where universities perform more than 25% of total national R&D during 1996-2004; indeed, the OECD data report that Canadian universities accounted for nearly 40% of national R&D performance by 2004. By contrast, U.S. and Japanese universities accounted for less than 15% of national R&D performance during the same period.

Cross-national data highlighting differences in the "division of labor" between universities and government laboratories in basic research indicate that the higher education sector's share of basic research performance is similar in most Western European economies and the United States, although higher than in most of the Eastern European and Asian countries for which data are available (OECD 2001b, Annex Table A.6.4.1). But a key difference between the United States and the European countries for which data are available is that a relatively low share of basic research outside the academic sector in the United States is performed by the government, and a relatively high share by industry[2].

---

[2] These data must be interpreted with caution, since the definitions used by the national statistical agencies whence they are drawn often differ. For example, in France CNRS is classified as part of the Higher Education Sector, whereas in Italy similar organizations are treated as part of the government sector. See OECD 2001b, Annex 2

Figure 1
University-performed R&D % of total national R&D, 1996 – 2004

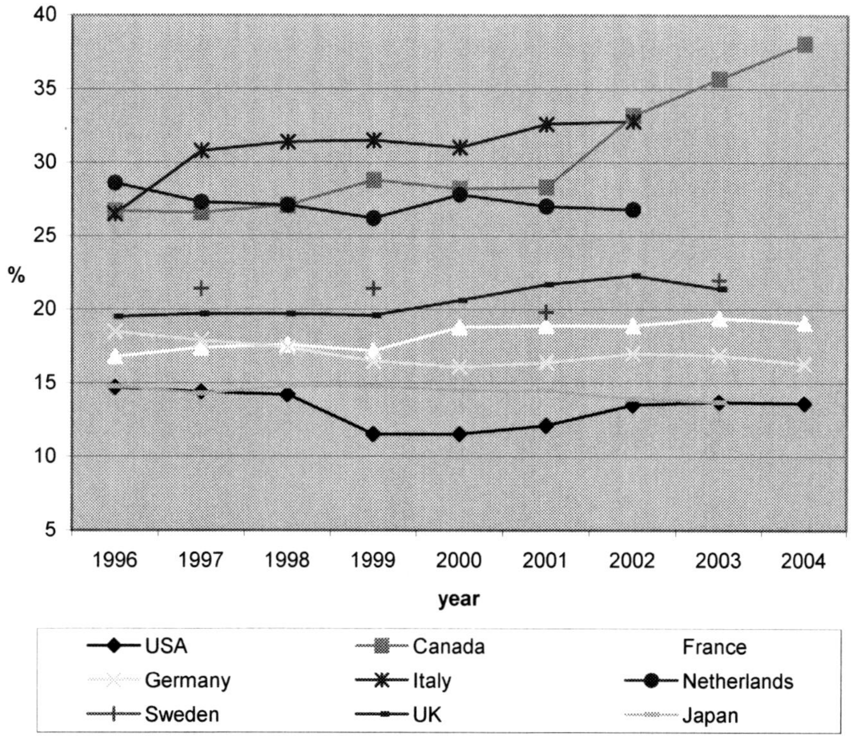

Figure 2 depicts the share of R&D funding within national higher education systems that is provided by industry. Despite the widely remarked closeness of U.S. university-industry research ties and collaboration (see Rosenberg and Nelson, 1994; Mowery et al., forthcoming), the share of R&D in higher education that is financed by industry is higher for Canada, Germany, and the United Kingdom than for the United States in the late 20th century.

Other qualitative data from the OECD 2002 study of "science-industry relationships" (2002, p. 37) compare the labor mobility and other "network relationships" linking universities and industry for Austria, Belgium, Finland, Germany, Ireland, Italy, Sweden, the UK, the US, and Japan. "R&D consulting with firms by university researchers" is greater than the EU average (the basis for these characterizations is not provided by the OECD study) in Austria, Germany, the UK, US, and Japan; such consulting is rated as "low" in Belgium, Finland, Ireland, and Italy. The annual flow of university researchers to industrial employment, another potentially important channel for knowledge exchange, is significantly higher than the EU average in Belgium, Finland,

Germany, Sweden, the UK, and the United States. Finally, the "significance of networks" linking universities and industry is rated as above the EU average for Finland, Germany, Sweden, the UK, the US, and Japan.

Figure 2
Share of higher ed R&D financed by industry, 1996-2004

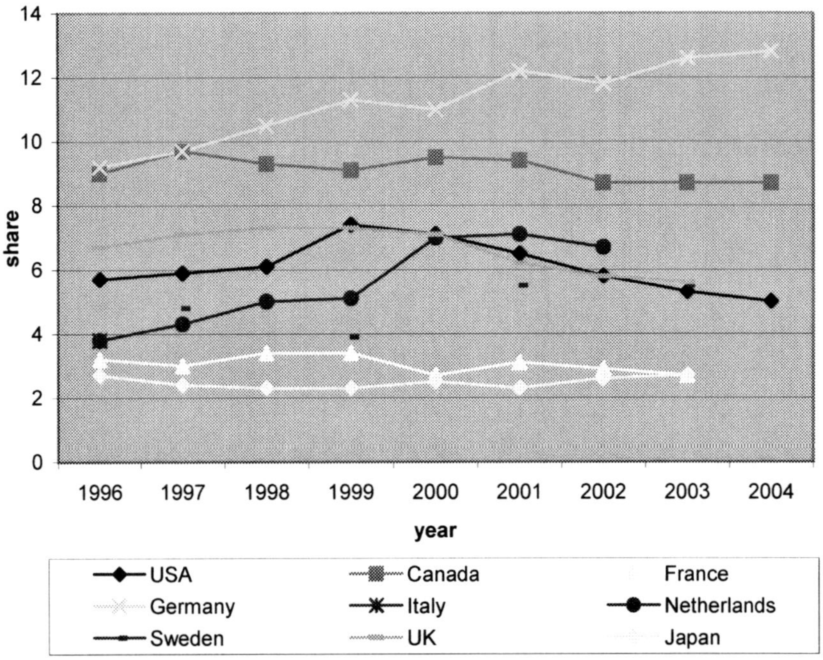

Surprisingly, in view of the frequency with which the United States is cited approvingly for the close links between university and industrial researchers, the evidence that university-industry relationships are "stronger" in the U.S. than elsewhere is mixed: the qualitative data on labor mobility support this characterization, while the data on industrial support of academic research do not. If the stereotypical view of U.S. universities as more closely linked with industrial research and innovation is indeed valid, it is striking that these indicators shed so little light on the structure of the linkages.

HOW DOES ACADEMIC RESEARCH INFLUENCE INDUSTRIAL INNOVATION?

A number of studies based on interviews and surveys of senior industrial managers in industries ranging from pharmaceuticals to electrical equipment have examined the influence of university research on industrial innovation. All of these studies (GUIRR, 1991; Mansfield, 1991; Levin et al., 1987; Cohen, Nelson, & Walsh, 2002) emphasize differences among industries in the relationship between university and industrial innovation. In particular, university research advances affect industrial innovation more significantly and directly in the biomedical field than is true of other sectors.

Table 1
The relevance of university science to industrial technology

Science	**# of Industries with "relevance" scores**		**Selected industries for which the reported "relevance" of university research was large (≥ 6)**
	≤ 5	≥ 6	
Biology	12	3	Animal feed, drugs, processed fruits/vegetables
Chemistry	19	3	Animal feed, meat products, drugs
Geology	0	0	None
Mathematics	5	1	Optical instruments
Physics	4	2	Optical instruments, electronics?
Agricultural science	17	7	Pesticides, animal feed, fertilizers, food products
Applied math/operations research	16	2	Meat products, logging/sawmills
Computer science	34	10	Optical instruments, logging/sawmills, paper machinery
Materials science	29	8	Synthetic rubber, nonferrous metals
Medical science	7	3	Surgical/medical instruments, drugs, coffee
Metallurgy	21	6	Nonferrous metals, fabricated metal products
Chemical engineering	19	6	Canned foods, fertilizers, malt beverages
Electrical engineering	22	2	Semiconductors, scientific instruments
Mechanical engineering	28	9	Hand tools, specialized industrial machinery

Source: Previously unpublished data from the Yale Survey on Appropriability and Technological Opportunity in Industry. For a description of the survey, see Levin et al. (1987).

The studies by Levin et al. (1987) and Cohen et al. (2002) summarize industrial R&D managers' views on the relevance to industrial innovation of various fields of university research (Table 1 summarizes the results discussed in Levin et al., 1987). Virtually all of the fields of university research that were rated as "important" or "very important" for their innovative activities by industrial respondents in both surveys were related to engineering or applied sciences, fields of U.S. university research with a long history of university-industry collaboration. Interestingly, with the exception of chemistry, few basic sciences appear on the list of university research fields deemed by industry respondents to be relevant to their innovative activities.

The absence of fields such as physics and mathematics in Table 1, however, should not be interpreted to mean that academic research in these fields does not contribute to technical advance in industry. Instead, these results reflect the fact that the effects on industrial innovation of basic research findings in such areas as physics, mathematics, and the physical sciences are realized only after a considerable lag. Moreover, application of academic research results often requires that these advances be incorporated into the applied sciences, such as chemical engineering, electrical engineering and material sciences.

The survey results summarized in Cohen et al. (2002) indicate that in most industries, university research results play a minor role in triggering new industrial R&D projects; instead, the stimuli originate with customers or from manufacturing operations. Pharmaceuticals is an exception, since university research in this field often triggers industrial R&D projects. A similar portrait of the relative importance of different outputs of university and public-laboratory research emerges from this study's summary of responses to questions about the importance to industrial R&D of various information channels (Table 2). Although pharmaceutical executives assign greater importance to patents and license agreements involving universities and public laboratories, even respondents from this industry rated research publications and conferences as a more important source of information. For most industries, patents and licenses involving inventions from university or public laboratories were reported to be of little importance, compared with publications, conferences, informal interaction with university researchers, and consulting.

The consistency in the findings of the Levin et al. (1987) study and the more recent survey conducted by Cohen and colleagues is striking—the "New Economy" notwithstanding, the late 1990s do not contrast sharply with the late 1970s. These studies also indicate that the relationship between academic research and industrial innovation in the biomedical field is different in other knowledge-intensive sectors. Academic research rarely produces "prototypes" of inventions for development and commercialization by industry, but academic research does inform the methods and disciplines employed by firms in their R&D facilities. Industrial R&D managers also rely on a variety of channels for learning about and exploiting the results of academic research, and the channels rated by industrial R&D managers as most important in this complex interaction between academic and industrial innovation rarely include patents and licenses.

Table 2
Importance to Industrial R&D of Sources of Information on Public R&D (including university research)

Information source	% rating it as "very important" for industrial R&D
Publications & reports	41.2%
Informal Interaction	35.6
Meetings & conferences	35.1
Consulting	31.8
Contract research	20.9
Recent hires	19.6
Cooperative R&D projects	17.9
Patents	17.5
Licenses	9.5
Personnel exchange	5.8

Source: Cohen et al. (2002).

Another important characteristic of the relationship between academic and industrial researchers is the dynamic and interactive character of this relationship. Industrial research may in fact "lead" and influence the agenda of academic research in some fields at some periods of time, as was the case in the early stages of research on light-emitting diodes and semiconductors. The movement of researchers between industry and academia is an important facilitating mechanism in this interactive relationship (e.g., William Shockley's move from Bell Labs to Shockley Semiconductor to Stanford University in the 1950s, or the move by Dr. Shuji Nakamura, a pioneering research in gallium-arsenide LEDs, from Nichia Chemicals in Japan to the University of California-Santa Barbara in 2000).

US UNIVERSITY-INDUSTRY COLLABORATION DURING THE 20TH CENTURY

U.S. university-industry research collaboration and technology transfer before the Bayh-Dole Act

University-industry collaboration in U.S. higher education has a long history, and was facilitated by the unusual structure of the U.S. higher education system, which was significantly larger than that of any other industrial economy for much of the 20th century. In addition, the U.S. system included a very heterogeneous collection of institutions (religious and secular, public and private, large and small, etc.), lacked any centralized national administrative control, and encouraged considerable interinstitutional competition for students, faculty, resources, and prestige (See Geiger, 1986, 1993; Trow, 1979, 1991, among other discussions). The reliance by many public universities on "local" (state-level) sources for political and financial support further enhanced their incentives to develop collaborative relationships with regional industrial and agricultural establishments. The structure of the U.S. higher education system thus strengthened incentives for faculty and academic administrators to collaborate in research and other activities with industry, and to do so through channels that included much more than patenting and licensing.

Although U.S. universities were patenting patent faculty inventions as early as the 1920s, few institutions had developed formal patent policies prior to the late 1940s, and a number of these policies embodied considerable ambivalence toward patenting. For example, despite the adoption by a growing number of universities of formal patent policies by the 1950s, many medical schools prohibited patenting of inventions. A number of universities active in patenting chose not to manage patenting and licensing themselves, in many cases because of concern over the political consequences of a visible role in profiting from faculty inventions, and in other cases because of fears that their nonprofit tax status could be jeopardized[3].

One important "independent" manager of the patenting and licensing activities for U.S. research universities from the 1930s through the 1960s was the Research Corporation, founded in 1912 by Frederick Cottrell, a University of California chemistry professor who wished to use the licensing revenues from his patents to support scientific research[4]. By the 1940s, the Corporation had become an important manager of patenting and licensing for such universities as MIT, Princeton, and Columbia.

World War II and the Cold War that followed transformed the structure of the U.S. national innovation system (Mowery and Rosenberg, 1998). Nowhere was this transformation more dramatic than in U.S. universities. Formerly funded largely by state governments and industry, academic research experienced a surge of federal funding, and many federal research sponsors required the development of a formal patent policy. Despite the adoption by a growing number of universities of formal patent policies by the 1950s, many of these policies, especially those at medical schools, prohibited patenting of inventions, and university patenting was less widespread than was true of the post-1980 period. Moreover, many universities chose not to manage patenting

---

[3] Etzkowitz's discussion of the debate within MIT over institutional patent policies during the 1930s (1994, p. 404) notes that "In 1936, the committee on patents of the institute put forward the view that: 'There is recognized to be danger in deriving any income whatever from inventions, first because of possible influence upon our tax exempt status, and second because of possible criticism of our methods leading to ill will among those upon whom we must depend for support. The first difficulty seems to be avoided if the actual handling of our affairs is delegated to some other organization".

[4] In addition to wishing to use the proceeds from licensing of his inventions for philanthropic purposes, Cottrell was concerned about the implications of having the University of California manage his patents: "A danger was involved, especially should the experiment prove highly profitable to the university and lead to a general emulation of the plan. University trustees are continually seeking for funds and in direct proportion to the success of our experiment its repetition might be expected elsewhere ... the danger this suggested was the possibility of growing commercialism and competition between institutions and an accompanying tendency for secrecy in scientific work" (Cottrell, 1932, p. 222).

and licensing themselves. The Research Corporation expanded its role as a manager of university patents and licensing during the 1950s and 1960s.

The number of universities establishing technology transfer offices and/or hiring technology transfer officers began to grow in the late 1960s, well before the passage of the Bayh-Dole Act of 1980. During the 1970s, U.S. universities expanded their patenting, especially in biomedical fields, and assumed a more prominent role in managing their patenting and licensing activities, supplanting the Research Corporation. Agreements between individual government research funding agencies and universities contributed to the growth of patenting during the 1970s. Private universities also expanded their patenting and licensing during this decade.

The Bayh-Dole Act of 1980

The Bayh-Dole Patent and Trademark Amendments Act of 1980 provided blanket permission for performers of federally funded research to file for patents on the results of such research and to grant licenses for these patents, including exclusive licenses, to other parties. The Act replaced a web of Institutional Patent Agreements (IPAs) that had been negotiated between individual universities and federal agencies with a uniform policy. The Act also expressed Congressional support for the negotiation of exclusive licenses between universities and industrial firms for the results of federally funded research. Of particular importance to U.S. universities, many of which lobbied actively for the passage of the Bayh-Dole Act was its reduction in the power of federal funding agencies to oversee the terms of licensing agreements between research performers and licensees.

The passage of the Bayh-Dole Act was one part of a broader shift in U.S. policy toward stronger intellectual property rights[5]. Among the most important of these policy initiatives was the establishment of the Court of Appeals for the Federal Circuit (CAFC) in 1982. Established to serve as the court of final appeal for patent cases throughout the federal judiciary, the CAFC soon emerged as a strong champion of patentholder rights[6]. But even before the establishment of the CAFC, the 1980 U.S. Supreme Court decision in *Diamond v. Chakrabarty* upheld the validity of a broad patent in the new industry of biotechnology, facilitating the patenting and licensing of inventions in this sector. The effects of Bayh-Dole thus must be viewed in the context of this larger shift in U.S. policy toward intellectual property rights.

---

[5] According to Katz and Ordover (1990), at least 14 Congressional bills passed during the 1980s focused on strengthening domestic and international protection for intellectual property rights, and the Court of Appeals for the Federal Circuit created in 1982 has upheld patent rights in roughly 80% of the cases argued before it, a considerable increase from the pre -1982 rate of 30% for the Federal bench.

[6] See Hall and Ziedonis (2001) for an analysis of the effects of the CAFC and related policy shifts on patenting in the U.S. semiconductor industry.

The effects of Bayh-Dole

A number of scholars have documented the role of Bayh-Dole in the growth of patenting and licensing by universities since 1980 (Henderson et al 1998). But Bayh-Dole is properly viewed as initiating the latest, rather than the first, phase in the history of U.S. university patenting. And this latest phase is characterized by a higher level of direct involvement by universities in management of their patenting and licensing activities, in contrast to the reluctance of many U.S. universities to become directly involved in patenting prior to the 1970s.

Keeping in mind that we are unable to separate the effects of Bayh-Dole from those of other influences, how has U.S. university patenting changed since 1980? Universities increased their share of patenting from less than 0.3% in 1963 to nearly 4% by 1999, but the rate of growth in this share began to accelerate before rather than after 1980 (Figure 3). Another issue of interest in academic patenting is the distribution among technology fields of university patents during the pre- and post-Bayh-Dole periods. Non-biomedical university patents increased by 90% from the 1968-70 period to the 1978-80 period, but biomedical university patents increased by 295%. The increased share of the biomedical disciplines within overall federal academic R&D funding, the dramatic advances in biomedical science that occurred during the 1960s and 1970s, and the strong industrial interest in the results of this biomedical research, all affected the growth of university patenting during this period.

Figure 3
US research univ. patents % of all domestic-assignee US patents, 1963 - 99

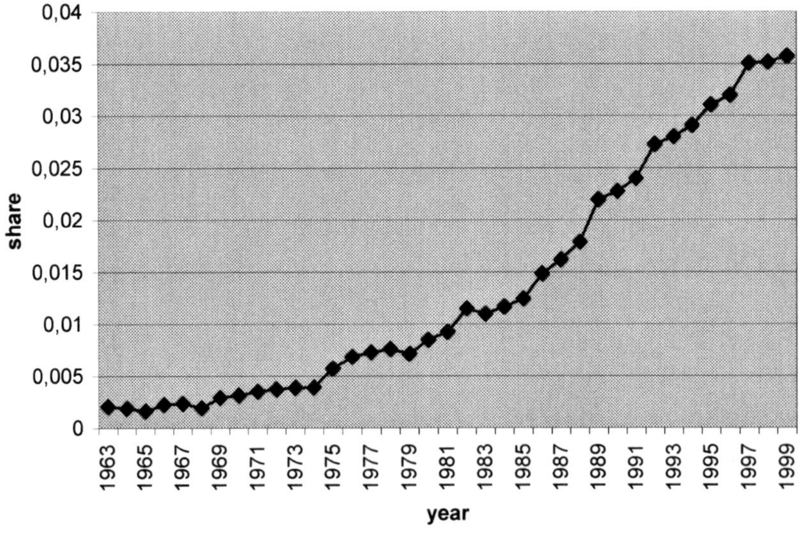

It is important to keep in mind the highly concentrated nature of this dimension of the "impact" of the Bayh-Dole Act. Far from affecting overall "university culture", or affecting the relationships among research, faculty, and industry in a great many fields, the Act appears to have affected a fairly narrow range of research fields. Moreover, the fact that Bayh-Dole's most significant effects on university patenting and licensing appear in a field in which the U.S. federal government has invested tens of billions in support for R&D should rebut any claims that university technology licensing can in some unspecified fashion compensate for a decline in public "mission-oriented" R&D support.

As I noted earlier, surveys of industry R&D managers and other evidence indicates that the relationship between academic research and industrial innovation in the biomedical field is unusual in both its closeness and the important role of patents as a channel for industry exploitation of scientific results. The dominant role of biomedical invention in post-Bayh-Dole university patenting and licensing therefore is not surprising, especially in light of the large sums of public funds invested in support for biomedical research. But the reliance on patenting and licensing that characterizes university-industry relationships in biomedical research may be less relevant to other fields. Indeed, a number of U.S. universities have begun to experiment with policies that recognize these differences among research fields (see below).

MANAGING UNIVERSITY PATENTING AND LICENSING

"Learning to patent" after the Bayh-Dole Act

The Bayh-Dole Act was followed by a wave of entry by universities into management of patenting and licensing. The share of U.S. research university patenting accounted for by institutions with at least 10 patents issued before 1980 declined from more than 85% during 1975-80 to less than 65% by 1992. Institutions with fewer than 10 patents issued before 1980, by contrast, increased their share of academic patents from 15% in 1981 to almost 30% in 1992. And institutions with no patenting activity during 1975-80 increased their share of overall academic patenting from zero in 1980 to more than 6% by 1992. These less-experienced entrant universities received less significant patents in the immediate aftermath of the passage of the Bayh-Dole Act, although the gap between the quality of their patents and those of experienced institutional patenters was narrowed by the end of the 1980s. The improvement in the signficance of their patents appears to have reflected the development at many of these entrant universities of a more selective approach to patenting during the 1980s.

This evidence concerning the relatively low quality of the early patents obtained by many "entrant" institutions also underscores the need for caution in using counts of patents (on their own, or relative to R&D spending) as a measure of the "productivity" of research universities. As is well known, patents vary widely in quality: like academic papers, a great many patents are never cited or actively "worked" by anyone, and the value of any portfolio of patents typically is dominated by a very small number of patents. Comparisons of patent productivity across universities, or (even more questionable) between universities and industry must use incorporate some adjustment for the "quality" of patents, e.g., through citation-weighting patents (or, possibly, papers)[7]. The very heterogeneous "value" of patents also underscores the inadvisability of using "patent productivity" as a criterion for evaluating the performance of faculty. Like published academic papers, the majority of patents are of minimal value. Unlike papers, however, the creation by universities of incentives for faculty to obtain patents may significantly raise the operating costs of the technology licensing offices charged with obtaining patents on faculty inventions, regardless of the value of these patents.

Evidence cited in Mowery et al. (2004) reveals that gross licensing revenues for Columbia University, Stanford University, and the University of California system were dominated by a small number of patents. For each of these three universities, the "top 5" patents accounted for more than 65% of gross licensing revenues. These "top 5" patents were mainly biomedical inventions. Universities lacking a major biomedical research program may not produce such "home run" patents and therefore may reap lower gross revenues.

Moreover, high levels of gross revenues do not necessarily translate into substantial net income for academic institutions. For example, average annual gross licensing revenues for the nine-campus University of California system during fiscal 2001 – 2004 were roughly $75 million. The net contribution to UC operating expenses, however, a figure that subtracts the operating expenses of the technology licensing office and payments to the faculty inventor, averaged slightly more than $15 million annually. This amount represents a small fraction (less than 1%) of the annual research budget for the UC system of more than $3 billion. Industry funding of academic research within the UC system in fiscal 2001 (the most recent year for which comprehensive data are available) amounted to $235 million, dwarfing both the average gross and net institutional revenues associated with licensing activities[8].

Revenues are of course not the only motive for university licensing activities. Other important motives include retention of faculty who wish to see their inventions patented and licensed; the transfer of university inventions to commercialization; regional or state-level economic development, and (in the wake of the 2003 *Madey v. Duke* decision of the Court of Appeals for the Federal Circuit, which eliminated the informal "experimental use" defense against claims of patent infringement) the preservation of freedom of academic scientists to conduct research.

---

[7] More generally, comparisons of the "cost-effectiveness" of R&D investments in universities and industry that rely on patents produced per R&D investment dollar are hazardous guides for policy. Such comparisons ignore the fact that research universities and industrial R&D performers fundamentally are pursuing different, complementary missions, and yield different outputs.

[8] http://www.ucop.edu/research/publications/pdf/resfund01.pdf, accessed February 20, 2006.

This array of potential goals for patenting and licensing activities, however, creates some challenges for management. First, these goals are not entirely compatible—for example, support for regional economic development may entail an acceptance of lower royalty rates on licenses for firms active in the vicinity of the university. Technology licensing thus will involve some tradeoffs among these goals. Second, in spite of these tradeoffs, as well as the evidence above on the relatively modest scale of net revenues at many university technology licensing offices, a recent survey of technology licensing officers (Jensen et al., 2001) indicates that these individuals cite licensing revenues as the most important goal of their activities.

Policy evolution at MIT, Stanford, and UC Berkeley

The Bayh-Dole Act of 1980 included few detailed guidelines for patenting and licensing policies, beyond a mandate that licensing revenues be used to support research. U.S. universities have considerable flexibility in the specific policies that they pursue for management of their patenting and licensing activities under the terms of the Act. In light of this flexibility, the lack of significant experimentation or innovation in most universities' policies during the 25 years since the passage of the Act is remarkable. Most universities rely on an institutional "technology licensing office" (or equivalent), rather than considering multi-institutional collaboration or some other form of outsourcing of management responsibility. Few if any universities have considered the development of field-specific licensing policies that might acknowledge the very different characteristics of research interactions and intellectual property protection across fields.

During the past 10 years, however, some research universities have altered the management and policies for their patenting and licensing activities. At least some of these initiatives have been undertaken in response to industry criticism (see below). Among the more interesting such changes are those undertaken by MIT, Stanford, and the University of California at Berkeley.

These three universities share a number of characteristics—all three have engineering colleges ranked among the top 5 in the United States, as well as strong physical-sciences research capabilities, and all three have long permitted patenting by faculty. All three institutions also have a history dating back to the early 20th century of collaborative relationships with industry that have contributed to the growth of regional information technology, electronics, and biomedical industrial complexes in northern California and eastern Massachusetts.

There are also significant contrasts among the three institutions. Alone among these three institutions, Stanford has a research-intensive medical school that has been an important source of licensed inventions. The university began direct management of patenting and licensing in 1970, having previously utilized the Research Corporation for these functions. MIT similarly reduced its reliance on the Research Corporation in the early 1960s (partly as a result of disputes over licensing policy—see Mowery and Sampat, 2001) and has managed its patenting and licensing activity through the Institute's Technology Licensing Office since the late 1960s. The University of California system has managed patenting and licensing since the 1940s, but the systemwide Office of Technology Transfer was strengthened and expanded in the 1970s. Since 1990, UC Berkeley has operated a campus-level technology licensing office, which shares responsibility for managing patenting and licensing activities with the systemwide OTT.

All three universities support an array of programs to support industry-university collaboration. For example, all three universities' colleges of engineering operate "Industrial Liaison Programs", which offer "memberships" to firms for a fee that allow firm employees to review research advances, visit campus laboratories, participate in regular meetings with researchers, and (depending on the structure of the particular program and the size of the annual fee paid by the firm) may include opportunities for firm employees to be work on a temporary basis in academic research facilities. All three universities also promote opportunities for industrial firms to support individual faculty research projects. Interestingly, however, the levels of industry-supported research differ significantly among these three institutions, all of which have roughly similar research budgets (ranging in fiscal 2003 from $486 million for MIT to $507 million at UC Berkeley and $603 million at Stanford). As Figure 4 shows, industry sources funded nearly 16% of MIT campus research in FY2003, more than twice the average for all U.S. universities (roughly 7.4%), but accounted for a substantially smaller share of the research budgets of UC Berkeley (4.4%) and Stanford (5.2%).

Figure 4
Industry-funded share of R&D: Stanford, UC Berkeley, MIT, FY 2003

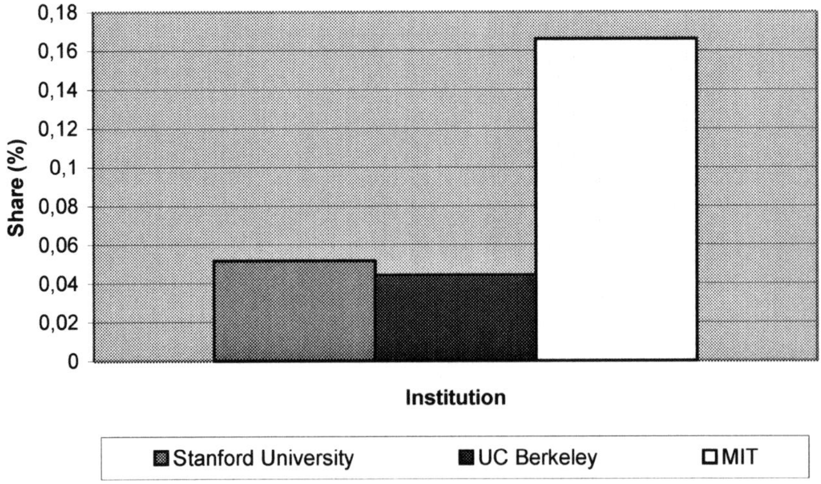

Additional comparative data on invention disclosures and licensing activity are displayed in Figures 5–7. The contrasts between the two private research universities, MIT and Stanford, and UC Berkeley are striking in all three Figures. Gross licensing income (which included some large "cashouts" of equity investments by the MIT and Stanford licensing offices in FY 2000 and 2001)[9] is substantially higher at both MIT and Stanford than at UC Berkeley. Indeed, the licensing activities of the UC Berkeley OTL yielded average annual net revenues to the campus of slightly more than $1 million during fiscal 2001 - 2004, after payments to the inventor and operating expenses are deducted.

Figure 5
Gross licensing royalties, FY2000 - 2004

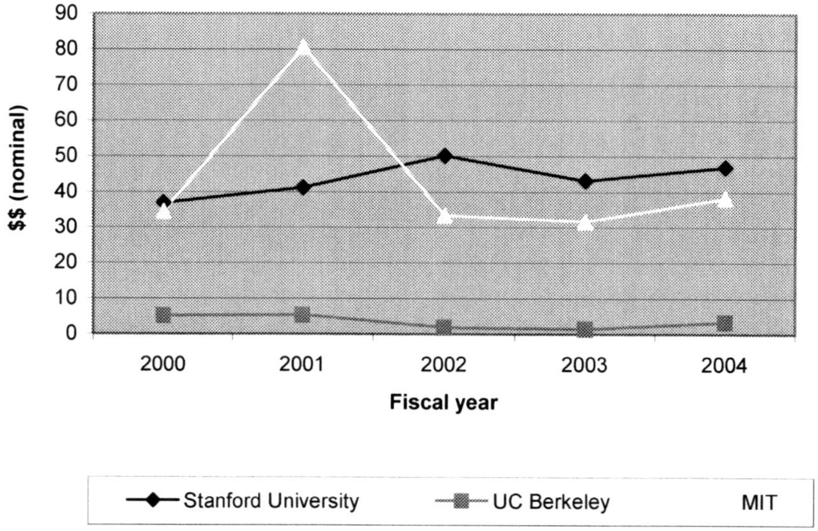

[9] All three universities' technology licensing programs are allowed to accept equity in startup licensees in lieu of patent prosecution costs or licensing fees. Only Stanford reports the annual number of transactions involving their acquisition of equity in licensees. In FY 2000 and 2001, MIT realized equity-based gains of $14.5 million and $55.6 million, respectively (these amounts are included in the data depicted in Figure 2). Stanford's OTL reported an equity-based profit of $2.1 million in FY 2001 and $336 million in FY 2005 (this bonanza was associated with the OTL sale of its equity stake in Google).

Figure 6
Licensing agreements, FY 2000 - 2004

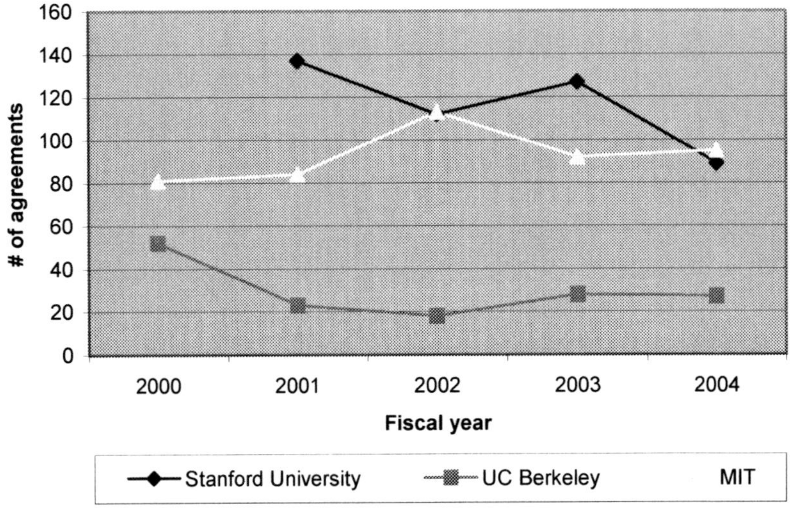

Figure 7
Annual invention disclosures, FY2000 - 2004

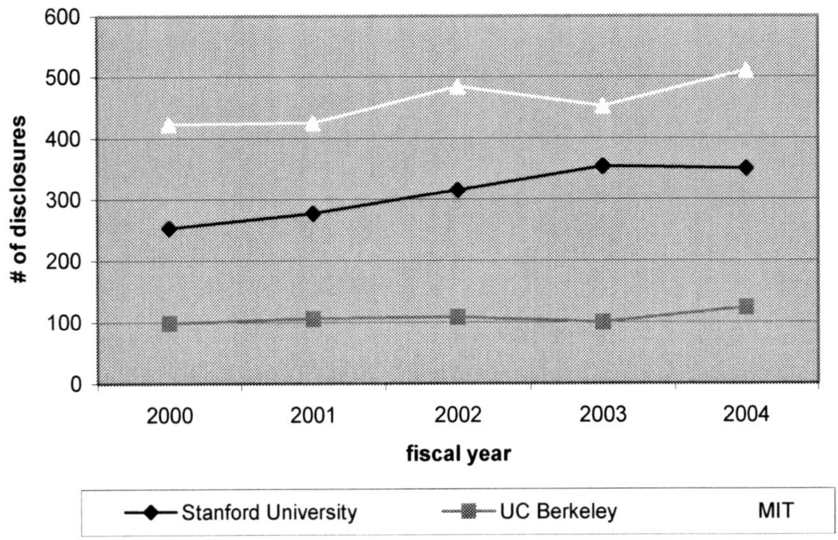

Some of these differences reflect the fact that its relatively recent establishment means that the UC Berkeley OTL has a smaller patent and license portfolio than the other two universities (licensing income associated with patents issued before the 1990 creation of the UC Berkeley OTL flows to the systemwide licensing office). But Figures 6 and 7 show that annual invention disclosures and licensing agreements also are higher at MIT and Stanford than at UC Berkeley, suggesting that the wide gap in gross licensing income between UC Berkeley and the other two institutions is not likely to vanish with the passage of time. The purely financial institutional benefits of patenting and licensing for UC Berkeley thus are modest, albeit positive, and are much smaller than those flowing to MIT or Stanford. Nonetheless, only Stanford University's average gross licensing income (netting out equity cashouts) during 2001-2004 exceeded its FY 2003 level of industry-funded research[10].

Although the patenting and licensing activities of all three universities are profitable on a "net institutional income" basis, these institutions now all manage their licensing activities as complements to broader programs promoting closer research relationships with industry. For example, the Stanford OTL Director also oversees the university's "Industrial Contracts Office", which manages sponsored-research agreements with industry, as well as overseeing materials transfer agreements (MTAs), which govern the transfer among researchers of research tools and materials. Industrial firms supporting campus research can receive licenses (in some cases, royalty-free licenses) to the results of this research.

A similar tradeoff between maximizing licensing revenues and obtaining industry research funding is apparent at the creation in 2003 of the "Intellectual Property & Industry Research Alliances" office at UC Berkeley, which absorbed the established Office of Technology Licensing and a new "Industry Alliances Office", charged with overseeing the negotiation of sponsored-research agreements with industry. Moreover, the UC Berkeley licensing office, along with other UC technology licensing offices, has implemented a new policy that recognizes the differences among industries in the value (and likely licensing income) of patents in different fields of research. In 2000, the UC President's Office authorized the negotiation of royalty-free licenses with industrial sponsors of campus research in electrical engineering and computer science. Another reflection of the changing priority assigned to licensing royalties relative to other goals at UC Berkeley is the "socially responsible licensing initiative", which negotiates royalty-free licenses on inventions sold in low-income economies[11].

---

[10] National Science Foundation data indicate that FY2003 industry-funded research amounted to $31 million at Stanford, $22 million at UC Berkeley, and $81 million at MIT.

[11] Yet another initiative of note on the UC Berkeley campus was the 1998 research agreement between the Novartis Corporation and the campus's Plant and Microbial Biology (PMB) department. The agreement involved the contribution by Novartis of $25 million to support PMB research over a five-year period. In exchange for this significant financial contribution, Novartis was granted rights to review all PMB department invention disclosures and exercise an option to negotiate a license to a share of these disclosures proportionate to its share of overall PMB research funding (roughly one-third). The initiative was controversial, in part because of excessive secrecy in its

The economic significance of these shifts in licensing strategy and policy is difficult to evaluate, and their implementation also raises challenges. Very few academic laboratories strictly separate research activities according to funding sources (the vast majority of academic research in the biomedical and nonbiomedical fields alike is financed by the federal government), and policies that seek to differentiate licensing terms based on the source of funding may prove to be infeasible. Similarly, the development and implementation of policies that promote differences in licensing terms according to characteristics of end-user markets will be difficult for the many inventions that serve both high- and low-income markets. Nonetheless, these initiatives, along with the assignment of responsibility for a broader set of relationships with industry to campus directors of technology licensing offices at both Stanford and UC Berkeley, suggest that these leading academic licensors are developing a more nuanced approach to the management of tradeoffs within their technology transfer strategies.

These policies seek to use licensing as a complement to policies that encourage industry-supported academic research, and create some tension with the popular goal at many university licensing offices of supporting the formation of new "spinoff" firms to commercialize university inventions. Even if one assumes that new firms are more effective at commercializing technologies and creating jobs (and the evidence for both of these propositions is mixed at best), newly established firms rarely are in a position to provide financial support for university research. Moreover, the new firm is likely to require an exclusive license to a relatively broad patent (or an exclusive license to a broad field of use for a patent) in order to attract the financing necessary to survive. As is often the case in university licensing policies, the tensions among these policies and goals have scarcely been articulated, let alone recognized or incorporated into management strategies.

INDUSTRY CRITICISM OF U.S. UNIVERSITY LICENSING POLICIES

Growth in U.S. university patenting and (to an even greater degree) licensing since 1980 was dominated by the biomedical sciences, where patents have considerable economic value and the number of patents associated with significant commercial innovations (e.g., a new pharmaceutical product) often is smaller than is true of commercial innovations in fields such as information technology. And in at least some of these nonbiomedical fields, anecdotal evidence suggests that the efforts of many universities to seek patent licensing income has been a source of friction, rather than facilitating collaboration, with industry. Dr. R. Stanley Williams of Hewlett Packard, a firm with a long history

---

negotiation and poor handling by campus administrators of its announcement, and Novartis elected not to renew the agreement. As of 2002, the firm had exercised its option on two of the department's disclosures, although the status of license negotiations is unclear. Given the controversy surrounding this undertaking, it seems unlikely that anything similar is likely to be negotiated by UC Berkeley officials in the near future.

of close research collaboration with U.S. universities, stated in testimony before the U.S. Senate Commerce Committee's Subcommittee on Science, Technology and Space that

> Largely as a result of the lack of federal funding for research, American Universities have become extremely aggressive in their attempts to raise funding from large corporations... Large US based corporations have become so disheartened and disgusted with the situation they are now working with foreign universities, especially the elite institutions in France, Russia and China, which are more than willing to offer extremely favorable intellectual property terms". (September 17, 2002; statement reproduced at http://www.memagazine.org/contents/current/webonly/webex319.html; accessed April 2, 2005).

In the biomedical field, the NIH Director's Working Group on Research Tools stated in its report that

> If there was one point on which virtually every private firm that we spoke to was in agreement, it was that universities take inconsistent positions on fair terms of access to research tools depending on whether they are importing tools or exporting them. Over and over again, firms, complained to us that universities "wear the mortarboard" when they seek access to tools developed by others, yet they impose the same sorts of restrictions when they enter into agreements to give firms access to their own tools. As one lawyer for a small biotechnology firm put it,
>
> "Universities want it both ways. They want to be commercial institutions when it comes to licensing their technology, but to be academic environments when it comes to accessing technology that others have developed...They throw the same things in the way of small companies". (National Institutes of Health, 1998, p. 15).

A more sweeping (and arguably exaggerated) assessment was presented at a 2003 conference organized by the Government-University-Industry Research Roundtable at the National Academy of Sciences:

> ...the universities' approach of securing iron-clad protection for intellectual property seems to be yielding diminishing returns, even within the narrow confines of the licensing activity itself...The requisite legal negotiations for IP-that-will-ultimately-prove-to-be-useless are laborious, individualized and negotiated between universities and companies on a case-by-case. The up-front legal negotiations can easily cost more than the total cost of the research project being conducted, and/or extent past the time when the company has interest in the technology path being pursued...In summary, the uncertainty of the true value of university-generated intellectual property, combined with a litigious culture, have combined to make the university-industry working relationship—one that has historically contributed greatly to graduate education—unaffordable and nearly unsustainable within the U.S. (Government-University-Industry Research Roundtable, 2003, p. 2).

These critical comments have triggered considerable discussion between large industrial firms (many of which are in the information technology sector) and U.S. research universities over intellectual property policies and licensing guidelines. In December 2005, four large IT firms (Cisco, Hewlett Packard,

IBM, and Intel) and seven universities—Carnegie Mellon University; Renssellaer Polytechnic; UC Berkeley; the Georgia Institute of Technology; Stanford University; University of Illinois at Champaign-Urbana; and University of Texas-Austin) agreed on a "statement of principles" for collaborative research on open-source software that emphasizes liberal dissemination of the results of collaborative work funded by industrial firms[12]. The Government-University-Industry Research Roundtable conference mentioned earlier is one of a series of meetings involving industrial firms, the Industrial Research Institute (representing R&D directors of large U.S. firms), and the National Conference of University Research Administrators that is supposed to produce a set of principles for patent and licensing policies.

In the nature of the case, most of the tensions that have received considerable press and some attention from policymakers involve relationships between established firms and universities—indeed, in some respects, the economic interests of established firms with large patent portfolios may differ from those of small startup firms that are owners or licensees of far fewer patents. Moreover, most of the major conflicts have involved firms outside of the biomedical sector, reflecting the fact that the value of individual patents in industries such as information technology typically is lower than is true of biomedical research. Nevertheless, the current controversies and discussions among U.S. industrial firms and U.S. research universities may result in a rethinking by universities of the value of patents in efforts to sustain collaborative research relationships with industry.

CONCLUSION

The relationship between U.S. university research and innovation in industry is a long and close one. Indeed, organized industrial research and the U.S. research university both first appeared in the late 19th century and have developed a complex interactive relationship. The unusual structure of the U.S. higher education infrastructure, which blended financial autonomy, public funding from state and local sources with federal research support, and

---

[12] The "Open Collaboration Principles" cover "…just one type of formal collaboration that can be used when appropriate and will co-exist with other models, such as sponsored research, consortia and other types of university/industry collaborations, where the results are intended to be proprietary or publicly disseminated". According to the "Principles", "The intellectual property created in the collaboration [between industry and academic researchers] must be made available for commercial and academic use by every member of the public free of charge for use in open source software, software related industry standards, software interoperability and other publicly available programs as may be agreed to be the collaborating parties…" (http://www.kauffman.org/pdf/open_collaboration_principles_12_05.pdf; accessed February 20, 2006). These principles originated in an August 2005 "University and Industry Innovation Summit" in Washington DC organized by the Kauffman Foundation of Kansas City and IBM (http://www.kauffman.org/items.cfm?itemID=662; accessed February 20, 2006).

substantial scale, provided strong incentives for university faculty and administrators to focus their efforts on research activities with local economic and social benefits. Rather than being exclusively concerned with fundamental scientific principles, much of U.S. university research throughout the late 19th and 20th centuries focused on problems of agriculture, public health, and industry.

U.S. universities have made important contributions to industrial innovation throughout the past century, not least through providing both advanced research and education. The strong links between education and research sustained a close relationship between the evolving scientific research agenda and problems of industry or agriculture, while at the same time providing an effective channel (in the form of trained students) for the transfer of this knowledge to industry and other economic sectors. In addition, many university researchers in engineering and medical schools maintained close ties with the users of their research and their graduates in industry, medical practice, and agriculture. The important role of U.S. universities in industrial innovation, particularly during the post-1945 period, also relied on institutions external to the university, including venture capitalists, equity-based financing of new firms, and high levels of labor mobility between academia and industry.

The Bayh-Dole Act of 1980, rather than transforming the nature of university-industry links and relationships in the United States, modified but did not fundamentally alter the longstanding structure of incentives and constraints that supported collaboration between university and industrial researchers in the United States. As the cases of MIT, Stanford University, and the University of California-Berkeley suggest, universities with long histories of close research links to industry appear to have shifted their priorities in managing patenting and licensing activities to accommodate a broader range of goals beyond those of maximizing royalty income. And claims by some critics within U.S. industry that the Bayh-Dole Act has added frictions to university-industry collaboration now appear to be triggering a broader debate over the appropriate management of research collaborations. It seems likely that further modifications in institution-specific policies will appear in the near future, reflecting the evolutionary nature of the century-old links between U.S. research universities and industrial innovation.

BIBLIOGRAPHY

Arrow, K., (1962), "Economic Welfare and the Allocation of Resources for Invention", [in:] R. R. Nelson (ed.), *The Rate and Direction of Inventive Activity*, Princeton University Press, Princeton.

Ben-David, J., (1968), *Fundamental Research and the Universities*, OECD, Paris.

Ben-David, J., (1971), *The Scientist's Role in Society*, New York, Prentice-Hall.

Cohen, W., R. Florida, L. Randazzese, and J. Walsh, (1998), "Industry and the Academy: Uneasy Partners in the Cause of Technological Advance", [in:]

R. Noll (ed.), *Challenges to the Research University*, Brookings Institution, Washington, D.C.

Cohen, W. M., R. R. Nelson, and J. P. Walsh, (2002), "Links and Impacts: The Influence of Public Research on Industrial R&D", *Management Science* 48, No. 1 (2002), pp. 1-23.

Etzkowitz, H., A. Webster, and P. Healey, (1998), "Introduction", [in:] H. Etzkowitz, A. Webster, and P. Healey (ed.), *Capitalizing Knowledge*, State University of New York Press, Albany.

Etzkowitz, H., and L. Leytesdorff, (1997), *Universities in the Global Economy: A Triple Helix of Academic-Industry-Government Relation*, Croom Helm, London.

Geiger, R., (1986), *To Advance Knowledge: The Growth of American Research Universities, 1900-1940*, Oxford University Press, New York.

Godin, B., and Y. Gingras, (2000), "The Place of Universities in the System of Knowledge Production", *Research Policy* 29, pp. 273-278.

Government University Industry Research Roundtable (GUIRR), *Industrial Perspectives on Innovation and Interactions with Universities*, National Academy Press, Washington, DC, 1991.

Graham, H. D., and N. Diamond, (1997), *The Rise of American Research Universities*, Johns Hopkins University Press, Baltimore.

Henderson, R., A. B. Jaffe, and M. Trajtenberg, (1998), "Universities as a Source of Commercial Technology: A Detailed Analysis of University Patenting, 1965-88", *Review of Economics & Statistics* 80, pp. 119-127.

Hicks, D., T. Breitzman, D. Olivastro, and K. Hamilton, (2001), "The Changing Composition of Innovative Activity in the U.S.: A Portrait Based on Patent Analysis", *Research Policy* 30, pp. 681-703.

Katz, M. L., and J. A. Ordover, (1990), "R&D Competition and Cooperation", *Brookings Papers on Economic Activity: Microeconomics*, pp. 137-192.

Levin, R. C., A. Klevorick, R. R. Nelson, and S. Winter, (1987), "Appropriating the Returns from Industrial Research and Development", *Brookings Papers on Economic Activity* 3, pp. 783-820.

Mowery, D., R. Nelson, B. Sampat, and A. Ziedonis, (2003), *Ivory Tower and Industrial Innovation: University-Industry Technology Transfer Before and After the Bayh-Dole Act*, Stanford University Press, Stanford, CA.

Mowery, D. C., R. R. Nelson, B. N Sampat, and A. A. Ziedonis, (2001), "The Growth of Patenting and Licensing by U.S. Universities: An Assessment of the Effects of the Bayh-Dole Act of 1980", *Research Policy* 30, pp. 99-119.

Mowery, D. C., and N. Rosenberg, (1989), *Technology and the Pursuit of Economic Growth*, Cambridge University Press, New York.

Mowery, D. C., and N. Rosenberg, (1993), "The U.S. National Innovation System", [in:] R. R. Nelson (ed.), *National Innovation Systems: A Comparative Analysis*, Oxford University Press, New York.

Nelson, R. R., (1959), "The Simple Economics of Basic Scientific Research", *Journal of Political Economy* 67, pp. 297-306.

OECD (Organization for Economic Cooperation and Development), *A New Economy,* (2000), OECD, Paris.

OECD (Organization for Economic Cooperation and Development), *Benchmarking Industry-Science Relationships*, (2002), OECD, Paris.

OECD (Organization for Economic Cooperation and Development), *Main Science and Technology Indicators: 2002*, (2002), OECD, Paris.

OECD (Organization for Economic Cooperation and Development), *Main Science and Technology Indicators: 2005*, (2005), OECD, Paris.

Rosenberg, N., and R. R. Nelson, (1994), "American Universities and Technical Advance in Industry", *Research Policy* 23, pp. 323-348.

Rosenberg, N., (1999), "American Universities as Economic Institutions", *Mimeo*.

Slaughter, S., and L. L. Leslie, (1997), *Academic Capitalism: Politics, Policies, and the Entrepreneurial University*, Johns Hopkins University Press, Baltimore.

Trajtenberg, M., R. Henderson, and A. B. Jaffe, (1997), "University Versus Corporate Patents: A Window on the Basicness of Inventions", *Economics of Innovation and New Technology* 5, pp. 19-50.

Trow, M., (1991), "American Higher Education: 'Exceptional' or Just Different", [in:] B.E. Shafer (ed.), *Is America Different? A New Look at American Exceptionalism*, Oxford University Press, New York.

3

Scientific Output, Social Infrastructure and Science Policies: the Case of Small and Medium-sized Countries

Hugo Horta
Francisco M. Veloso

INTRODUCTION

National innovation systems are characterized by particular developmental trajectories (Nelson, 1991). In each country, the national innovation system has specific institutional and systemic characteristics that make its development unique and dissimilar from any other country. Therefore, each country typically follows a particular developmental path influenced by past history, institutional arrangements, and specific conditions of its own national context. An important part of this developmental trajectory is associated with the evolution of the science and technology systems in which national innovation systems are rooted (Lundvall and Christensen, 2003). Trajectories explain asymmetries in a country's performance because they contain the history of knowledge accumulation through learning processes (Johnson, 1992). Although this knowledge resides significantly in the cognitive ability of individuals and institutions (Bach and Matt, 2005), it is also expressed in codified form and diffused as information contained in scientific and technical publications, reflecting the advancements of both individual and collective knowledge. A country's scientific output is, therefore, a good indicator of its knowledge accumulation and the cumulative learning of its individuals and institutions (Bach and Georghiou, 1998). In fact, statistics focusing on scientific outputs and outcomes are in great demand (Weingart, 2005). Countries with developed science and technology systems, such as the United States, have long started to gather data on Science and Technology (S&T) funding, human resources, and scientific output, aiming to understand the evolution and performance of their scientific systems[1]. S&T statistics are also increasingly used to complement peer review analysis as part of evaluation assessments (Weingart, 2005). The analysis of scientific production, with the aim of understanding the scientific development and structure of nations, has also received attention in the specialized literature. A number of papers have been published in recent years looking at the scientific structure of large countries (Bayers, 2005) or countries with significant scientific output (Adams, 1998; King, 2004). In addition, research has also focused on specific scientific fields (Gu, 2004), scientific fields within a given country (Guimarães, 2002), scientific collaboration issues (Basu and Aggarwal, 2001; Bordons *et al.*, 1996) and, less often, on the scientific output of developing countries (Osareh and Wilson, 2000).

This paper contributes to previous research results by looking at a sample of small and medium sized countries. We compare the scientific output of each country and analyze their scientific structure, aiming to identify patterns of development and the influence of each country's specific national institutional system. The paper is structured as follows: Section 2 provides the theoretical framework, where we discuss the evolution of knowledge as critical for enabling sustained social and economic development. In this Section we also examine the potential role of institutions, policies, and resources on the scientific structure of a nation. Section 3 presents the data and methodology used in our study. Section 4 encompasses the data analysis. The first part of the analysis focuses on the relationship between scientific output and both scientific and economic intensity. The second part classifies the scientific structure of the sample countries, as well as their scientific impact, by disciplinary area. The last section is comprised of discussion and conclusions.

KNOWLEDGE, LEARNING AND THE EVOLUTION OF SCIENTIFIC SYSTEMS

For several decades now, we have come to recognize the relevance of knowledge for economic growth. Arrow (1962) was among the first to explore how increases in the stock of capital and labor were not sufficient to explain economic growth, and that learning and knowledge accumulation also play a key role. This idea led to further research on why and how knowledge relates to economic development. One of the notions, proposed in Nelson and Romer's

---

[1] The National Science Technology has compiled S&T funding data since the 1950s (NSB, 2005) and the Higher Education Research Institute (HERI) has applied a biannual survey to higher education institutions since 1966 that include individual data concerning faculty scientific publications.

(1996) taxonomy, is to distinguish between skills and ideas, or using the author's nomenclature, between 'wetware' and 'software'. Skills refer to each individual's knowledge as it is stored in the brain. People express their skills through convictions, talents and abilities. In contrast, ideas are codified knowledge stored outside the human brain. The basic difference between skills and ideas lies on the level of codification. Ideas can be formalized into symbols, languages and codes, while skills always remain in a tacit form (Nonaka *et al.*, 2003).

Tacit knowledge is difficult to learn and transfer, furthermore, it is both rival (cannot be used at the same time by other users) and exclusionary (it can only belong to the individual). Nevertheless, tacit knowledge is not only utterly necessary to understand ideas, but also to create new ideas (Nelson and Winter, 1982; Polanyi, 1966). Ideas, on the other hand, are easily reproduced and diffused as long as there are individuals with sufficient skills to understand them and use them. It is the dynamic interaction between skills and ideas that enables the accumulation of knowledge (for a detailed analysis on this subject, see Conceição and Heitor, 1999; Freeman and Soete, 1997), which in turn ultimately contributes to economic development.

In Nelson and Romer's (1996) framework, formal learning processes such as education (Lucas, 1988) and scientific research (Romer, 1990) are seen as processes that create ideas and develop skills. Education and research are thus critical to sustaining economic growth. Landes (1998) went further and argued that it is the growth of knowledge capacity that explains the per capita income increase in the most developed countries.

The role and importance of a country's knowledge capacity in explaining economic growth goes beyond the notion of accumulation only. Hall and Jones (1999) demonstrate that social infrastructure, including government policies and national institutions, are central factors explaining the existing differences among countries regarding knowledge creation as well as income per capita. Thus learning processes, or the ability to continuously generate skills and ideas, are not only central to sustaining economic growth (Bruton, 1998) but are also path-dependent, interactive, and influenced by the institutional setup of the economy and of the science and technology system (Johnson, 1992; Lundvall and Johnson, 1994). These interdependencies are especially critical because emphasis is increasingly being placed not on individual learning, but on collective and institutional learning processes (Wright, 1999).

In the knowledge accumulation process, the research and training provided by public institutions such as universities is central (Rosenberg, 2002; Pavitt, 1998). The importance of basic research produced by public institutions is also recognized as crucial to many industrial innovations (NAE, 2003). Moreover, these traditional suppliers of knowledge are joined by firms, not only as users of knowledge (Brooks and Randazzese, 1998; Cohen *et al.*, 1998; Nonaka and Takeuchi, 1995), but also increasingly as knowledge producers, a role they perceive as critical in their "race to innovate first" (Conceição *et al.*, 2004). This leads to the idea that a co-evolution does exist between scientific performance, technology, and the economy (Pavitt, 1996), reinforcing the rationale that academic research is needed to foster a country's capacity for technical change (Mowery and Rosenberg, 1998).

In this context, understanding the evolution of the scientific systems of nations is of critical importance. According to Price (1961) science, measured by scientific papers and journals, has been growing more rapidly than the population or the economy. Price's (1956) theory also suggests that the larger the size of the science system, the faster is its growth. However, Price (1961) notes that a rapid expansion of science will demand resources to such an extent that its development will eventually reach a state of saturation, thereby slowing down to a steady level. In this perspective, the evolution of a scientific system is characterized by three phases: a first phase with few increments to knowledge (infant phase); followed by one of purely exponential growth (adolescent phase), and a third or adult phase when it reaches a state of saturation (for a detailed account of this process, see Price, 1961; Ruivo, 1995). Scientific production can thus be suitably used to understand the evolution of science (Price, 1961). Moreover, since a relationship can be established between resources (researchers, S&T funding), national policies, and institutions (Lundvall and Johnson, 1994), it can also be employed to perceive countries' scientific structures (King, 2004). This postulate will be the focus of this work.

DATA AND METHODS

Although most papers dealing with cross-national scientific productivity compare large, medium, and small-sized countries at the same time (Doré and Ojasoo, 2001; Grupp *et al.*, 2001), this article deals with small and medium-sized countries alone. We take the term 'small and medium-sized countries' to mean countries with a population not larger than 25 million persons. We also assume the condition proposed by Price (1965; 1961), i.e. that the production of science is strongly associated with the size of a country's population. In addition to population we added another factor that enables us to identify countries with greater similarities: the human development index reported by the United Nations Development Program. This index, expressed as a value between 0 and 1, results from the computation of gross domestic production per capita, adjusted for national purchasing power, life expectancy at birth, gross enrollment ratio and adult literacy ratio (for a detailed explanation see UNDP, 2004).

In this context, a sample of countries was chosen based on their high human development index, with a mark above 0.850 for 2002 (UNDP, 2004). This sample is composed of eight European Union countries and two Asian countries, with differing economic performances as well as different science and technology structures and investment levels (OECD, 2004). These countries are Belgium, Czech Republic, Denmark, Finland, Ireland, Netherlands, Portugal, Slovenia, Singapore and Taiwan. Taiwan was the only country not chosen based on the United Nations Human development index, but added instead due to the size of its population.

In order to understand the scientific productivity and scientific structures of these countries, we used the ISI Thomson National Science Indicators On Diskette (NSIOD-2003) database, which includes the number of published research papers and reviews and respective citations attributed to a country (so long as one of the authors has an address in that country). The Thomson ISI databases are recognized as having several shortcomings (see May, 1997 for a discussion on this subject). But despite their flaws, they have a good coverage of most scientific fields at a worldwide level. Thomson ISI databases are used extensively in both governmental reports and scientific literature (BIE, 1996).

The analysis presented in this paper is based on the work of the Australian Bureau of Industrial Economics (BIE) in their analysis of the reported decline of the impact of Australian scientific production (BIE, 1996), as well as research by May (1997) and, more recently, by King (2004). This paper applies methodological concepts and analysis instruments similar to those used in the aforementioned studies. For example, concepts such as wealth intensity, referring to GDP per capita, are based on the paper by the latter author. Other concepts, such as specialization of science structure or impact relative to field, used in the BIE (1996) report, are explained below.

The specialization of a country's scientific structure by disciplinary field not only reveals the scientific development of a country, but also indicates how the scientific structure is associated with the country's economic activity. Firstly, countries with a developed science base typically have broader scientific structures (King, 2004). This provides a wide knowledge base, which is an important aspect when it is hard to foresee where new scientific developments will come from (BIE, 1996; Gibbons *et al.*, 1994). Secondly, the BIE (1996) results demonstrate that scientific structure is associated with the main economic activities of a country. For example, the scientific structures of Australia, Canada or Chile are clearly natural resource based, focusing on agricultural sciences, while the scientific structures of India, Singapore or Taiwan are associated with an industry-based economy focusing on engineering and natural sciences – as this paper confirms. The specialization of a country is calculated first by dividing the number of publications of a country in a given year in a disciplinary field by the average of the total publications of the sample countries in the same year for the same disciplinary field. Then, the variance of this measure is calculated across the disciplinary fields of each country, providing an overall measure of specialization. In examining the fields, we further understand where such specialization lies.

The Impact Relative to Field (IRL) is an important indicator of the visibility and quality of a given field of science. State agencies (BIE, 1996) as well as researchers (May, 1997) often use this measure when performing bibliometric studies. The impact relative to a field refers to the citation impact for each country in a given field, divided by overall citation impact for the field worldwide. Whenever the value achieved by a country is higher than 1, then that country in that specific scientific field has an impact above the world average; if it has a value lower than 1, then its impact is below the world average. This measure for country i and disciplinary field j is calculated through the following formula, where P*ij* refers to the number of papers in country *i* in a disciplinary field *j*, C*ij* refers to citations, and *N* refers to all the countries of the world:

$$IRL_{ij} = (C_{ij} / \sum P_{ij}) / (\sum^{N}_{j=1} C_{ij} / \sum^{N}_{j=1} P_{ij})$$

The disciplinary fields used in this paper are based on the classification of fields of science and technology developed by UNESCO and OECD. In order to facilitate comparative analyses, UNESCO and OECD developed standard procedures to analyze and compare scientific areas. OECD (2002) uses six broad areas: natural sciences, engineering and technology, medical sciences, agricultural sciences, social sciences, and the humanities. In this study we use these scientific fields, but aggregate the natural sciences with agriculture (in order to facilitate reading however, we only use the term natural sciences throughout the paper).

EVOLVING SCIENTIFIC PRODUCTION: COUNTRY PROFILES

Scientific production at the world level is growing at a very high rate. In the last 12 years, the number of publications produced per year in the world has increased 30%. According to King (2004), the largest countries are showing mixed growing trends regarding their publication and their world share of citations. As Figure 1 shows, small and medium-sized countries, chosen as a group for our analysis, provide a growing contribution to the overall world publication output. Their share of world publications has almost doubled since 1990, meaning that their annual average rate growth has been far superior to the overall world annual average rate growth. The increased share in citations attained by the sample group of countries also indicates that the international impact of scientific production produced in these countries also evolved considerably during the past decade.

The contribution made by our sample of small and medium-sized countries to overall world publication and citations is an important indicator of the increasing scientific relevance of these countries. As expected, some of the countries in the group have contributed more than others to this increasing world relevance. Table 1 shows that, in 2002, half of the sample countries, including the Netherlands, Belgium, Denmark, Finland and Taiwan, contributed disproportionately to this increasing share of worldwide publications and citations (77% of publications and 83% of citations). This difference reflects the considerable country specificities that led to the high concentration in the share of papers produced and citations received in the aforementioned five countries out of the total group of ten.

Figure 1
Percentage relative to world in publication output and number of citations

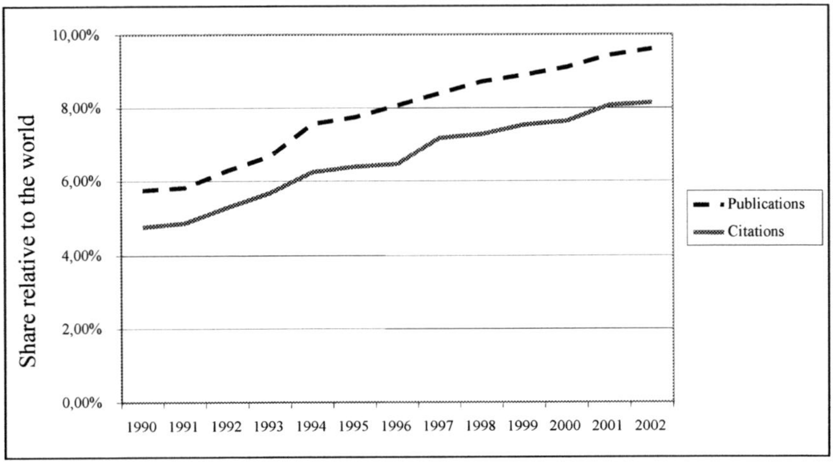

Source: Thomson's 2003 NSIOD – Deluxe database (ISI)

Table 1
Share of publications and citations in the total world output of the sample countries, 2002

	BEL	CZ REP	DK	FIN	IRL	NL	PT	SING	SLOV	TAI
Publications	14%	6%	11%	10%	4%	27%	5%	6%	2%	15%
Citations	16%	4%	13%	12%	4%	34%	4%	4%	1%	8%

The differences between the sample countries can also be perceived by looking at a set of economic performance indicators. Figure 2 compares a country's GDP per capita with publication and citation intensity, or in other words, the total papers or citations per million GDP. The analysis shows that, as the level of development improves, nations tend to improve their scientific impact, as measured by citations. Yet there is no distinguishing characteristic as to the placement of countries either above or below the average relative to the two indicators, neither whether using citations or publications as an output measure. As shown in King's (2004) paper, the Scandinavian countries and the Netherlands perform well in this measure, while the Asian countries and Portugal demonstrate the poorest performances.

Figure 2
Country GDP per capita and citation and publication intensity

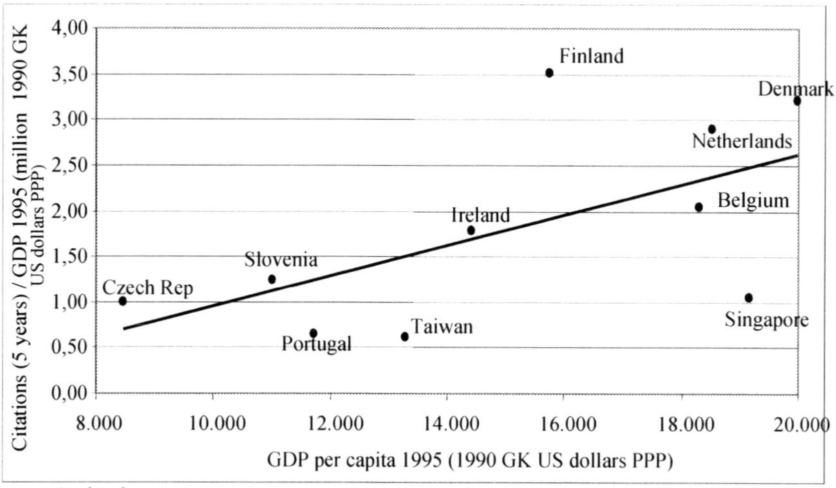

a) citations

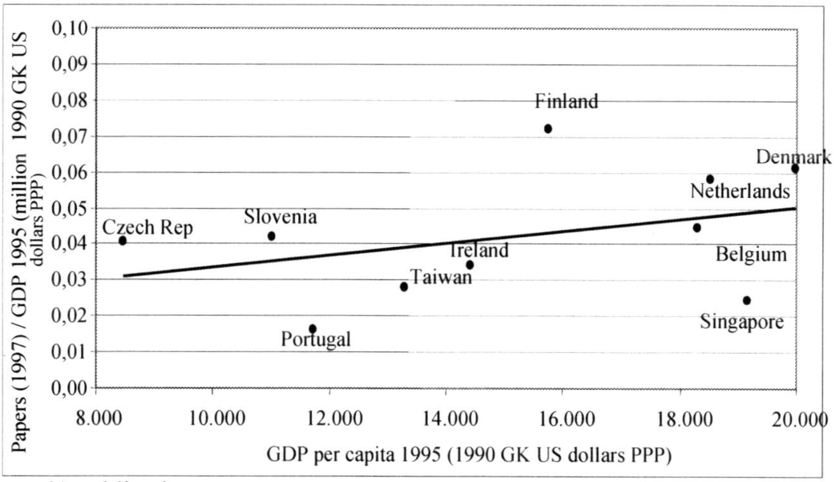

b) publications

Source: Groningen Growth and Development Centre and The Conference Board, Total Economy Database, January 2005, http://www.ggdc.net and Thomson's 2003 NSIOD – Deluxe database (ISI)

Notes: The citations refer to the overall citations in all papers during the years 1998-2002. Both GDP and GDP per capita refer to 1995, harmonized by the Geary-Khamis method of aggregation in order to achieve power purchase parity between the countries. The economic indicators refer to 1995, so that a minimum time lag of 4 to 6 years exists between the period of investment (1995) and the period of publication and exposition to peers reflected in the number of citations. The same rationale is used for the publications, using a two year lag, and therefore the publication data refers to 1997.

Another possible way to look at a country's scientific performance is to measure the total number of citations that the papers published by researchers in that nation received in a given year, and compare it against the country's investment in research and development. Following the recommendation of Grant *et al.* (1997) to have a time lag between the input measure and the time the papers produced and citations obtained are counted, Figure 3 shows the gross R&D investment per capita for 1995, and its return in terms of total citations per total papers produced during the period between 1998 and 2002. This time lag assumes two to four years from the date of investment until papers to be published, and two more years for them to reach their citation peak. Unlike Grant *et al.* (1997), we use GERD per capita as an indicator of scientific investment, instead of mere GERD (Gross Expenditure in Research and Development) as the input. The advantage of this measure is that it combines two input measures in one: the investment in R&D and the size of the population. On the output side, citations and paper production are also combined to provide a relative measure of the quality of the codified knowledge produced. Nevertheless, the figure still provides only a crude measure of the scientific return on investment in R&D potential

Figure 3
Relation between scientific intensity (GERD per capita) and outcome

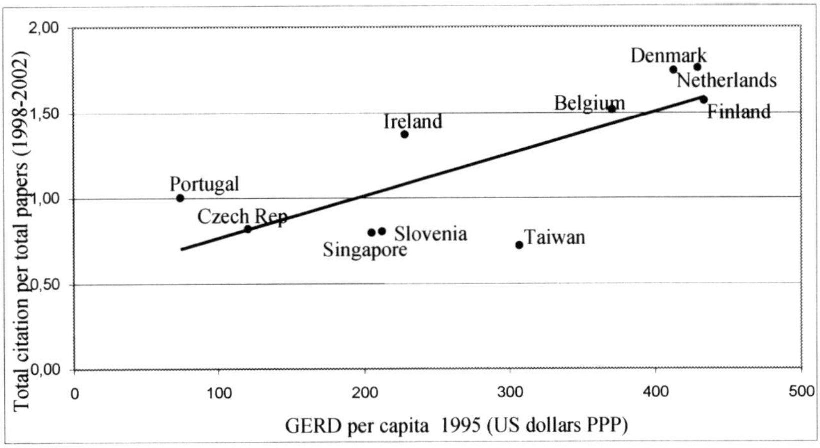

Source: OECD, MSTI (2003-1); Thomson's 2003 NSIOD – Deluxe database (ISI)

Note: total citations per paper refer to the division of the sum of all yearly citations by the sum of all yearly papers produced from 1998 until 2002, divided by the number of years.

The analysis of this figure leads to two observations. First, countries at the top in terms of GERD per capita also tend to have a higher citation rate (Netherlands, Finland, Denmark, Belgium). Yet in the case of the countries with a lower GERD per capita, such a positive correlation is less evident, with countries such as Taiwan, Singapore or Slovenia showing below average citations, even though they have larger investment in R&D per capita than countries such as the Czech Republic or Portugal. This suggests either that there is a minimum level of critical mass which makes a difference in terms of impact, or other factors are influencing the countries' scientific performance and create the variation around the overall positive relation between investment and impact.

Going beyond the above results, a deeper analysis can be conducted using scientific input, output, and quality indicators, aiming to better understand how different countries in the sample compare in terms of performance and efficiency. Figure 4 shows a radial graph with five distinct indicators. Two indicators can be considered as input indicators: Business Expenditure on Research and Development (BERD) as a percentage of GDP, and Government-financed Gross Expenditure on Research and Development (GERD) as a percentage of GDP. The use of these two indicators is associated with the idea that the production of codified knowledge assumes different forms and characteristics according to whether the funding sources are public (government financed) or private (BERD as a percentage of GDP) (David, 2000). The average number of papers published per researcher is an output indicator. It measures the productivity of the researchers of a country across all scientific fields. Finally, the figure includes two quality indicators: the average number of citations attained by a researcher and the impact relative to the world. The latter measure is the average of the impact relative to field (IRL) for all the scientific fields for a country in a given year (see the Section on Data and Methods).

The first observation concerning Figure 4 is that there are important asymmetries among countries. Since there are many countries in the graph, we decided to begin the analysis of Figure 4 by comparing two Scandinavian countries, Finland and Denmark, in order to illustrate the performance asymmetries among nations. Finland has two times more and Denmark 23% more business expenditure in R&D than the average of the sample countries. When comparing these two countries in terms of government funded GERD as percentage of GDP, Finland has a 46% larger investment than the average, while Denmark is just 12% higher than average. Although these two countries present different levels of public and private investment, these are not directly translated into outputs or outcomes as one might expect. Finland presents a low volume of publications per researcher (0.72), a value for citations close to average (0.92) and a result considerably above the mean (1.38) as far as world impact is concerned. On the other hand Denmark, with lower investments than Finland in both BERD and government funded GERD, shows a much better performance in all output and outcome indicators[2]. The lower performance of Finland when compared to Denmark in terms of scientific production cannot be related to a low investment in R&D. In 2002, Finnish GERD was 26% larger than the Danish, and it has been larger than Denmark's GERD since 1981.

---

[2] (publications: 1.41; citations per researcher: 1.89; impact relative to world: 1.46).

Figure 4
Performance measures of the S&T system for ten countries (2001 – 2003)

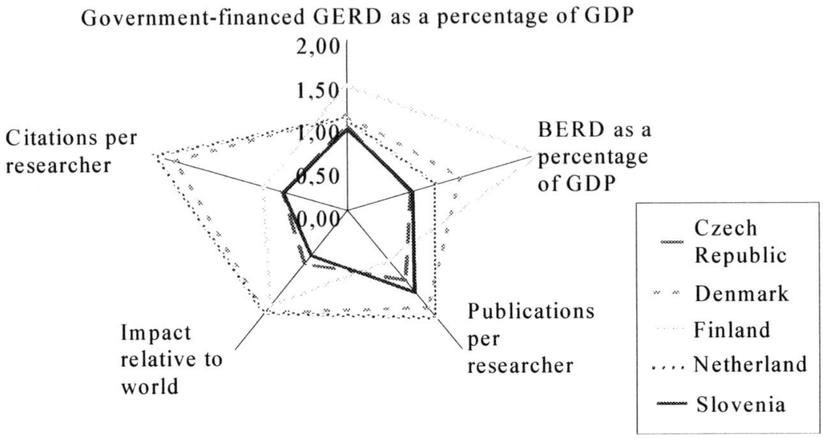

a) countries above sample countries' average for Figure 1 part b (publications)

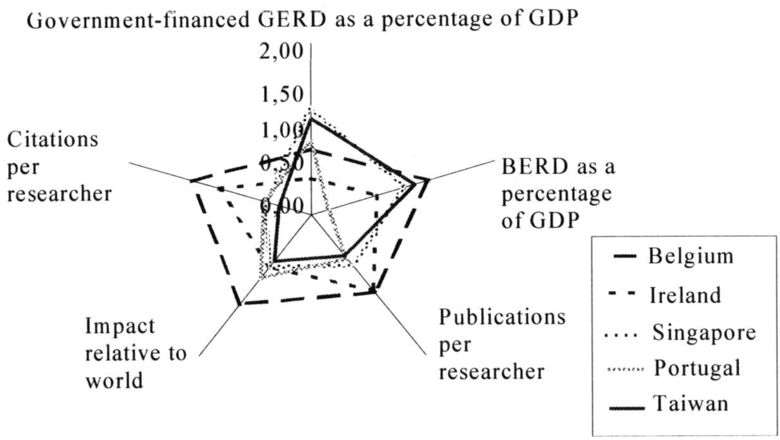

b) countries below sample countries' average for Figure 1 part b (publications)

Source: Thomson's 2003 NSIOD – Deluxe database (ISI).

Note: The average mean for the overall countries presented in the figure is 1. BERD as percentage of GDP: Finland: 2,08; Citations per researcher: Netherlands: 2,11.

In terms of GERD, both countries show a very similar system, dominated by the business sector, while the non-business sector share is roughly equally distributed among the universities and state laboratories. This suggests that differences in scientific production and quality are likely to be associated with each country's social infrastructure, namely their S&T policy, as suggested by Hall and Jones (1999).

Although Finland has a larger government funding structure in support of R&D when compared to Denmark, much of it is channeled to industry through the Finish Funding Agency for Technology and Innovation (TEKES). This suggests a strategy much more directed towards technological development for direct economic purposes rather than scientific production aimed at the international scientific community (TEKES, 2006). Taiwan is similar, with a relatively large GERD, and above average BERD as well as government financed GERD. Yet Taiwan shows poor performances on all scientific output measures. The case of Taiwan is much more striking than Finland, because Taiwan's scientific production and quality results are below the performance of countries with low levels of R&D investment such as Portugal or the Czech Republic (OECD, 2004). The emphasis given by Taiwan and Finland towards technological development is reinforced by their patenting activity. In 2000, from the 9574 patents granted by the United States Patent and Trademark Office (USPTO) to the sample group of countries in all fields of knowledge, 54% were given to Taiwan and 10% to Finland, respectively the first and the third countries with the largest share of patents granted by the USPTO to the sample countries for that given year[3] (OECD, 2004).

Portugal, Ireland, Slovenia, and the Czech Republic are the countries with the lowest investment in R&D. But the scientific performance within this group is still varied. Ireland has values for publications and citations per researcher around the mean of the group average, while the Czech Republic and Slovenia only reach the mean in publications per researcher. These two countries overlap in R&D investment, outputs, and quality indicators, suggesting a similar development in the overall S&T system for both countries. This perception is confirmed by the similarity as regards GERD (OECD, 2004). Portugal has lower values than this group of countries in both production of papers and citations per researcher, while its scientific production impact is higher than Ireland, Slovenia, or the Czech Republic. Portugal is the only country where the state, rather than the business sector, is the main source of R&D funding. It is also the only country in the group where most of the R&D is performed in universities instead of firms.

The analyses of the previous figures already provide some insights into the different countries' scientific structures. It should be noted, however, that they only provide a static view of their systems. Since R&D systems are dynamic and continuously evolving (Ziman, 1994; Price, 1965), a longitudinal analysis of scientific production and output is important in order to understand their developmental path. Figure 5 includes paper production evolution divided by the GDP (A), and citation evolution divided by the GDP (B), from 1981 until 2003.

---

[3] The Netherlands had the second largest share of patents granted by the USPTO in 2000.

Figure 5
Publications and citations evolution for the sample countries, 1981-2003

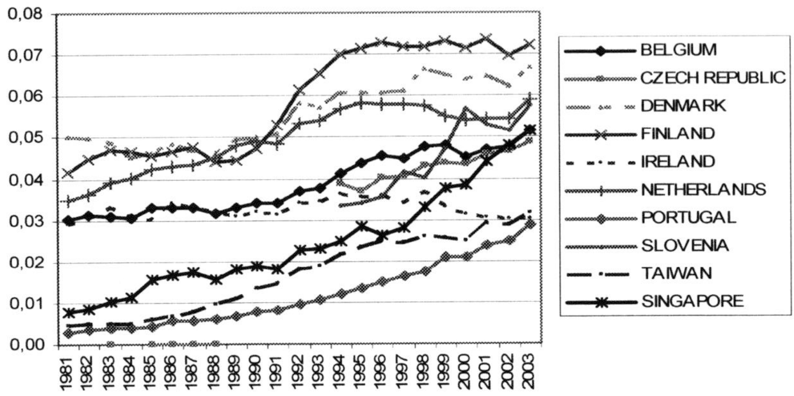

A - Total papers produced divided by GDP, 1981-2003

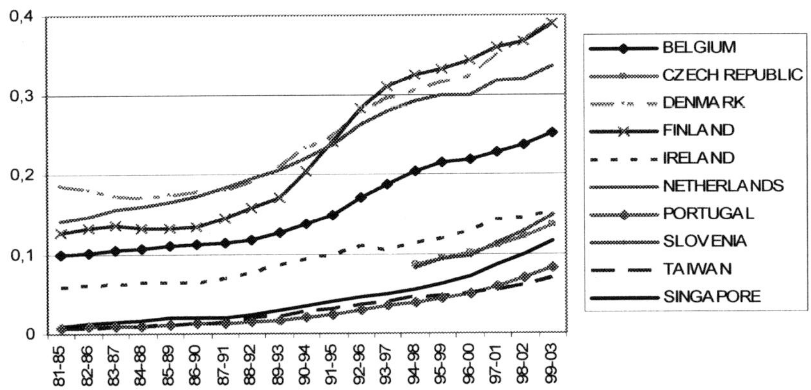

B – Total citations received for overall fields in 5 year overlapping periods, divided by GDP, for 1981-2003

Source: Groningen Growth and Development Centre and The Conference Board, Total Economy Database, January 2005, http://www.ggdc.net and Thomson's 2003 NSIOD – Deluxe database (ISI)

Figure 5 reinforces the argument that small and medium-sized countries' scientific production is on the rise. Finland and Denmark are the countries that, since the early 1990s, have the highest performance in terms of both total publications and citations per GDP, although the Netherlands also performs well on both indicators. Yet it is Singapore's growth in publications that stands out the most. This is especially relevant because its scientific production per GDP started out within that group of countries with lowest volume of scientific production, and evolved into placement among countries such as Belgium or the

Czech Republic, with an intermediate level of productivity. Other countries with a substantial growth rate (measured in CAGR since 1981)[4] in terms of this indicator are Portugal (10.37%) and Taiwan (8.41%), although both still exhibit a below average performance. The growth rate of these countries is much more moderate in terms of total citations divided by GDP, where they continue to demonstrate a poor performance when compared with most scientifically developed countries such as the Scandinavian countries.

As was the case in Figure 4, the Czech Republic and Slovenia present similar trends both in terms of publications and citations divided by GDP. They are clearly above average and growing fast in the former category, while remaining below average in citations divided by GDP. Ireland is the only country that declined in terms of publications divided by GDP. Nevertheless, it is important to note that this is mostly driven by its much stronger GDP growth when compared to the evolution in publications. Yet Ireland's citations grew dramatically in this period. In 99-03 an Irish paper was cited on average almost 5 times, while in 86-90, an Irish paper was cited only twice.

Using Figure 5B as a reference in terms of number of citations, two distinct groups of countries can be identified. The above average group of countries, when considering their 81-85 trend, can be considered as the group of the most scientifically advanced countries. This group is composed of Finland, Denmark, the Netherlands and Belgium. All the other countries form the below average group and thus are still developing their scientific position.

So far, we have examined the sample countries' scientific production and quality in an aggregated form. Disaggregating the scientific areas provides a more detailed view of the countries' scientific production and quality in specific areas. This will be presented in the following sections.

SCIENTIFIC STRUCTURES AMONG SMALL AND MEDIUM-SIZED COUNTRIES

The analysis of the structure of scientific production by disciplinary fields provides insights into the scientific strengths of each country's scientific system (King, 2004; BIE, 1996). Figure 6 once again displays important asymmetries among countries. Firstly, we can observe that the publication structure of Asian countries is specialized. Taiwan presents a highly specialized publication structure, focused on natural sciences and engineering, while Singapore shows a less specialized publication structure based on engineering. The Netherlands and Belgium also present a specialized publication structure (more specialized than Singapore), but with different areas of strength. Belgium's strength is the humanities, while the Netherlands is focused on medicine and social sciences.

---

[4] The compound annual growth rate is a method of assessing the average growth of a value over a specific period of time. It is calculated by taking the nth root of the total percentage growth rate, where n refers to the number of years in the considered period. The Compound annual growth rate can be calculated using the following formula: CAGR = $(A/B)^{(1/T)}$ -1, where A refers to the ending value, B to the beginning value, and T is the number of years between the beginning and the ending values.

Figure 6 a & b
Scientific production structure and levels of specialization, 1994 and 2003

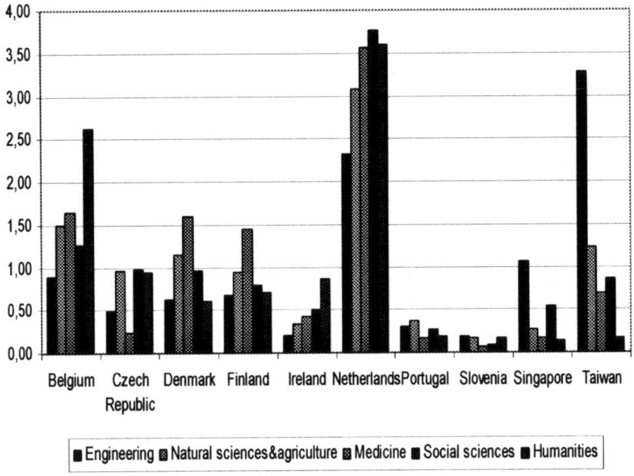

a) scientific production structure 1994

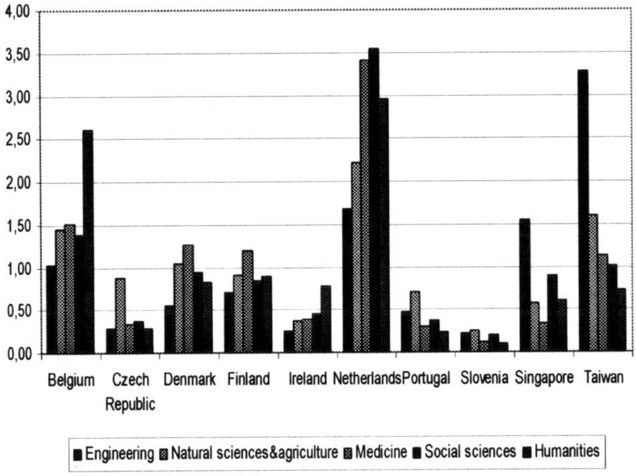

b) scientific production structure 2003

Figure 6 c
Scientific production structure and levels of specialization, 1994 and 2003

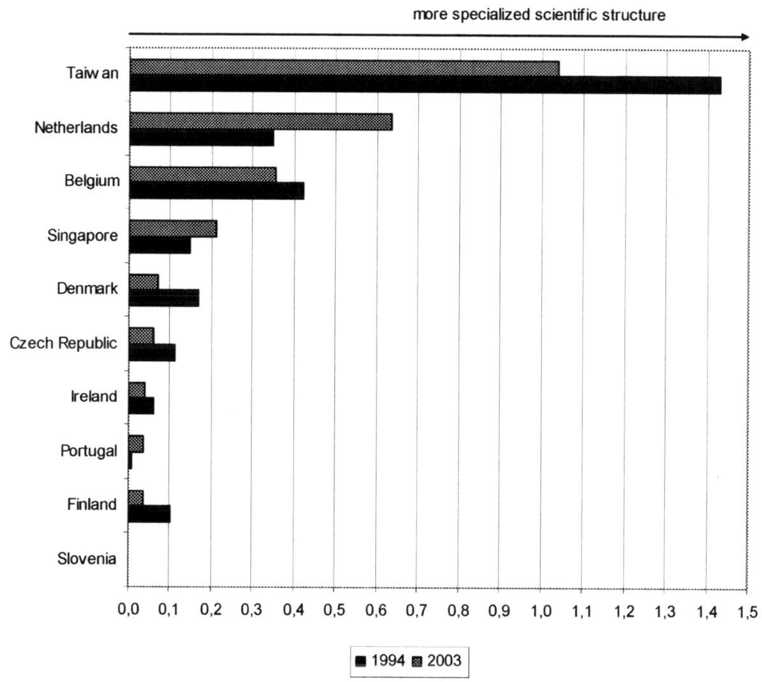

c) scientific structure specialization

Source: Thomson's 2003 NSIOD – Deluxe database (ISI)

All other countries present broader structures, albeit with different focuses. Denmark's paper production is particularly intensive in medicine, natural sciences and social sciences, while Finland's strong point is in medicine; Ireland is strong in the humanities and social sciences, Portugal in natural sciences and engineering, the Czech Republic in natural sciences, and Slovenia in natural sciences, engineering, and social sciences.

In terms of systemic evolution, only Slovenia demonstrates a stable scientific production structure when comparing 2003 with 1994. All other countries have witnessed changes in their scientific production structures. The Netherlands, Singapore, and Portugal had less specialized scientific production structures in 1994, while all other countries had more specialized scientific structures in that year. The scientific structure in terms of the prominence of scientific fields has remained practically unchanged in most of the countries. The exception is the Czech Republic, where the level of paper production in social sciences and the humanities was considerably reduced, and natural sciences became the most prominent field in terms of scientific production. The Dutch researchers publish the most in all scientific fields except for engineering, where Taiwan leads by far.

Scientific impact: who is who in terms of scientific quality?

The scientific structure of a country indicates its ability to produce scientific output, and concomitantly, its strength in producing codified knowledge. However, an analysis of scientific production which focused only on the number of published papers would be incomplete without taking into consideration the issue of quality. Although it is known that in order for a paper to be published in a journal it must undergo a process of peer-review and acceptance by peers – a central process in the scientific system since at least the seventeenth century (Zukerman and Merton, 1971) – it is the citation of work by colleagues and peers that usually confirms the validity and usefulness of the produced codified knowledge (Small, 1978). In this context, we measured the quality of the papers produced by a country in a given disciplinary field by using the impact relative to field (IRL) measure, which has been earlier described in the data and methods section.

Figure 7 shows the countries' IRL for the period 1991-1995 and 1999-2003. The first observation that one can draw is that the quality of the knowledge produced by the small and medium-sized countries, measured in terms of citations, is growing *vis* a *vis* the world[5]. Comparing the 1999-2003 period with the 1991-1995 period in terms of aggregate impact, we see an increase of the sample countries in relation to the world impact per scientific field. If the impact of the group is added together and compared in the two periods, one can note that the impact of these countries increased by 71% in the humanities, 46% in the social sciences, 39% in medicine, 23% in the natural sciences and agriculture, and 22% in engineering[6].

When comparing the two time periods chosen, two additional trends can be identified. The first relates to the path towards a more even IRL among all scientific fields (following a similar trend towards a broader scientific structure by most countries as reported in the analysis of Figure 6); while the second refers to the differing impact between countries. In the period from 1995 to 2003, there is a clearer delineation, in terms of impact, between that group of countries that includes Netherlands, Belgium, Denmark and Finland and all the other sample countries, all of which demonstrated a much weaker impact across scientific fields.

---

[5] This result was already implicit in the analysis of Figure 1, which demonstrated that the share of citations of the sample of small and medium-sized sample countries was growing at a faster rate than the overall world growth in citations.

[6] These values are over estimated due to the fact that the Czech Republic and Slovenia only started to compile their scientific production and citation data in 1994 and 1993 respectively.

Figure 7
Impact Relative to Field, 1991-1995 and 1999-2003

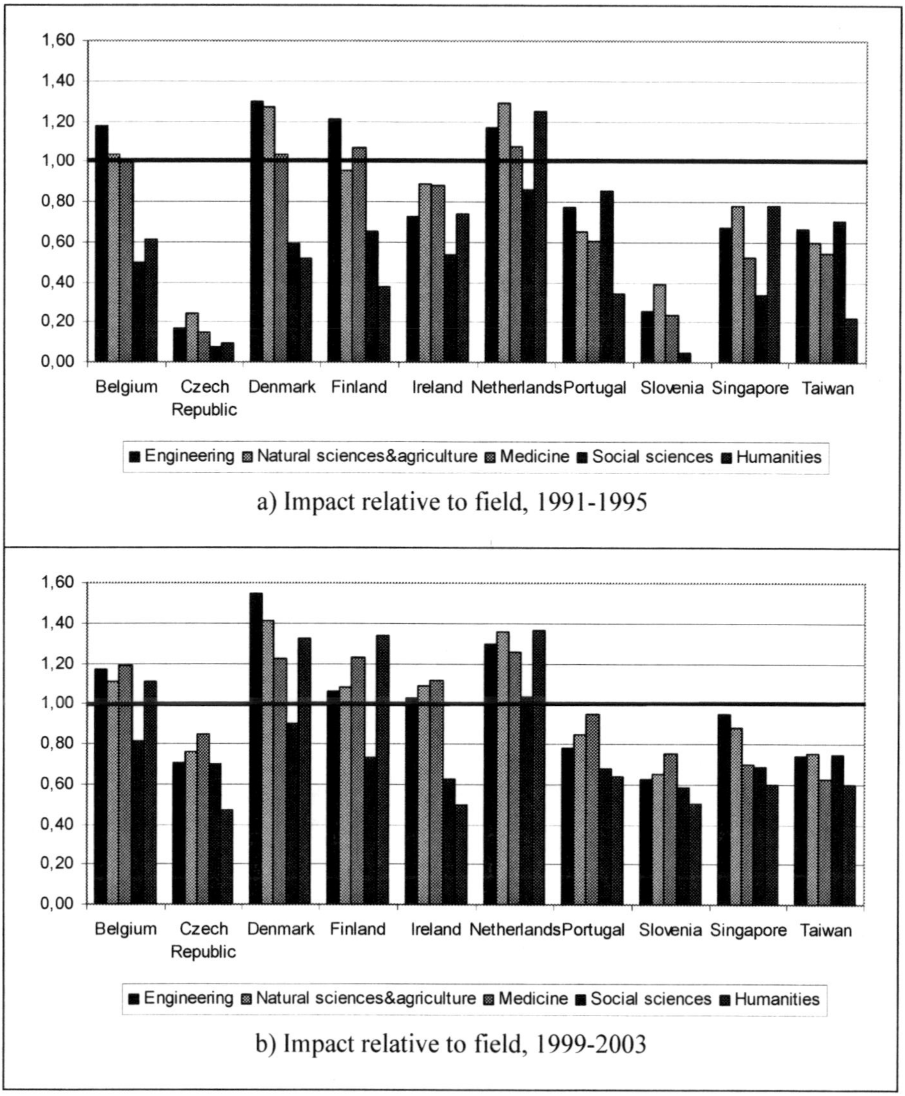

a) Impact relative to field, 1991-1995

b) Impact relative to field, 1999-2003

Source: Thomson's 2003 NSIOD – Deluxe database (ISI)

When examining the sample countries as a group, the Netherlands demonstrates a greater impact, compared to the world base impact, in humanities (1.36), medicine (1.26) and social sciences (1.04), while Denmark has a better performance in engineering (1.55) and the natural sciences (1.41). Netherlands is the only country in the group that has an above average IRL in all the scientific fields. Since the 91-95 period all scientific fields increased their IRL, especially in the social sciences, which evolved from 0.86 (below average) to 1.04. Denmark has an above average IRL in all scientific fields except the social sciences, although it has a near average value (0.90) therein. All scientific fields also quite substantially increased their impact, especially the social sciences and the humanities. Like Denmark, Belgium's IRL impact is above average in all scientific fields except the social sciences. In terms of fields, the humanities (0.61 to 1.11) and the social sciences (0.50 to 0.82) increased their impacts substantially since the 1991-95 period, while Medicine and the natural sciences slightly increased their impacts.

Finland's strongest IRL is in the humanities (1.34), where its value is similar value to that obtained by the Netherlands. Like Belgium and Denmark, Finland has an above average IRL in all scientific fields except the social sciences. Since 91-95, the Finnish impact in humanities has tripled, which makes it the major growing scientific field in terms of impact. All of the other scientific fields reported growth (all around 13%), with the exception of engineering, which registered a decrease. Ireland presents a variation among the so-called "hard" and "soft" sciences. Scientific fields such as medicine (1.12), the natural sciences (1.09) and engineering (1.03) attain values slightly above the IRL average, while the social sciences and the humanities have impact values only around half of the IRL's world average. Since the 91-95 period, its impact across all scientific fields grew, with the exception of the humanities.

All the other small and medium-size countries that comprise the sample group had a below average IRL in all scientific fields. Engineering (0.95) is Singapore's strongest scientific, while it is weakest in the humanities (0.60). Compared to the earlier sample period, all scientific fields increased their IRL except for the humanities. Like Singapore, Taiwan has a below average IRL in all scientific fields, albeit a more uniform one, as the variation between all the scientific fields in Taiwan is no more than 0.15. Surprisingly the social sciences, which is the least productive scientific field, has an impact similar to that of the natural sciences and engineering. Thus for some scientific fields in Taiwan there seems to be no relation between the increasing scientific production and the IRL obtained. All scientific fields improved their IRL, but the humanities registered the most significant growth (almost tripling).

Portugal's highest IRL is in medicine (0.95). Other scientific fields with considerable impact are the natural sciences and engineering, 0.85 and 0.78 respectively. The impact of the social sciences and the humanities impact is weaker, and similar in size (0.68 and 0.64). Some changes were noted since 91-95. At that time, the social sciences was the field with the highest IRL in Portugal, but in the intervening years this was the only scientific field in which Portugal's impact decreased. Others such as engineering maintained their IRL while some, such as the humanities, increased their IRL.

The two Eastern European countries also present a below average IRL in all scientific fields. But in their case we must limit our analysis to the 99-03 period, because data was not collected for these countries prior to 1993. In the Czech Republic, medicine, the natural sciences, engineering, and the social sciences present a similar impact range, between 0.85 (medicine) and 0.70 (social sciences). The only exception is the humanities (0.47). Slovenia has a similar IRL structure to Portugal, although with a weaker impact in all scientific areas. Medicine is the scientific area with most impact (0.75), followed by the natural sciences and engineering (0.65 and 0.62), while the social sciences (0.59) and the humanities (0.50) reveal weaker impact indicators.

DISCUSSION

Small and medium-sized countries are making valuable contributions to the world's overall scientific production. This is demonstrated by their growing shares in terms of both papers and citations. However, the contributions of these countries differ. Countries with the highest relative GERD tend to have higher citation rates, but funding efforts do not uniformly correlate to scientific production or quality. The empirical evidence gathered in this paper supports this conclusion. Of all countries studied, Taiwan has had the highest GERD since the 1990s, yet it demonstrates a comparatively poor performance in the overall indicators of scientific knowledge output and quality. Countries with a low GERD, such as Portugal or Ireland, reflect better knowledge production and quality performance than Taiwan. This suggests that the institutional structure is also important and that countries with both high and low R&D funding levels may demonstrate a wide range in terms of performance (Conceição and Heitor, 2005). Although investment into science is important, it is the process of knowledge production and accumulation, or in other words learning, that matters the most. The structure of the innovation and scientific system seems to be as important as the investment in science itself. It is not only important to accumulate knowledge, but it is also critical to understand how that knowledge is accumulated and used to lead to new knowledge (Kuhn, 1970). The scientific evolution of countries and their institutional policies and research framework, discussed below, reinforces this argument.

Taiwan is in a state of scientific system adolescence (Price, 1965), characterized by its impressive gross investment in R&D and rapidly growing scientific production, while demonstrating a lower citation impact. But Taiwan favors technology over science policy, which may be the reason for its lower citation rate. Taiwan's science and technology policies have been mainly directed towards technological development (Jayakar, 2003) of an industry-based economy (BIE, 1996). This is further confirmed by Taiwan's highly specialized scientific structure in favor of the natural sciences, and especially the engineering, fields. The accent on these scientific fields is associated with Taiwan's patenting activity, which is mainly focused on technology areas such as semiconductors and electronics, electrical appliances and components, and industrial machinery and tools, all of which require strong knowledge in the natural sciences and engineering fields (Hu and Mathews, 2005).

Finland has a different profile. This country has the highest BERD and government financed GERD among the sample group, yet its scientific production and quality indicators are lower than that of countries such as

Denmark, which have a lower GERD. This relatively lower performance in terms of scientific knowledge production and quality can also be explained by a S&T policy directed towards technological development (Kivinen and Rinne, 1996), similar to Taiwan. However unlike Taiwan, Finland has a relatively broader scientific structure, not only it terms of its scientific production, but also with regard to its citation impact in the various scientific fields. This aspect, together with Finland's consistent impact in the natural sciences and decrease in engineering, may indicate a change in its science policy, going from applied to basic fields, namely the Biotechnology sector (Smaglik, 2004; Reiss *et al.*, 2004). Nevertheless, due to the above average impact relative to field in all (or nearly all) scientific fields Finland, along with Netherlands, Belgium and Denmark, can be considered to belong to the group of scientifically mature countries within the sample. These countries are also those with the strongest economies. This comes as no surprise, since economic progress is closely related to scientific development – no scientific system can develop itself without resources (Price, 1961). It is the continuous funding of science that sustains the consolidation of new research institutions, the reform of old research institutions, the building up of infrastructures[7], and the refinement of research and development incentives (Conceição *et al.*, 2003).

Singapore has a GDP per capita similar to Taiwan, Finland and Denmark, but it is clearly an outsider in terms of the nature of its science system. Its relatively weak GERD, at the level of countries such as the Czech Republic or Portugal, as well as its below average impact relative to all scientific fields, distinguishes Singapore from the other countries in the sample with the highest economic intensity. Nonetheless, Singapore is in a rapid catching-up phase, both in terms of its GERD as well as its scientific production and impact, which are both growing at very fast rates.

The Netherlands, Belgium, and Denmark are the most economically and scientifically mature countries in the study. Their R&D performance structures illustrate the differences between them. In the Netherlands, the business sector does not have such a dominant position as it has in Belgium or Denmark[8], while the higher education sector in the Netherlands has a strong position. Higher education institutions are the pillars of basic research and the main source of scientific production in today's knowledge society (Rosenberg, 2002). The higher education sector is also the R&D performing sector that produces most of the codified knowledge, through scientific articles. In the Netherlands, higher education institutions perform 27% of all research, while higher education institutions perform only 19% in Belgium and Denmark. However, more scientific production does not necessarily translate into a better impact relative to field. Denmark, for instance, has a higher impact relative to field than Netherlands in the natural sciences and engineering fields. The impact of

---

[7] In this context, infrastructure refers to the potential to create capabilities that arise from the general technical structure, human capital and physical assets (see Lall, 1992)

[8] The business expenditures in research and development (BERD) are 58% in the Netherlands, 74% in Belgium, and 69% in Denmark.

Denmark is reflected in its relatively weak production yet high citation rates, suggesting that Danish researchers do not produce so much, but what they do produce in these scientific fields is of high scientific quality as measured via citations.

In the scientifically developing countries such as Portugal, the Czech Republic, and Slovenia, scientific performance and structures are also closely associated with their scientific institutional systems. The Czech Republic and Slovenia are countries with a similar scientific structure in both business and government sectors. Their scientific production structure is also rather similar. The Czech scientific production grew at a faster rate than Slovenia, but the citation impact is similar for both countries. Both nations also reflect a broad scientific production structure, although Slovenia has several scientific fields that are quite strong. These similarities suggest a particular pattern of scientific development that is typical for Eastern European countries, which are emerging from years of scientific closure and therefore carrying out a process of institutional change (Radosevic and Auriol, 1999). Portugal is the only country in the sample where the higher education system is the main performer of R&D. However, unlike the Netherlands, its S&T and higher education system are both still in a process of capacity building in all scientific fields. The system is still hampered by uneven funding and low investment in S&T (Heitor and Horta, 2004). This has not prevented Portugal from achieving reasonable results in some scientific fields. In particular, medicine in Portugal is approaching the world average in terms of relative impact.

In summary, continuous support for R&D activities is important, but not enough to assure research competitiveness in terms of scientific publication and quality. Infrastructure, resources and institutions play a central role (Conceição *et al.*, 2003). The scientific production of countries is also associated with the structure of the science and technology system. Taiwan's specialized structure, towards the engineering and the natural sciences, reflects a science structure built up to support an industry-based economy and a specific technological paradigm. The same is true for Finland. Although this country has a much broader scientific structure than Taiwan, Finland is now favoring biotechnology and medicine as potential economic drivers of the country's overall economy. The divergent patterns for the scientific structure of these countries, towards either specialized or broader structures, also suggest that they are adapting their scientific structures to their socio-economic needs and S&T policies.

BIBLIOGRAPHY

Adams, J., (1998), "Benchmarking International Research", *Nature*, Vol. 396, pp. 615-618.

Arrow, K., (1962), "The Economic Implications of Learning by Doing," *Review of Economic Studies*, 28, pp. 155-73.

Bach, L., and L. Georghiou, (1998), "The Nature and Scope of RTD Impact Measurement", discussion paper for the International Workshop on Measurement of RTD Results/Impact, Brussels, 28-29 May 1998.

Bach, L., and M. Matt, (2005), "From Economic Foundations to S&T Policy Tools: a Comparative Analysis of the Dominant Paradigms", [in:] P. Lerena, and M. Matt, (eds), *Innovation Policy in a Knowledge-Based Economy*, Berlin, Springer.

Basu, A., and R. Aggarwal, (2001), "International Collaboration in Science in India and Its Impact on Institutional Performance", *Scientometrics*, Vol. 52, pp. 379-394.

Bayers, N. K., (2005), "Using ISI Data in the Analysis of German National and Institutional Research Output", *Scientometrics*, Vol. 62, 155-163.

Bordons, M., I. Gomez, M. T. Fernandez and A. Mendez, (1996), "Local, Domestic and International Scientific Collaboration in Biomedical Research", *Scientometrics*, Vol. 37, pp. 279-295.

Brooks, H., and L. P. Randazzese, (1998), "University-Industry Relations: The Next Four Years and Beyond", [in:] L. M. Branscomb and J. H. Keller, (eds.), *Investing in Innovation*, Cambridge, MIT Press.

Bruton, H. J., (1998), "A Reconsideration of Import Substitution", *The Journal of Economic Literature*, XXXVI (June), pp. 903-936.

Bureau of Industry Economics (BIE), (1996), *Australian Science: Performance from Published Papers*, Canberra, Australian Government Publishing Service.

Cohen, W., R. Florida, L. Randazzese, and J. Walsh, (1998), "Industry and the Academy: Uneasy Partners in the Cause of Technological Advance", [in:] R. Noll, (ed.), *Challenges to Research Universities*, Washington DC, Brookings Institution Press.

Conceição, P., and M. V. Heitor, (2005), *Innovation for All? Learning from the Portuguese Path to Technical Change and the Dynamics of Innovation*, London: Praeger.

Conceição, P. and M. V. Heitor, (1999), "On the Role of the University in the Knowledge Economy", *Science and Public Policy*, 26(1), pp. 37-51.

Conceição, P., M. V. Heitor, G. Sirilli, and R. Wilson, (2004), "The 'Swing of the Pendulum' from Public to Market Support for Science and Technology: Is the U.S. Leading the Way?", *Technological Forecasting and Social Change*,71, pp. 553-578.

Conceição, P., M. V. Heitor and F. Veloso, (2003), "Infrastructures, Incentives and Institutions: Fostering Distributed Knowledge Bases for the Learning Society", *Technological Forecasting and Social Change*, 70, pp. 583-617.

David, P., (2000), "A Tragedy of the Public 'Commons'? Global Science, Intellectual Property and The Digital Technology Boomerang", Stanford Institute for Economic Policy Research (SIEPR) Discussion Paper No. 00-02, October.

Doré, J.-C. and T. Ojasoo, (2001), "How to Analyze Publication Time Trends by Correspondence Factor Analysis: Analysis of Publications by 48 Countries in 18 Disciplines over 12 Years", *Journal of the American Society for Information Science and Technology*, 52, 9, pp. 763-769.

Freeman, C. and L. Soete, (1997), *The Economics of Industrial Innovation*, Cambridge, MIT Press.

Gibbons, M., C. Limoges, H. Nowotny, P. Scott, S. Schwartzman, M. Trow, (1994) *The New Production of Knowledge – the Dynamics of Science and Research in Contemporary Societies*, London, Sage Publications.

Grant, J., G. Lewison and R. M. May, (1997), "Government Funding of Research and Development", *Nature*, Vol. 278, pp. 878-880.

Grupp, H., U. Shmoch and S. Hinze, (2001), "International Alignment and Scientific Regard as Macro-indicators for International Comparisons of Publications", *Scientometrics*, 51, 2, pp. 359-380.

Gu, Y., (2004), "Global Knowledge Management Research: a Bibliometric Analysis", *Scientometrics*, Vol. 61, pp. 171-190.

Guimarães, P., (2002), "The State of Portuguese Research in Economics: an Analysis Based on Publications in International Journals", *Portuguese Economic Journal*, 1, pp. 3-25.

Hall, R. E. and C. I. Jones, (1999), "Why do Some Countries Produce so Much More Output per Worker than Others?", *Quarterly Journal of Economics*, 114(1), pp. 83-116.

Heitor, M. V., and H. Horta, (2004), "Engenharia e desenvolvimento científico: o atraso estrutural português explicado no contexto histórico", [in:] M. V. Heitor, J. M. B. Brito, M. F. Rollo, (eds), *Momentos de Inovação e Engenharia em Portugal no seculo XX*, Lisboa, D. Quixote.

Hu, M.-C., and J. A. Mathews, (2005), "National Innovative Capacity in East Asia", *Research Policy*, 34, pp. 1322-1349.

Jayakar, K. P., (2003), "Cross-sectoral Differences in Intellectual Property Enforcement in Developing Countries: the Role of State-industry Linkages", *Information Society*, 19, pp. 155-169.

Johnson, B., (1992), "Institutional Learning", [in:] B.-A. Lundvall (ed.), *National Systems of Innovation – Towards a Theory of Innovation and Interactive Learning*, London, Pinter Publishers.

Landes, D., (1998), *The Wealth and Poverty of Nations: Why Some are so Rich and Some so Poor*, New York, W. W. Norton & Company.

Lucas, R. E., (1988), "On the Mechanics of Economic Development", *Journal of Monetary Economics*, 22, pp. 3-42.

Lundvall, B.-A. and J. L. Christensen, (2003) "Broadening the Analysis of Innovation Systems – Competition, Organisational Change and Employment Dynamics in the Danish Model", [in:] P. Conceição, M. V. Heitor and B.-A. Lundvall (eds), *Innovation, Competence Building, and Social Cohesion in Europe-Towards a Learning Society*, London, Edward Elgar.

Lundvall, B. –A. and B. Johnson, (1994), "The Learning Economy", *Journal of Industry Studies*, 1/2, pp. 23-42.

King, D. A., (2004), "The Scientific Impact of Nations", *Nature*, Vol. 430, pp. 311-316.

Kivinen, O. and R. Rinne, (1996), "The Problem of Diversification in Higher Education: Countertendencies Between Divergence and Convergence in the Finnish Higher Education System Since the 1950s", [in:] V. L. Meek, L. Goedegebuure, O. Kivinen, and R. Rinne (eds), *The Mockers and Mocked:*

Comparative Perspectives on Differentiation, Convergence and Diversity in Higher Education, Oxford: Pergamon.

Kuhn, T. S., (1970), *The Structure of Scientific Revolutions*, Chicago, The University of Chicago Press.

May, R. M., (1997), "The Scientific Wealth of Nations", *Science*, Vol. 275, pp. 793-796.

Mowery, D. C., and N. Rosenberg, (1998), *Paths of Innovation - Technological Change in the 20th Century America*, Cambridge, Cambridge University Press.

National Academy of Engineering, (NAE), (2003), *The Impact of Academic Research on Industrial Performance*, Washington, NAE Press.

National Science Indicators, (NSB), (2005), *Science and Engineering Indicators*, Arlington, National Science Foundation.

Nelson, R., (1991), *National Innovation Systems*, Oxford, Oxford University Press.

Nelson, R. R. and P. Romer, (1996), "Science, Economic Growth, and Public Policy", [in:] B. L. R. Smith and C. E. Barfield, (eds), *Technology, R&D, and the Economy*, Washington, D.C., Brookings.

Nelson, R. R. and S. G. Winter, (1982), *An Evolutionary Theory of Economic Change*, Cambridge, Harvard University Press.

Nonaka, I., S. Sasaki and M. Ahmed, (2003), "Continuous Innovation in Japan: The Power of Tacit Knowledge", [in:] L. V. Shavinina (ed.), *International Handbook on Innovation*, Oxford, Elsevier.

Nonaka, I. and H. Takeuchi, (1995), *The Knowledge Creating Company*, Oxford, Oxford University Press.

OECD, (2004), *Main Science and Technology Indicators 2004*, Paris, OECD.

OECD, (2002), *Frascatti Manual – Proposed Standard Practice for Surveys on Research and Experimental Development*, Paris, OECD.

Osareh, F. and C. S. Wilson, (2000), "A Comparison of Iranian Scientific Publications in the Science Citation Index: 1985-1989 and 1990-1994", *Scientometrics*, Vol. 48, pp. 427-442.

Pavitt, K., (1998), "The Social Shaping of the National Science Base", *Research Policy*, 27(8), pp. 793-805.

Pavitt, K., (1996), "National Policies for Technical Change: Where Are the Increasing Returns to Economic Research?", *Proceedings of the National Academy of Sciences*, Issue 23, Nov. 12, Washington, DC.

Polanyi, M., (1966), *The Tacit Dimension*, New York: Doubleday.

Price, D. S., (1965), "The Scientific Foundations of Science Policy", *Nature*, Vol. 206, pp. 233-238.

Price, D. S., (1961), *Science Since Babylon*, New Haven, Yale University Press.

Price, D. J.,(1956), "The exponential curve of science", *Discovery*, 17, 240-243.

Radosevic, S. and L. Auriol, (1999), "Patterns of Restructuring in Research, Development and Innovation Activities in Central and Eastern European Countries: an Analysis Based on S&T Indicators", *Research Policy*, 28, pp. 351-376.

Reiss, T., S. Hinze and D. Lacasa, (2004), "Performance of European Member States in Biotechnology", *Science and Public Policy*, Vol. 31, No. 5, pp. 344-358.

Romer, P., (1990), "Endogenous Technological Growth", *Journal of Political Economy*, 98(5), pp. 71-102.

Rosenberg, N., (2002), "Knowledge and Innovation for Economic Development: Should Universities be Economic Institutions?", [in:] P. Conceição, D. V. Gibson, M. V. Heitor, G. Sirilli and F. Veloso, *Knowledge for Inclusive Development*, Westport, Quorum.

Ruivo, B., (1995), *As políticas de ciência e tecnologia e o sistema de investigação*, Lisboa, INCM.

Smaglik, P., (2004), "Boosting Biotech in Finland", *Nature*, Vol. 431, J1020-J1021.

Small, H. G., (1978), "Cited Documents as Concept Symbols", *Social Studies of Science* 8, pp. 327-340.

TEKES, (2006), Tekes Annual Review 2005, Helsinki, TEKES.

United Nations, (2004), *Human Development Report 2004 – Cultural Liberty in Today's Diverse World*, New York, United Nations Development Programme.

Weingart, P., (2005), "Impact of Bibliometrics Upon the Science System: Inadvertent Consequences?", *Scientometrics*, Vol. 62, pp. 117-131.

Wright, G., (1999), "Can a Nation Learn? American Technology as a Network Phenomenon", [in:] N. Lamoreaux, D. M. G. Raff, and P. Temin (eds.), *Learning by Doing in Markets, Firms, and Countries*, Chicago, The University of Chicago Press.

Ziman, J., (1994), *Prometheus Bound: Science in a Dynamic State*, Cambridge, Cambridge University Press.

Zuckerman, H. and R. K. Merton, (1971), "Patterns of Evaluation in Science: Institutionalisation, Structure, and Functions of the Referee System", *Minerva*, Vol. 9, No. 1, pp. 66-100.

4

Intellectual Capital Approaches within Higher Education Institutions: Lessons from the Autonomous University of Madrid

Susana Elena

INTRODUCTION

There is a high degree of consensus around the idea of considering universities and research centres as critical institutional actors in national innovation systems within the knowledge-based economy.

Since the end of the last decade, European Higher Education (HE) institutions have become a major issue in the political agenda of the European countries and they are now immersed in strong transformation processes which aspire to make universities and research institutes more comparable, flexible, transparent and competitive. Given the new societal demands and the new mode of knowledge production (Gibbons et al., 1994), HE institutions are forced to change their internal organisational structures, governing bodies, and management systems. Indeed, there is a high degree of consensus in the specialised literature about the fact that the governing modes of today's universities are not completely able to fulfil the new roles of these organisations and to operate with efficiency and effectiveness, applying the criteria of excellence. Accordingly, new managerial instruments, skills and capacities are required (Amaral *et.al.*, 2003).

In this changing context, Intellectual Capital (IC) models are rapidly gaining importance as an approach for measuring intangibles, not only at firm level but within universities and research organisations.

Given this climate, the purpose of this paper is to analyse the growing interest in IC approaches within European HE institutions as a new response to the need to improve internal management and attain transparency, and to present the analytical framework developed within the Observatory of European University (OEU), focusing on the difficulties encountered at the Autonomous University of Madrid (UAM) when applying the common model.

The paper is structured as follows. Section 2 presents the current role of universities in the knowledge-based economy from an evolutionary perspective, taking into account the main structural transformations that are reshaping the European HE scenario as a result of the so-called Bologna process. In Section 3, the importance of IC approaches within universities, as an instrument to face these new challenges of improving internal management and transparency, is explored. In the realm of practice, Section 4 is devoted to highlighting the university reform which occurred in Austrian universities during the 90s decade. Austrian universities are the first HE institutions worldwide obliged by law (University Organisation Act, 2002) to submit Intellectual Capital Reports (ICRs) annually. In addition, other voluntary initiatives implemented in European universities and research centers are described. Section 5 describes, firstly, the Observatory of European University (OEU) project which aims at designing a tool to manage intangibles in universities; and secondly presents the UAM experience in the process of data gathering and analysis. Finally, Section 5 draws some preliminary conclusions and poses new questions for further research.

THE ROLE OF UNIVERSITIES IN THE KNOWLEDGE BASED ECONOMY

There is a high degree of consensus around the idea that the transformation of the knowledge production mode is one of the core processes that characterises the prevailing socio-economic context. Changes in universities are directly influenced by this new conceptualisation of knowledge production. The linear model of innovation is not useful anymore in the new scenario, where the differences between basic and applied research have become blurred, where universities are not the only institutions creating new knowledge, and where the traditional disciplines are not able to provide the solutions demanded by society to the present problems.

In this scenario, some authors point out that knowledge creation modes turn to more comprehensive and open approaches responding to socioeconomic, institutional and corporate needs. In contrast with the `Mode 1′, which operates within the linear paradigm, a new `Mode 2′ is defined by dynamic trans-disciplinary links between academia and industry and, in accordance with `solution-focused´ and `design-oriented´ models, is characterised by a "constant flow back and forth between the fundamental and the applied, between the theoretical and the practical" (Gibbons et al., 1994; p. 19).

In this new production mode, the main change regarding universities is that knowledge production and dissemination are considered "no longer self contained activities, carried out in relative isolation; they now involve interaction with a variety of other knowledge producers" (Gibbons, 1998; p. 1).

Despite the fact that this notion of Mode 2 is generally accepted in the specialised literature, there are some critics which argue that it would be more precise and accurate to recognise that it is not a completely new context, "but rather a shift in the balance between the already existing forms of Mode 1 and Mode 2" (Martin, 2003; p. 12). Indeed, argues Martin, applied research was already developed by universities at the end of the nineteenth century, mainly in the United States and Germany. Thus the distinction made today is a matter of considering Mode 2 more central nowadays than a century earlier (Martin, 2003).

Another approach directly linked with "Mode 2" is the Triple Helix Model proposed by Etzkowitz & Leydesdorff (1996). In it University-Industry-Government relationships are analysed "in terms of three interlocking dynamics: institutional transformations, evolutionary mechanisms and the new position of the university" (Etzkowitz & Leydesdorff, 2001; p. 6). It is suggested that not only the relationships among actors are changing, but also that there are internal transformations in each institution. In this sense, hybrid institutions, alliances with companies, trilateral networks and new joint centers are being created to produce and disseminate knowledge (see Figure 1).

Figure 1
Triple Helix Model

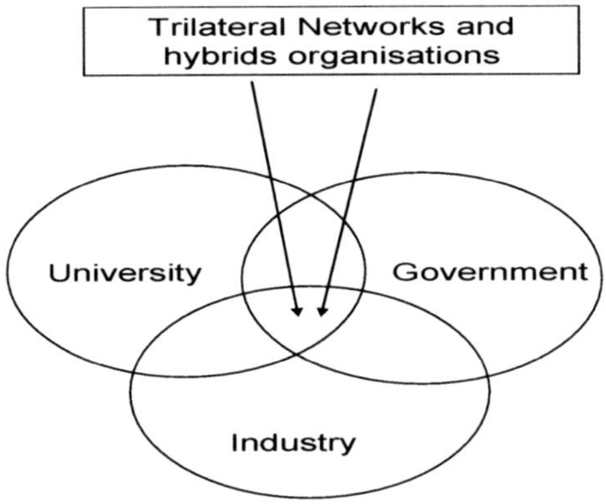

Source: Etzkowitz & Leydesdorff (2000).

While in the National Systems of Innovation approach (Lundvall, 1992 & Nelson 1993) firms are considered to have the leading role in innovation processes, and in the "Triangle Model" (Sabato, 1975) the State is contemplated as the privileged agent[1], the Triple Helix postulate considers the three spheres as having equal importance in the innovation network (Etzkowitz & Leydesdorff, 1996).

As regards universities, Etzkowitz & Leydesdorff (2001; p. 1) argue that the Triple Helix thesis assumes "that the university can play an enhanced role in innovation in increasingly knowledge-based societies". There was an academic revolution in the late nineteenth century in which research was incorporated into the university's mission, making it basically compatible with teaching activities. In this sense, there is a tension between the two activities, but nevertheless they co-exist "because it has been found to be both more productive and cost effective to combine the two functions" (Etzkowitz & Leydesdorff, 2001; p. 11). In any case, this process of change has led to a "re-evaluation of the mission and the role of the university in society" (Etzkowitz & Leydesdorff, 2001; p. 2).

These analytical frameworks are quite useful for conceptualizing the role of universities within the innovation process in knowledge-based economies, and emphasize the importance of strong links between universities and other institutional actors in OECD economies (Mowery & Sampat, 2004). However, as these authors point out, "what is not clear in these frameworks is a set of criteria to assess the strength of such linkages and a set of indicators to guide the collection of data" (Mowery & Sampat, 2004, p. 8).

Under this new paradigm for conceptualizing the role of universities within the knowledge-based society, European HE and research systems are immersed in strong transformation processes, which aspire to make our universities and research institutes more comparable, flexible, transparent, and competitive, in both teaching and research. As Goldsmith & Berndtson (2002)[2] note "higher education is affected today by a number of new challenges, which already have changed our way of teaching and research".

The European Commission considers that "Europe must strengthen the three poles of its knowledge triangle: education, research, and innovation. Universities are essential in all three. Investing more and better in the modernisation and quality of universities is a direct investment in the future of Europe and Europeans" (European Commission, 2005; p. 2).

On the one hand, the so-called *Bologna process*, which now involves more than 40 countries, began at the end of the last decade with the Sorbonne Joint Declaration. According to this Declaration (1998), higher education plays a central role in the development of modern societies as it enhances the "intellectual, cultural, social and technical dimensions of our continent" (Sorbonne Joint Declaration, 1998; p. 1). In other words, the declaration stressed the idea that universities are a pivotal element for reaching the main aspiration

[1] In Etzkowitz & Leydesdorff, 2001.

[2] In Berndtson (2003, p. 7).

of the European Union: to be "the world's leading knowledge-based economy by 2010" (Lisbon European Council, 2000).

All the endeavours and initiatives undertaken since the Sorbonne Joint Declaration[3] have been aimed at the establishment of a European Higher Education Area (EHEA) by the end of the decade. This means making European educational and research systems more comparable, increasing their level of quality, improving their flexibility, and making higher education and research activities more transparent and competitive.

On the other hand, and although there is no doubt about considering research as an essential driving force for economic and social growth, indicators reveal that European countries' investments in this regard are still inferior to those of our major competitors. Indeed, the European Union's share of overall public R&D effort in relation to GDP has been steadily declining for ten years (1988-1997) and, what is worse, the gap in relation to the United States and Japan during that period has been widening (European Commission, 2000c).

With the aim of tackling this lack of R&D investment, since 1998 the EU has developed several Framework Programmes in order to boost research cooperation among European countries. However, while some progress has been attained, it seems that the main research framework remains national in scope, overlapping in some fields with European Union policies (European Commission, 2000c). Having regard for the diversity of national frameworks of innovation, the Trend Chart on Innovation in Europe[4], in effect since January 2000, summarises information and statistics on innovation policies and trends in Europe in an annual report (Innovation Policy in Europe).

In order to make progress towards a better organisation of research in Europe, the European Research Area (ERA) is now being developed. This initiative aims to create the necessary conditions to increase the impact of European research efforts "by strengthening the coherence of research activities and policies conducted in Europe" and "offers a new horizon for scientific and technological activity and for research policy in Europe" (European Commission, 2000b; p. 3).

These two European initiatives were conceived based on the premise that the quality of higher education and research is, and will be, a crucial element of Europe's international attractiveness and competitiveness (Prague Declaration, 2001).

These policy initiatives increase the pressure on universities to re-think their role in national innovation systems, and should result in a structural change in order to contribute to the Lisbon goal of becoming the "most competitive and dynamic knowledge-based economy in the world by 2010" *(*Lisbon European Council, 2000).

---

[3] Bologna Declaration (1999), Prague Conference (2001), Berlin Conference (2003), and Bergen Conference (2005).

[4] Joint initiative between the Innovation Policy Directorate of the Enterprise and the Industry Directorate General of the European Commission.

In accordance with this aim, it is suggested that, in order to succeed, efforts should be undertaken to secure links between higher education and research. Both systems should benefit from their combined results and experiences, and synergies should be identified and exploited. These symbiotic relationships are deemed to configure a crucial pillar of the foreseen knowledge-based society in Europe.

THE IMPORTANCE OF MANAGING INTELLECTUAL CAPITAL IN PUBLIC UNIVERSITIES AND RESEARCH ORGANIZATIONS AS A NEW RESPONSE FOR IMPROVING TRANSPARENCY AND INTERNAL GOVERNANCE

Since the second half of the 20th century, mainstream economic theories have recognised, to a greater or lesser degree, the existence of intangible elements that underlie part of economic growth (Solow, 1957; Deninson, 1962; Arrow, 1962; Shultz, 1969; Kendrick, 1974; Becker, 1975; Freeman, 1982; Nonaka & Takeuchi, 1995; Gorey & Dobat, 1996; OECD, 1996; European Commission, 2000a). In a knowledge-based economy intangible assets and investments are seen as essential elements to value creation in companies and, hence, to economic wealth (Cañibano *et. al.*, 2000). According to this view, intangibles and Intellectual Capital (IC) have become a major issue, not only for academics but also for governments, regulators, enterprises, public organisations, investors and other stakeholders.

In accordance with Lev (2000), we can consider that the terms intangibles, knowledge assets and intellectual capital are interchangeable. Although there are different classifications in the specialised literature for the terms 'intangibles' and 'intellectual capital', there is a high degree of consensus around most conceptual aspects. The most relevant classification of Intellectual Capital distinguishes between human, structural, and relational capital (Cañibano and Sánchez, 2004), but it is important to point out that IC is more than the sum of these three elements: "it is about how to let the knowledge of an organisation work for it and have it create value" (Roberts, 1999).

As pointed out in the previous section, given the new pressures and demands of the rapidly changing environment, contemporary universities across Europe are force to be multi-mission oriented[5] and to work within a more competitive environment. Consequently, HE institutions have begun to re-consider their organizational structures and to incorporate management systems to run their affairs, since traditional governance modes are not fully able to manage these organisations with the effectiveness and efficiency criteria demanded by the new socio-economic context (Amaral *et al.*, 2003). In this sense, since the mid-80s the reform of HE systems has became a priority on the

[5] A university's main functions are: the generation of knowledge (research), the transmission of knowledge (teaching), and contributing to local and regional wealth and economic development.

political agenda of Western countries and has created an intense debate about how these institutions should be managed (Deem, 2001).

As a response to the new demands, a trend called *new manageralism* (Amaral *et al.*, 2003; Deem, 2001) emerged. This concept is used to refer to the application in universities of concepts, techniques and management tools originally implemented only in the business world. This new managerial view implies a greater autonomy at the institutional level in order to allow universities to manage not only their academic aspects, but also their financial and administrative affairs.

In this context, IC approaches are acquiring growing importance within universities and research centers as an answer to the need to improve both internal management and transparency at the global level[6], and especially in the European Union. Proof of this can be found in the recent interest for IC shown by the European Commission, which set up a High-Level Expert group in December 2004 to recommend a set of measures aimed at stimulating small and medium sized companies (SMEs) to report on their intangibles. This effort resulted in compilation of the document RICARDIS (Reporting Intellectual Capital to Augment Research, Development and Innovation in SME's), which will be published in the upcoming months.

By doing so, the Expert Group has also taken into consideration the relationship between SME's and Universities and the need of these latter institutions to report on their Intellectual Capital. The main assumption that supports the recommendations for reporting is that Intellectual Capital is the hidden driver of the knowledge-based economy, and that diffusing information on IC may stimulate and make the R&D efforts of these organizations more visible. The document produced by the Group includes among its policy recommendations the need for both the Commission and the member states "to promote the reporting of IC by universities and research and technology organizations" (European Commission, 2006).

Although most of the knowledge management and IC analyses developed during the last decades refer to private companies (See the Balanced Scored Card (Kaplan & Norton, 1992); Navigator of Skandia (Edvinsson & Malone, 1997); Technology Broker (Brooking, 1996); West Ontario University (Bontis, 1996); Canadian Imperial Bank (Saint-Onge, 1996); Intellectual Asset Monitor (Sveiby, 1997); INTELECT Model, (Instituto Universitario Euroforum, 1998);

---

[6] Supranational organisations such as the OECD, the World Bank, the European Commission, the World Intellectual Property Organisation, the European Patent Office and the European Investment Bank are involved in different activities to identify, measure, manage and disclose information about intellectual assets following an IC scheme, and in supporting and participating in conferences and events. See for instance, the first Workshop on "Visualising, Measuring and Managing Intangibles and Intellectual Capital" and the OECD International Policy Conference entitled "Intellectual Assets and Innovation: Value Creation in the Knowledge Economy", both held in Ferrara in October 2005. The next OCDE Conference on the subject, entitled "Intellectual assets-based management toward innovation & sustainable growth", will take place in Tokyo on December 7th and 8th.

Intellectual capital (Dragonetti & Ross, 1998); The Value Explorer (Andriessen, 2000); MERITUM Project (2002)[7]), there is a growing interest in public organisations[8] such as universities and research centres (see Leitner, 2005; Altenburger & Novoty-Farakas, 2005; Sanchez & Elena, 2005), not only in terms of management, but also in terms of the informational disclosure on intangibles to external stakeholders.

The main reasons underlying the premise that IC models could constitute significant management and reporting tools for universities can be summarised as follows:

1. Even though the main inputs and outputs of universities are basically intangibles (mostly knowledge and human resources), only a small portion of these are identified, and very limited instruments exist to measure and manage them (Cañibano & Sánchez, 2004).
2. Because of the new pressures for accountability within public institutions, universities and research centers are forced to be more transparent and to disseminate more information to stakeholders (students, public authorities, funding organisations, the labour market, and society as a whole). As asserted by the European Commission (2003a; p. 13) "universities have a duty to their stakeholders to maximise the social return of their investment". However, despite the increase in external demands for greater information and transparency on the use of public funds (Warden, 2003), an empirical study carried out throughout different universities in Australia, Canada, France, Germany, New Zealand, Spain, the United Kingdom and the USA confirms that most of them have not yet assumed a general practice of elaborating external information reports (Campos, Noverto & Villanueva, 2003).
3. The globalisation processes, the increasingly competitive environment, and the European Union's efforts to create the European Higher Educational and Research Area are pushing universities to improve their attractiveness in order to attract the best students, researchers and professors, and to compete for public and private funds.
4. Finally, because of changes in the method and rationale for distributing public funds in Western countries, governments and universities are forced to implement mechanisms and tools to assess research and teaching activities in order to elaborate criteria for the allocation of resources, linking funding rationale with academic performance (Geuna & Martin, 2003).

[7] The MERITUM Project developed 77 case studies in six European countries (Spain, France, Finland, Sweden, Denmark and Norway).

[8] Compare the experiences within health care organisations (Vagnomi & Castellini, 2005), cultural institutions (Donato, 2005), and the efforts to identify, measure, manage and report on intangibles and IC at meso-level in certain regions (Gallardo & Castilla, 2005).

PRACTICAL EXAMPLES OF THE APPLICATION OF INTELLECTUAL CAPITAL APPROACHES WITHIN EUROPEAN UNIVERSITIES AND RESEARCH CENTERS

As described in the previous section, intangibles and IC within HE institutions have became a major issue, not only for academics but also for practitioners and politicians, during the 90s decade. This growing interest is being translated into different initiatives in the realm of practice, with varying implications and impacts on the governing structures of universities and on their transparency.

Undoubtedly one of the most significant initiatives regarding the introduction of Intellectual Capital Reports (ICRs) within the HE sector is the university reform which has been taking place in Austria since the beginning of the 90s decade. The University Organisation Act of 2002 (hereafter UG2002) was the culminating legal step aimed at providing universities with more institutional autonomy and new governing bodies in order to make them more competitive. Austrian public universities are the first higher education institutions in the world which are compelled to elaborate and diffuse ICRs (UG2002, art.13). As with other relevant university documents, ICRs should be available to the public (art 20.6).

Accordingly, each Austrian university is obliged to submit annually its own ICR, which should include, at a minimum, the following features:

- The university's activities, social goals, and self-imposed objectives and strategies.
- Its Intellectual Capital, broken down into human, structural, and relational capital.
- The processes set out in the performance agreement, including outputs and impacts.

The model on which the University ICR is based has been developed by the Austrian Research Center (ARC), one of the pioneer research institutions in Europe in the application of IC models to manage intangibles and in the reporting of this information (see figure 2).

Figure 2
ARC Intellectual Capital Model

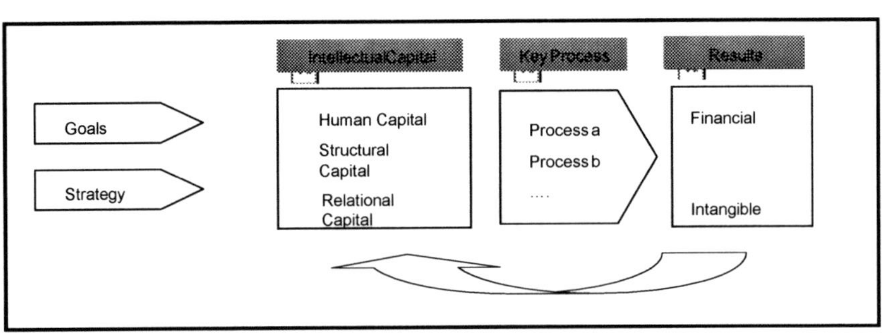

Source: Leitner & Warden (2003).

Table 1
Voluntary IC Measurement in European Public Universities and Research Centres

Organisation	**Description of the Initiative regarding IC management**
Austrian Research Centres (ARS)[9]	ARS is the largest publicly funded research organisation in Austria and performs an important function as a link between basic research in universities and applied research carried out by companies. It was the first research centre in Europe to publish an Intellectual Capital Report (1999). The model it proposed incorporates the main categorizations of Intellectual Capital: Human Capital, Structural Capital and Relational Capital, and it uses an indicator-based approach.
INGENIO Project[10]	An Innovation and Knowledge Management Institute created in 1999 by CSIC and the Polytechnic University of Valencia. This joint initiative developed a research project concerning the use of knowledge management technologies to improve quality management in Spanish universities (2002).
Higher Education Institutions and Research Organisations (HEROs)	This is an initiative led by the members of the European Association of Research Managers and Administrators (EARMA)[11] in collaboration with the European Centre for the Strategic Management of Universities (ESMU)[12]. Its main goal is "*raising awareness and disseminating good practices in the fields of managing and reporting intellectual capital among universities and research organisations*" (Leitner & Warden, 2003).
PCI Project	This Project developed an Intellectual Capital Indicators Programme which was applied to the research activities of three universities[13] and two public research institutions[14] in Madrid (Spain). The starting methodological framework was the Intellect Model (Instituto Universitario Euroforum, 1998), in which a panel of indicators for each public institution and university was proposed.
University of the Basque Country (UPV)	This university developed a knowledge management case-study project in a key and strategic cross-organisational process (Research-Development-Transfer of Knowledge) in order to spread the knowledge management process to the entire university.

Source: Own elaboration based on Austrian Research Centres (1999), INGENIO (2001); Warden, (2003); Comunidad de Madrid (2002); Araujo, (2000).

Moreover, the increasing awareness about the importance of measuring, managing, and reporting on intangibles has led some other individual

[9] http://www.arcs.ac.at/publik/fulltext/wissensbilanz

[10] http://www.ingenio.upv.es

[11] http://www.earma.org

[12] http://www.esmu.be

[13] Autonomous University of Madrid, Carlos III University, and Polytechnic University of Madrid.

[14] National Biotechnological Centre (CNB) and CIEMAT.

universities and research institutions across Europe to construct their own model voluntarily. Although it is not the objective of this work present an in-depth analysis of each of these initiatives, Table 1 briefly presents the most relevant experiences.

The analysis of these practical experiences demonstrates that, in general, there is a significantly increasing interest in HE institutions in adopting management tools from the private sector, in particular with regard to IC approaches.

The Austrian university reform indicates that it is possible to create a radical shift within the HE sector via legislation. The establishment of a legal obligation for universities to submit ICRs on an annual basis is a crucial step for the proliferation of IC models world-wide, not only for management purposes but also in order to disclose information to stakeholders. The use of homogeneous ICRs across the country will facilitate benchmarking analysis and comparative studies, which will help the decision-making process and the articulation of public policies.

On the other hand, while the proliferation of various models on a voluntary basis emphasizes the increasing awareness of universities about their internal governing modes, the existence of different approaches and different sets of indicators in each institution creates the problem of comparability.

MANAGING RESEARCH ACTIVITY IN EUROPEAN UNIVERSITIES: AUTONOMOUS UNIVERSITIES CASE STUDY

Creating a common framework of analysis: The Observatory of European Universities

The Observatory of European University (OEU) is an on-ongoing European project[15] which is being developed within the PRIME Network of Excellence (Policies for Research and Innovation in the Move towards the European Research Area), supported by the VI Framework Programme. The project started in June 2004 and is composed of fifteen universities and research institutes from eight different European countries[16]. Its main goal is to contribute to a better understanding of the importance of managing intangibles assets and activities in public universities. Taking into account the necessity to improve the level of quality, competitiveness and transparency of European universities, the Observatory aims to provide these organisations with the tools, instruments and techniques which will enable them to identify, measure and manage their research activity.

[15] The project will finish in November 2006.

[16] Germany, Spain, France, The Netherlands, Hungary, Italy, Portugal and Switzerland.

Figure 3
OEU Analytical Framework: Strategic Matrix

	Funding	Human Resources	Academic Outcomes	Third Mission	Governance
Autonomy	Key Questions Indicators				
Strategic capabilities					
Attractiveness					
Differentiation Profile					
Territorial Embedding					

Source: Observatory of European University (2005).

Given the complexity of universities and the degree of heterogeneity which characterise HE institutions across Europe (European Commission, 2003a), the OEU has decided to start with an analysis of research activity, leaving teaching and administrative activities for future stages. Accordingly, the project is generating a common analytical framework and defining a battery of indicators to measure and compare research activities among institutions. It is important to note that the main purpose of the project is not to create a ranking of universities, but to help them improve their internal management systems and to provide the information to funding institutions, students, professors, and other stakeholders which they need in order to make decisions and implement public policies.

The analytical structure proposed by the Observatory combines five thematic dimensions and five transversal questions into a Strategic Matrix (see Figure 3).

The "Thematic Dimensions" refer to:

- Funding: all budget elements are analysed, including revenues and expenses.
- Human Resources: administrative staff, researchers/teachers and PhDs.
- Academic Production: the results of research activities in all the fields: articles, academic publications, non-written results, and the knowledge embodied in PhDs.
- Third Mission: all the activities and relations between a university and non-academic partners: firms, non-profit organisations, public authorities, local government, and society as a whole.
- Governance: the process by which the university converts its inputs (funding and human resources) into research outputs (academic outcomes and third missions).

Giving its qualitative profile, this should be approached mainly as a characterization issue.

The "Transversal Issues" considered are:

- Autonomy: the university's margin for manoeuvre - in other words its degree of freedom to allocate resources or to use funds - is analysed.
- Strategic Capabilities: the university's real ability to implement its own strategic choices.
- Attractiveness: the university's capacity to attract resources (money, people, equipment, collaboration, etc.) within a context of scarcity.
- Differentiation Profile: the main features of a university which distinguish it from the other strategic actors.
- Territorial Embedding: the geographical distribution of university involvement, contacts, collaborations, etc.

Each cell of the Matrix contains various key questions and a set of indicators. The process of developing and applying the above mentioned Strategic Matrix has provided universities with the opportunity to think about and discuss the way they manage and account for research, and to understand better the problems of governance. In order to complete the Matrix with indicators for each institution, three steps have been followed:

1. "*State of the art*" analysis in order to become acquainted with the internal structure, the management systems and the current situation of each university. In order to do this, a preliminary battery of indicators (both financial and non-financial), aimed at describing and measuring the research activities in all disciplines and Faculties, was proposed. This step was possible due to the close collaboration between the OEU research teams and the governing and research staff of the universities involved.
2. Homogenization of the experiences with other university participants in the Project. Since one of the main objectives of the Observatory is to facilitate benchmarking analysis, it was crucial to develop a framework able to provide homogeneous information about the research activities carried out by a range of European universities and, simultaneously, to be useful as management tool for individual organisations.
3. The project is now in its last step, in which the selected indicators are being tested empirically. A series of interviews are being conducted with the main decision makers at the UAM and other research and teaching positions which are considered strategically important based on the amount of information that they receive and their decision making capacity.

As regards quantitative indicators, according to the analysis developed by the MERITUM Project (2002) they should be checked against the following set of characteristics:

- Useful: an indicator is useful if it facilitates decision making both among internal and external users.

- Relevant: an indicator is relevant when it provides information that can modify or reassure the expectations of decision makers. In order to accomplish this, it should be:

  - Significant: related to issues critical for universities,
  - Understandable: presented in a way it is easily understood by potential users,
  - Timely: available when it is required for analysis, comparison, or decision making purposes.

- Comparable: an indicator should follow generally accepted criteria, so that users may make comparisons over time and across institutions; and
- Reliable: trustworthy. This requires that an indicator be:

  - Objective: its value should not be affected by any bias arising from the interests of the parties involved in the preparation of the information,
  - Truthful: the information it provides should reflect the real situation,
  - Verifiable: it should be possible to assess the credibility of the information an indicator provides.

- Feasible: the information for the elaboration of an indicator should be available from the University's information system, or alternatively the cost of modifying the information system in order to obtain the required information should be lower than the benefits (private or social) arising from the use of the indicator.

It should be noted however that despite the importance of quantitative measures, the Observatory is aware of the necessity to provide a narrative to describe the indicators and to supply additional further information.

Finally, in addition to the above-mentioned analytical framework and the learning process created within the HE institutions, one of the main outputs of the Observatory will be the development of specific guidelines for each thematic dimension. These general guidelines will be crucial for a better understanding of the research activity, and provide the governing bodies of universities with an instrument to improve their internal management.

Some lessons from the Autonomous University of Madrid (UAM)

One of the most pro-active institutions within the Observatory is the Autonomous University of Madrid (UAM), Spain[17]. It is a relatively young (founded in 1968) Spanish public University, in which teaching and research activities are considered of equal importance. UAM is comprised of 8 Faculties made up of 60 Departments, 8 Research Institutes, and 20 additional Centres, which conducts 34 undergraduate programmes, 75 Masters' programmes, and 89 PhD programmes.

The research group has made an important effort to get the university involved in the Project and has obtained the strong commitment of the Vice-

[17] http://www.uam.es

rector of Research, whose participation is crucial for the successful development of the Project.

In the process of applying the OEU principles and filling out the strategic matrix, the UAM has encountered some difficulties. The situations described below seem to be common in universities with similar structures and governing bodies, hence the analyses presented might be useful to others in the search for possible solutions that could be applied to various universities:

– Sources of information. The university does not operate with an integrated database. Quite the contrary, UAM works with different information systems for academic management, human resources, economic resources, etc. This situation slows down the data collection process. Consequently, much of the needed data is scattered throughout the University: in various departments, institutes, administrative offices, etc. This creates a situation whereby, on the one hand, comparability between areas and issues is not always possible, and on the other hand managers have only a partial notion of the university's activities. In order to address these problems, the UAM is in the process of integrating all the databases into a common data warehouse.

In addition, despite the fact that much data is public and issued in different reports, still some crucial information for management purposes is internal. As a result it is critical to select the right person in the organisation to interview and obtain the necessary new data.

– Time frame. A dynamic analysis is essential in order to examine the evolution and the trends in research activity over time. Inasmuch as most universities have been immersed in an intense transformation process throughout the last decade, and their aims and goals are changing, not all the information needed is available for all the years required.
– Related to the above-mentioned issues, it is important to consider the dynamic notions implicit in indicators. In other words, while the indicators are a tool for measurement, if the priorities and strategic objectives of the organisation change, the indicators need enough flexibility to add or eliminate them. Thus any framework used should incorporate this dynamic perspective.
– When applying a particular IC model, there is usually a tendency to produce a long list of indicators. Indeed, university managers have singled out this issue as one of the main management difficulties. At the beginning of the process, everything looks as if it can be and should be measured. However, the process of gathering information and producing relevant indicators is costly, and it can happen that they turn out to be useless for management purposes. Accordingly, the measurement process should allow for the efficient elimination of those indicators that do not reasonably fulfil the characteristics mentioned above
– Although some strategic questions may be answered using quantitative indicators, some will require qualitative analysis. In order to obtain comparable results in a qualitative analysis it is crucial to develop a multiple-choice questionnaire. In this sense one of the most difficult areas to analyse is probably the so-called "Third Mission", which has many elements in common with the Relational Capital of the IC framework. The heterogeneity stemming from the multitude of possible ways to interact with society and to assess performance makes the construction of

quantitative indicators very complicated. Description and qualitative arguments are therefore needed, which in turn makes comparisons much less accurate.

CONCLUSIONS AND FURTHER RESEARCH

The intense transformation process in which European universities are involved and their increasing role in the knowledge-based economy have evidenced the need to introduce new management system, both to improve internal management and to provide better information to stakeholders. Considering the inherent importance of intangible assets for this type of organisation, and taking into account the fact that intangibles and IC have became a major issue for academics, practitioners and politicians, an analysis of the approaches used seems to be crucial in the new HE scenario.

Not surprisingly, different initiatives regarding IC management are taking place across European HE institutions. On one hand, the Austrian experience shows that it is possible to create a radical shift within universities through changes in legislation. The UG2002 aims to increase universities institutional autonomy by creating new governing bodies and introducing new management systems. The establishment of a legal obligation for universities to submit an ICR every year is a crucial step for the spread of IC models world-wide, not only for management purposes but also in order to disclose information to stakeholders. Having homogeneous ICRs across the country will enhance benchmarking analysis and facilitate the articulation of public policies. On the other hand, the increasing awareness about the importance of measuring, management and reporting on intangibles has also led some universities and research institutions to create their own model voluntarily. Despite the importance of these experiences, the proliferation of different models with different sets of indicators creates a problem of comparability across Europe.

Also on a voluntary basis, the Observatory of European University aims to provide a management tool to universities and research centers. From the conceptual point of view, the OEU is elaborating a management instrument with which to characterise research activities. The so-called Strategic Matrix was designed in two dimensions regarding five thematic dimensions and five transversal issues, with two main goals in mind: facilitating internal management within individual institutions, and benchmarking analysis across universities. The main results of the project are to be the creation of strategic matrix including all the quantitative and qualitative indicators of each university, as well as general guidelines for each thematic dimension - Funding, Human Resources, Academic Outcomes, Third Mission and Governance - both of which are designed to explain how to effectively apply the analytical framework.

Although numerous endeavours and projects are trying to harmonise European educational and research systems, the development of a common framework under different institutional conditions is revealing several methodological problems within the data collection process.

The UAM experience was presented in this paper, mainly focusing on the identification of problems and barriers in the data gathering process which could be faced in every university with similar characteristics when dealing with similar issues. In other words, the analysis of the individual problems encountered in the UAM might aid in the formulation of best practices in terms of managing and reporting Intellectual Capital.

Having homogeneous IC models across European universities will facilitate benchmarking analysis and comparative studies, which in turn will aid in the decision making process and in the articulation of public policies. However, despite the great advances have been achieved during the OEU project in terms of measuring intangibles within HE universities, the analytical framework developed does not include IC terminology. In this sense, further research about the impact of the strategic matrix, which coexists with other initiatives which follow IC standards, is required.

Moreover, as pointed out in previous sections, all the universities involved in the project are working voluntarily. Accordingly, there is a need to create proper incentives for the application of the analytical framework to internal management and to the distribution of information to stakeholders.

Finally, further research needs to be conducted to analyse if these new management systems within universities, such as the IC models analysed in this paper, can be compatible with traditional university decision-making structures based on collegial models. In this sense, although some authors claim that universities should professionalize their internal management, the application of managerial ideas to HE institutions has created enormous controversy within academia itself. Additional research is required in order to provide empirical evidence which would quantify the long-term effects of introducing professional governance and management systems into academic research.

BIBLIOGRAPHY

Altenburger, O. & Z. Novotny-Farkas, (2005), “Intellectual Capital Reports for Universities – A Trial Intellectual Capital Report at the University of Viena”, Working paper submitted to the 1st Workshop on “Visualising, Measuring, and Managing Intangibles and Intellectual Capital”. Ferrara 18-20 October.

Amaral, A., V. L. Meek, and I. M. Larse, (eds.) (2003), *The Higher Education Managerial Revolution?,* Kluwer Academic Publishers, Kluwer.

Andriessen, D., (2001), “Weightless Wealth. Four Modifications to Standard Intellectual Capital Theory”, Paper presented at the 4th World Congress on the Management of Intellectual Capital. Hamilton, Ontario.

Araujo, A, (2000), “Gestión del Conocimiento, Universidad y Empresa”, Paper presented at Jornadas Universitarias sobre la gestión del conocimiento en empresas y organizaciones, Centre for University Knowledge Management, Bilbao, pp. 13-15.

Arrow, K. J., (1962), “The Economic Implication of Learning by Doing”, *Review of Economic Studies,* Vol. 29 June, pp. 155-173.

Austrian Research Centers, (1999), “Intellectual Capital Report 1999”, Austrian Research Centers, Seibersdorf.

Becker, G. S., (1975), *Human Capital*, 2nd Edition, Chicago University Press, Chicago.

Berndtson, (2003), "The European Higher Education Area: to Change or not to Change?", The EpsNet General Conference, Paris, June 13-14, 2003.

Bologna Declaration, (1999), "The European Higher Education Area", Joint Declaration of the European Ministers of Education, Bologna, 19 June 1999, http://www.bologna-berlin2003.de/pdf/bologna_declaration.pdf

Bontis, N., (1996), "There's a Price on Your Head: Managing Intellectual Capital Strategically", *Ivey Business Quarterly*, Vol. 60 (4), pp. 40.

Brooking, A., (1996). *Intellectual Capital: Core Asset for Third Millennium Enterprise*, International Thomson Business Press, London.

Campos Fernández, M., M. C. Noverto Laborda & E. Villanueva García, (2003), "Reporting de las Instituciones Universitaria", *Revista AECA*, N° 62, Enero-Abril, http://www.aeca1.org/revistaeca/revista62/revista62.htm

Cañibano, L., M. García-Ayuso & M. P. Sánchez, (2000), "Accounting for Intangibles: A Literature Review", *Journal of Accounting Literature*, Vol. 19, pp. 102-130.

Cañibano, L., P. Sánchez, (2004), "Measurement, Management and Reporting on Intangibles: State of the Art", *Accounting and Business Review,* 3 (1), pp 37-62 .

Communiqué of the Conference of European Ministers Responsible for Higher Education, (2003), "Realising the European Higher Education Area", Berlin, 19 September.

Communiqué of the Conference of European Ministers Responsible for Higher Education, (2005), "The European Higher Education Area. Achieving the Goals", 19-20 May, Bergen.

Comunidad de Madrid, (2002), "Capital Intelectual y producción Científica", Dirección general de Investigación, Consejería de Educación, Comunidad de Madrid, Madrid.

Davenport, T. y Prusak, (1998), *Working Knowledge: How Organizations manage GAT They Know*, Harvard Business School Press, Boston, Massachusetts.

Deem, R., (2001), Globalisation, New Managerialism, Academic Capitalism and Entrepreneurialism in Universities: is the Local Dimension Still Important?, *Comparative Education*, Vol. 37, N°1, pp. 7-20.

Deninson, E. F., (1962), *The Sources of Economic Growth in the United States and the Alternatives before us*, Committee for Economic Development, New York.

Donato, F., (2005), "The Orientation Towards the Management and the Reporting of the Intangibles in Cultural Organisations: a Starting Point or an Escape Route?", Working paper submitted to the 1st Workshop on "Visualising, Measuring, and Managing Intangibles and Intellectual Capital", Ferrara, 18-20 October.

Dragonetti, N. C. & G. Ross, (1998), "Efficiency and Effectiveness in Government Programmes: An Intellectual Capital Perspective", 2nd World Congress on Intellectual Capital I, January 1998, McMasters University, Hamilton.

Edvisson, L., M. S. Malone, (1997), *Intellectual Capital. Realizing your Company's True Value by Finding its Hidden Brainpower*, Harper Collins Publishers, Inc., 1ª ed.

Etzkowitz, H. & L. Leydesdorff, (2001), The Dynamics of Innovation: from National Systems and "Mode 2" to a Triple Helix of University-Industry-Government Relations", *Research Policy* 29(2), pp. 109-123, http://users.fmg.uva.nl/lleydesdorff/th2/index.htm.

Etkowitz, H. & L. Leydesdorff, (2000), "The Dynamics of Innovation: From National Systems and "Mode 2" to a Triple Helix of University-Industry-Government Relations", *Research Policy*, Vol. 29, pp. 109-123.

Etzkowitz, H. & L. Leydesdorff, (1996), "The Triple Helix as a Model for Innovation Studies", *Science and Public Policy* 25(3) (1998) 195-203, http://users.fmg.uva.nl/lleydesdorff/th2/index.htm.

European Commission, (2000a), "Innovation Policy in a Knowledge-Based Economy", A Merit Study Commissioned by the European Commission Enterprise Directorate General, EUR 17023.

European Commission, (2000b), "Making a Reality of The European Research Area: Guidelines for EU Research Activities (2002-2006)", Brussels, 04/10/2000, COM (2000)612 Final.

European Commission (2000c), "Towards a European Research Area", Brussels, 18/01/2000, COM (2000) 6.

European Commission (2003a), "The role of the Universities in the Europe of Knowledge". Brussels 05/02/2003, COM (2003) 58 Final.

European Commission (2005), "Mobilising the Brainpower of Europe: Enabling Universities to Make Their Full contribution to the Lisbon Strategy". Brussels, 20/04/2005, COM (2005) 152 Final.

European Commission (2006), "RICARDIS (Reporting Intellectual Capital to Augment Research, Development and Innovation in SME's)"

Federal Ministry of Education, Science and Culture of Austria (2002), "University Organisation and Studies Act – University Act 2002" Nº 120/2002, http://www.bmbwk.gv.at.

Freeman, C., (1982), *The Economics of Industrial Innovation*, Pinter Publishers, 2nd Edition, London.

Gallardo & Castilla, (2005), "Qualitative Analysis of The Presence of Intellectual Capital in Less Developed Zones: Implications for Business Management". Paper submitted to the 1st Workshop on "Visualising, Measuring, and Managing Intangibles and Intellectual Capital", Ferrara, 18-20 October.

Geuna, A. & R. B. Martin, (2003), "University Research Evaluation and Funding: an International Comparison", *Minerva* 41, pp. 277-304.

Gibbons, M., (1998), "Higher Education Relevance in the 21st Century", Paper presented at UNESCO World Conference on Higher Education, 5-9 October, Paris.

Gibbons, M., C. Limoges, H. Nowotny, S. Schwartzman, P. Scott, & M. Trow, (1994), *The New Production of Knowledge: the Dynamics of Science and Research in Contemporary Societies*, Sage Publications, London.

Goldsmith & Berndtson, (2002), "Teaching Challenges for Political Science in Europe", *European Political Science,* Vol.1 (3), pp. 69-80.

Gorey R. M. & D. R. Dovat, (1996), *Managing on the Knowledge Era*, New York.

INGENIO (Instituto de la Gestión de la Innovación y del Conocimiento) (2001), http://www.ingenio.upv.es/4_1_1b.html,

Instituto Universitario Euroforum El Escorial (1998), Medición del Capital Intelectual, Madrid.

Kaplan, R., & D. Norton, (1992), "The Balanced Scorecard - Measures That Drive Performance", *Harvard Business Review*, 70(1), pp. 71-79.

Kendrick, J. W., (1974), "The Accounting of Human Investment and Capital", *Review of Income and Wealth*, Vol. 20, December, pp. 439-468.

Leitner, K-H. & C. Warden, (2003), "Managing and Reporting Knowledge-based Resources and Processes in Research Organisations: Specifics, Lessons Learned and Perspectives".

Leitner, K. H., (2005), "Managing and Reporting Intellectual Capital in Public Research Organisations and Universities: Background, Development and Experiences of Austrian Organisations", Working paper submitted to the 1st Workshop on "Visualising, Measuring, and Managing Intangibles and Intellectual Capital", Ferrara, 18-20 October.

Lev, B., (2000), "Intangibles: Management, Measurement and Reporting", http//www.baruch-lev.com.

Lisbon European Council, (2000), 23-24 March 2000, http://www.bologna-berlin2003.de/pdf/PRESIDENCY_CONCLUSIONS_Lissabon.pdf

Lundvall, B. A. (ed.), (1992), *National Systems of Innovation*, Pinter, London.

Martin, B. R., (2003), "The Changing Social Contract for Science and the Evolution of the University", [in:] A. Geuna, J. A. Salter, & W. E. Steinmueller, (eds.), (2005), *Science and Innovation. Rethinking the Rationales for Funding and Governance*, Ed. Edward Elgar, pp. 7-29.

MERITUM (2002), Guidelines for Managing and Reporting on Intangibles (Intellectual Capital Statements). Vodafone Foundation. Madrid.

Mowery, D. C. & B. N. Sampat, (2004), "Universities in National Innovation Systems", Paper presented in Globelics Academy, 25 May – 04 June, Lisbon, 2004, http://www.globelicsacademy.net/pdf/DavidMowery_1.pdf

Nelson, R. (ed.), (1993), *National Innovation Systems: a Comparative Study*, Oxford University Press, New York.

Nonaka, I. & H. Takeuchi, (1995), *The Knowledge-Creating Company*, Oxford University Press, Boston.

Observatory of European Universities (2005), PRIME Network of Excellence, work in progress.

OECD (1996), "The Knowledge-Based Economy", Paris, OECD. OCDE/GD(96)102.

Prague Declaration (2001), "Towards the European Higher Education Area. Policy and Reform of Higher Education, Inter-University Cooperation, Academic Recognition and Mobility", Prague, May 19th.

Roberts, H., (1999), "The Control of Intangibles in the Knowledge-Intensive Firm", Communication presented at 22th Annual Congress European Accounting Association, Burdox.

Sabato, J., (1975), *El Pensamiento Latinoamericano en la Problemática Ciencia-Tecnología-Desarrollo-Dependencia*. Ed. Paidós, Buenos Aires.

Saint-Onge, H., (1996), "Tapping into the Tacit Knowledge of the Organisation", paper presented at the Knowledge Challenge Conference, MCE. Brussels, 30-31, May.

Sánchez, M. P. and S. Elena, (2005), "Managing Intellectual Capital in Public Universities. The Autonomous University of Madrid example", Working paper submitted to the 1st Workshop on "Visualising, Measuring, and Managing Intangibles and Intellectual Capital", Ferrara, 18-20 October.

Schultz, T. W., (1969), "Investment in Human Capital", [in:] E. S. Phelps (ed.), *The Goal of Economic Growth*, Norton, New York.

Solow, R., (1957), "Technical Change and the Aggregate Production Function", *Review of Economics and Statistics*, Vol. 39, pp. 312-20.

Sorbonne Joint Declaration, (1998), *Joint Declaration on Harmonisation of the Architecture of the European Higher Education System*, Paris, May 25.

Sveiby, K. E., (1997), "Measuring Intangibles Assets" (1997), http://sveiby.com.au/IntangAss/MeasureInatgibleAssets.html.

Vagnomi, E. & M. Castellini, (2005), "Designing Intellectual Capital Report: A Study on Health Care Organizations", Paper submitted to the 1st Workshop on "Visualising, Measuring, and Managing Intangibles and Intellectual Capital", Ferrara 18-20 October 2005.

Warden, C., (2003), "Managing and Reporting Intellectual Capital: New Strategic Challenges for HEROs" in IP Helpdesk Bulletin, nº 8, April-May 2003, http:// www.ipr-helpdesk.org/newsletter/8/pdf/EN/N08_EN.pdf.

5

Supporting High-Tech Start-Ups: Lessons from Italian Technology Policy

Massimo G. Colombo
Luca Grilli

INTRODUCTION

Economic analyses have identified two main rationales for governments to offer public direct subsidies to young high-tech firms. First, the social optimal level of R&D expenditures may be higher than the private optimal level due to the presence of R&D spillovers. Firms may invest less than the social optimum because they are unable to defend innovation and extract sufficient profits in the product market (Griliches 1992; Jaffe, 1996. On the difficulty to defend innovation by small and young high-tech firms, see also Teece 1986). Second, a large body of empirical literature on entrepreneurship has pointed to the presence of important financial constraints suffered by new firms (Evans and Jovanovic 1989, Evans and Leighton 1989, Meyer 1990, Holtz-Eakin *et al.* 1994a and 1994b, Black *et al.* 1996, Blanchflower and Oswald 1998). Access to the credit market is also quite problematic, especially for new technology-based firms (NTBFs) (Westhead and Storey 1997, Storey and Tether 1998, Carpenter and Petersen 2002. For the Italian case, see Colombo and Grilli 2005a and Grilli 2005). Many obstacles to access debt financing by this typology of enterprises stem from the inability of banks and other similar financial institutions to discern good projects from "lemons" in sectors usually characterised by highly skewed returns, *ex-ante* (i.e. hidden information) and *ex-post* (i.e. hidden information), asymmetric information and a lack of internal collateral to secure debt (see Carpenter and Petersen 2002). On the other hand, even though external private equity financing has advantages over debt for NTBFs (see again Carpenter and Petersen 2002), this mode of financing may still present problems related to *ex-ante* asymmetric information (Myers and Mayluf 1984) and high transaction costs (Asquith and Mullins 1986, Lee *et al.* 1996), which inhibit access to seed and start-up equity capital for most new high-tech ventures.

Both arguments lead to the same conclusion: innovative high value-added projects may be disregarded and never be implemented. Naturally, from a social point of view this constitutes a missed opportunity and thus calls for public intervention. However several contributions have emphasized the distortions that may result from government subsidies. As articulated by Holtz-Eakin (2000) and Santarelli and Vivarelli (2002) failure rates are naturally high among new small firms, and public support may disturb and delay the competitive selection process, thus subsidising inefficiencies. Firstly, screening difficulties may be magnified for policy makers. Secondly, politicians may seek to direct subsidies in a manner that creates political consensus and benefits themselves rather than citizens (see Lerner 2002 and the literature mentioned there).

As recently suggested by Siegel *et al.* (2003), the assessment of the effectiveness of public policy measures designed to promote innovation in NTBFs and to solve market inefficiencies has become a key policy issue, given the potentially crucial role of high-tech enterprises in terms of economic growth and job creation. If there is great debate within the scientific community on the most appropriate policy evaluation methodologies (see European Commission Joint Research Centre 2002 for a comprehensive treatment of this issue), there is also little doubt on what makes a policy scheme effective. First of all, it has to avoid the generation of *substitution effects* in the sector. These occur when beneficiaries of the subsidy manage to stay in the market at the expense of more efficient potential entrants. Secondly, it has to avoid creating a situation whereby subsidies flow towards beneficiaries that do not need any public aid in order to realize their potential. If this occurs, the policy measure does not produce any truly positive additional result with respect to the *ex-ante* situation and generates only *deadweight effects.* Since most of the works evaluate the effectiveness of public programs mainly by looking at the performance of subsidized and non-subsidized firms (see Lerner and Kegler 2001, and Lerner 1999 and 2002 for contributions that refer to specific NTBFs-targeted schemes), we want to make a step back and focus on those characteristics which enable firms to obtain access to direct financial public subsidies. More precisely, the objective of this paper is to investigate whether the characteristics of subsidized NTBFs are compatible with a scenario in which public funds are efficiently allocated. Specifically, while focusing on NTBFs' access to national direct public subsidies, we aim at verifying whether existing policy measures with regard to high-tech sectors are well designed to provide assistance to businesses which are likely to succeed. In other words, we investigate whether or not existing policy measures are likely to have generated important *substitution effects* in high-tech sectors.

For this purpose, the Italian experience is very interesting. In fact Italy, as many other EU countries, has never had any national financial support scheme exclusively targeting young high-tech firms (see Storey and Tether 1998), though these firms have benefited from public assistance through measures that were also available to broader typologies of firms. Furthermore, there has been no systematic indirect support to NTBFs. So it is legitimate to question whether direct general-purpose policy measures have been effective in supporting NTBFs. Recognition of the fact that horizontal public effort has been largely inefficient would urge a re-consideration of the Italian technology policy towards this typology of firms and would call for the implementation of more specific and customised measures (i.e. a vertical approach) than those presently in existance.

In order to accomplish our goal, we conduct an empirical analysis based on a sample of 507 Italian firms that were established in the '80s and '90s and operate in the high-tech manufacturing and service industries. The sampled firms were independent at the time of their foundation and have remained so as of 1/1/2004 (i.e. they were not controlled by another business organization, even though other organizations may hold minority shareholdings in the new firms). Very detailed firm-level data is available on the firms' activity, structure, and performance, as well as on the characteristics of the founding teams, such as education level and prior working experience. And also on NTBFs' access to national public direct support measures. The data was provided by the RITA database developed at Politecnico di Milano.

The paper proceeds as follow: in the next, second, section we rely on the rich literature studying NTBFs' growth performances in order to identify those characteristics that distinguish the successful archetype among new high-tech firms. Section 3 defines a taxonomy of all direct support measures implemented by the Italian national authorities that have been accessed by NTBFs. In Section 4 we present the data set. Section 5 provides some descriptive statistics on NTBFs' access to national subsidies. In Section 6 we run an econometric analysis in order to test whether public funds have prevalently flown towards those NTBFs whose characteristics are typical of "gazelles" in high-tech sectors. The final section discusses our main findings and the paper concludes with policy implications.

NTBFS WITH HIGH-GROWTH PROSPECTS

Any policy measure that aims at being effective and maximising social welfare should primarily target those NTBFs that are potentially successful and are characterised by high-growth prospects. If this is the case, the risk of generating a *substitution* effect is alleviated. The second prerequisite for a successful policy intervention is to target those firms with high-growth prospects that suffer from financial and other market imperfections and need public aid in order to realize their potential. If this principle is observed, the policy measure has a very limited risk of generating a *deadweight* effect. While studying NTBFs' access to direct national subsidies, we investigate whether the first prerequisite is met, although little can be said as regards the second one. Clearly, an eventual failure of the existing horizontal policy measures to designed provide aid to NBTFs with high-growth prospects, and conversely evidence that public money largely flows towards NTBFs characterised by low-growth prospects should urge, of its own accord, a re-consideration of the Italian technology policy towards young high-tech firms.

Let us now focus on what makes a business successful in high-tech sectors. Several studies have indentified a precise set of determinants of the growth of young high-tech firms. Following Storey (1994) and Almus and Nerlinger (1999), we can separate them into three different categories.

The first category is represented by *founder-specific* characteristics. Founders' human capital is reputedly a primary asset for the competitive advantage of a NTBF[1]. The available empirical evidence generally, although not unanimously, lends support to the positive relationship between founders' competencies, measured by their level of educational attainments and professional experience, and firms' growth (see for example Bates 1990, Brüderl *et al.* 1992, Brüderl and Preisendörfer 2000, Almus 2002; for a survey see Storey 1994), and it is widely documented as well for NTBFs' sectors (Cooper and Bruno 1977, Feeser and Willard 1990, Westhead and Cowling 1995, Dahlstrand 1997, Nerlinger 1998, Almus and Nerlinger 1999; see Colombo and Grilli 2005b for the Italian case). As regards founders' educational level, Westhead and Cowling (1995) find that it has a positive and statistically significant impact on firms' growth, based on a sample of UK NTBFs. Colombo and Grilli (2005b) analyse a large sample of Italian NTBFs (the ones used in the present study) and provide evidence that post-graduate education, gained both in economic-managerial fields and to a lesser extent in technical-scientific subjects, has a positive impact on firms' growth. But in particular, it is the *specific* component of founders' human capital that exerts a positive influence on firms' performance. *Specific* human capital is commonly defined in the literature (see Becker 1975) as encompassing all those capabilities of individuals that can directly and profitably be applied to the entrepreneurial task in the newly created firm. It embraces the *industry-specific* skills that founders learned in the organization by which they were formerly employed and the *leadership-competencies* gained by entrepreneurs, either through a managerial position in another firm or in prior self-employment (see Preisendörfer and Voss 1990, Brüderl *et al.* 1992, Brüderl and Preisendörfer 2000). In this respect, Cooper and Bruno (1977) considered young high-tech firms in the '70s that were located in the San Francisco Peninsula. By comparing high-growth and terminated firms, they demonstrated that the former were more likely than the latter to have been founded by individuals who came from incubating organizations that operated in the same industry of the new firm. Later studies in the '80s and '90s generally found similar results as well for firms operating in low- and medium- tech

[1] See Cooper and Bruno (1977, p. 21): "for a new, high-technology firm, the primary assets are the knowledge and skills of the founders. Any competitive advantage the new firm achieves is likely to be based upon what the founders can do better than others".

industries (see for instance Cooper 1985, Dunkelberg *et al.* 1987, Chandler and Jansen 1992, Siegel *et al.* 1993; for an exception see Storey 1994). Furthermore, Feeser and Willard (1990), in comparing 39 high-growth computer producers with a matching set of low-growth firms, demonstrated that the former firms are more likely than the latter to have products, markets and technologies closely related to those of their founders' incubating organization. Similar results are found by Roberts (1992) and Dahlstrand (1997). Roberts considered a sample of NTBFs located in the Greater Boston area and showed that a rapid transfer of technology from an advanced technology-based incubating organization, together with a deep understanding of customer needs and market conditions, are key factors in determining a firm's performance. Dahlstrand, using a sample of new Swedish high-tech firms, documented that after a short initial period, corporate spin-off firms grow more rapidly than non spin-off ones. Lastly, Colombo and Grilli (2005b) show that founders' work experience, in particular when it is gained in the same sector of the start-up and in technical departments (R&D, design, engineering and production), has a positive impact on NTBFs' growth. The same applies for the level of entrepreneurial experience gained by founders before a firm's creation. The last *founder-specific* characteristic usually considered in the empirical literature concerns the number of founders of a firm. Larger founding teams are reputed to have access to more resources of both a tangible (financial funds) and intangible (competencies) nature, which stimulates more rapid growth in firms. Generally, the available empirical evidence lends support to such contentions (Teach *et al.* 1986, Feeser and Willard 1990, Eisenhardt and Schoonhoven 1990, Colombo and Grilli 2005b; but see Almus and Nerlinger 1999 for different results)[2].

The second group of factors that affect NTBFs' performance is represented by *firm-specific* characteristics. In particular, in several studies the ability of a NTBF to implement and maintain links with external firms and public research institutions is found to positively affect its performance (Nerlinger 1998, Almus and Nerlinger 1999, Maskell 2001). Similarly, Chiesa and Piccaluga (2000) and Colombo and Piva (2005) show that new Italian high-tech firms created by individuals with previous professional experience in academia perform better than the rest of Italian NTBFs over a range of dimensions, including innovative activity. Similar results were found in Sweden (Dahlstrand 1997). Furthermore, the fact that a newly-created firm has received valuable tangible and/or intangible resources from a "mother" company (e.g. complementary technologies, access to distribution channels, after-sale services, support for entry into international markets, financing) are likely to positively influence a firm's performance, especially in its infancy stage. Accordingly, Colombo *et al.* (2004) finds that it positively affects NTBFs' start-up size. The location in a technology incubator also has a positive impact on NTBFs' growth (see Colombo and Delmastro 2002, Sternberg 1990 and Westhead and Storey 1994

[2] Almus and Nerlinger (1999) do not detect any significant relation between team foundation and the growth of new high-tech firms, even though they do find a positive effect of team foundation for low-and medium-tech firms.

for Italy, Germany and the United Kingdom respectively; for opposite conclusions see Westhead and Cowling 1995, and Westhead 1997 on NTBFs located in the United Kingdom). Finally, a firm's legal status is found to affect NTBFs' growth. In particular, Almus and Nerlinger (1999) and Hamilton *et al.* (2002) show that NTBFs with a corporate limited liability status realize *ceteris paribus* higher growth rates than other typologies of new high-tech firms.

As concerns the third category, NTBFs' performance is likely to be influenced by the socio-economic characteristics of the area in which firms are located. More specifically, Colombo and Grilli (2005b) provide evidence that Italian NTBFs benefit from operating in well-infrastructured counties. Analogously, Audretsch and Dohse (2004) document that being located in a region characterised by a high availability of knowledge resources in terms of human capital is more conducive to a firm's growth than being located in less endowed areas. For more on the benefits that firms receive from being embedded in an innovative high-technology *milieux,* see also Capello (1999) and Breschi and Lissoni (2001), as well as Feldman (1999) for a survey on the topic.

The question thus arises: Do these determinants of NTBFs' growth performance also drive the allocation of direct public funds granted by national authorities? Before investigating this issue, the following sections are devoted both to defining a taxonomy of Italian national schemes providing direct support to NTBFs, and to illustrating some descriptive statistics on NTBFs' access to such measures.

TAXONOMY OF ITALIAN NATIONAL DIRECT SUPPORT SCHEMES

Table 1 presents a taxonomy of all Italian national policy measures that have provided direct financial aid to NTBFs. Italy lacks of any direct policy measure that explicitly targets NTBFs. Furthermore there has not been any systematic indirect support to this typology of firms. Nonetheless, Italian young high-tech firms may benefit from those national horizontal schemes implemented in order to sustain entrepreneurship. Overall, 28 national laws that have provided some type of financial support to Italian NTBFs have been classified using the following criteria.

Main objective. The main purpose of the scheme is identified, and a distinction is made between measures that aim at providing support to: (a) investments in R&D ; (b) general-purpose investments (i.e. machineries, new plants, employment, etc.); and (c) depressed areas (i.e. in particular the South of Italy and the area defined as "Mezzogiorno d'Italia")[3].

---

[3] The area defined as "Mezzogiorno d'Italia" includes the following regions: Abruzzo, Basilicata, Calabria, Campania, Molise, Puglia, Sardegna and Sicilia.

Table 1
Taxonomy of national direct support schemes covering Italian NTBFs.

Number of national schemes		**1329**	**1089**	**902**	**46**	**696**	**26**	**44**	**64**	**113**	**318**	**346**	**181**	**317**	**215**
Year of implementation		‘65	‘68	‘76	‘82	‘83	‘86	‘86	‘86	‘86	‘87	‘88	89	‘91	‘92
MAIN OBJECTIVE	R&D Investments		X		X							X		X	
	General-purpose Investments	X		X		X				X	X		X		X
	Depressed Areas						X	X	X						
EVALUATION METHOD	Automatic	X				X					X			X	X
	Discretionary		X	X	X		X	X	X	X		X	X		
MAIN INSTRUMENT	Fiscal Benefits or Incentives		X				X							X	X
	Financial Contributions	X	X	X	X	X	X	X	X	X	X	X	X		X
MAIN TARGET GROUP	All Firms	X	X	X	X		X			X		X	X		
	New Entrepreneurship or SMEs					X		X	X		X			X	X

Number of national schemes		**488**	**17**	**236**	**451**	**95**	**341**	**140**	**196**	**266**	**449**	**297**	**185**	**388**	**383**
Year of implementation		‘92	‘93	‘93	‘94	‘95	‘95	‘97	‘97	‘97	‘97	‘99	‘00	‘00	‘01
MAIN OBJECTIVE	R&D Investments							X				X			
	General-purpose Investments		X	X	X	X			X	X	X			X	X
	Depressed Areas	X					X						X		
EVALUATION METHOD	Automatic		X		X		X		X	X	X	X	X	X	X
	Discretionary	X		X		X		X							
MAIN INSTRUMENT	Fiscal Benefits or Incentives		X		X		X	X	X	X	X	X		X	X
	Financial Contributions	X		X		X	X					X	X		
MAIN TARGET GROUP	All Firms	X	X	X	X		X	X				X		X	X
	New Entrepreneurship or SMEs					X			X	X	X		X		

Evaluation method. National schemes are distinguished based on an evaluation method of applicants. In particular we distinguish between (a) automatic and (b) discretionary measures. An automatic scheme gives financial assistance to all applicants provided that the applicants fulfil all the requirements specified in the law. A discretionary scheme provides financial support to selected applicants. In this latter case, applicants compete for receiving a financial subsidy and they (or their projects) are judged most of the time by committees formed by experts who are nominated by the national authority.

Main instrument. Laws are classified on the basis of the financial instrument used in order to sustain entrepreneurship. In particular, we identify the following two categories: (a) fiscal benefits or incentives; (b) financial contributions.

Main target group. National schemes may be devoted to support: (a) all firms; (b) new entrepreneurship and small and medium-sized enterprises (SMEs).

Our analysis of the taxonomy reveals some interesting tendencies. First of all, policy makers have shown in the past a neat preference for implementing measures that support general-purpose investments (i.e. machineries, new plants, employment, etc.) rather than schemes with a more focussed objective, such as the support of R&D activities or aid to economically depressed areas. Secondly, it is not possible to derive a strong emphasis of the regulator on the basis of a particular evaluation method: both automatic and discretionary modalities are equally pursued (15 and 13, respectively) even if there is a clear tendency in recent years to favour the automatic scheme. Thirdly, legislators have historically preferred to grant public direct subsidies more as financial contributions than as fiscal benefits or incentives (19 versus 14 schemes). However, also in this case the tendency seems to be reversing itself and most of the recent policy measures provide tax credits. Finally, the policy makers' propensity to implement very general-purposes measures is confirmed by the analysis of the typology of the main beneficiaries of the schemes: 18 policy measures are directed to support all firms while the number of schemes devoted to stimulate the creation of new entrepreneurship or to sustain SMEs (whose NTBFs usually represent a subset) is 11.

DATA

In this paper we consider a sample composed of 507 Italian NTBFs. The sampled firms were established in 1980 or later, were independent at the time of creation and have remained so of 1/1/2004 (*i.e.* they were not controlled by another business organization even though other organizations may hold minority shareholdings in the new firms), and operate in the following high-tech sectors, in manufacturing and services: computers, electronic components, telecommunication equipment, optical, medical and electronic instruments, biotechnology, pharmaceuticals, advanced materials, robotics and process automation equipment, multimedia content, software, Internet (e-commerce, ISP, web-related services), and telecommunication services.

The sample of NTBFs was extracted from the RITA (Research on Entrepreneurship in Advanced Technologies) database (see also Colombo *et al.* 2004, Colombo and Grilli 2005b), developed at Politecnico di Milano. The development of the database went through a series of steps.

First, Italian firms that complied with the above mentioned criteria relating to age and sector of operations were identified. For the construction of the target population a number of sources were used. These included lists provided by national industry associations, on-line and off-line commercial firm directories, and lists of participants in industry trades and expositions. Information provided by the national financial press, specialized magazines, other sectoral studies, and regional Chambers of Commerce was also considered. Altogether, we collected information in the public domain for 1974 firms. For each firm, a contact person (i.e. one of the owner-managers) was also identified. Unfortunately, data provided by official national statistics did not allow us to obtain a reliable description of the universe of Italian NTBFs[4]. Secondly, a questionnaire was sent to the contact person of the target firms either by fax or by e-mail. The first section of the questionnaire provides detailed information on the human capital characteristics of firms' founders. The second section comprises further questions concerning the characteristics of the firms, including their eventual access to public support measures granted by the national authority. Lastly, answers to the questionnaire were checked for internal coherence by educated personnel. In several cases, phone or face-to-face follow-up interviews were made with firms' owner-managers. This final step was crucial in order to obtain missing data and ensure that the data used was reliable[5].

The sample used in the present work consists of 507 NTBFs, all of which are RITA firms for which we were able to create a complete data set[6]. The only exceptions concern data on previous entrepreneurial experiences of firms' founders and NTBFs' corporate status, which was available only for a sub-sample of 437 firms. Note that only firms having survived up to the survey date could be included in the sample. In principle, attrition may generate a sample selection bias that is difficult to control. As long as the failure rates of new firms decrease in proportion to the firms' access to public funds, our data might overestimate the use of national support schemes by Italian NTBFs. This notwithstanding, the sample is quite large and it exhibits considerable heterogeneity as to the variables which are of interest to us. In addition, the information relating to NTBFs, and in particular to the human capital of their founders, is more fine-grained than in previous datasets of similar size.

To our knowledge this is the first large scale empirical study to evaluate Italian NTBFs' recourse to public direct support measures granted by the national authority.

---

[4] The main problem is that in Italy most individuals who are defined as "self-employed" by official statistics actually are salaried workers with atypical employment contracts. Unfortunately, on the basis of official data such individuals cannot be distinguished from entrepreneurs who created a new firm.

[5] Note that for only 3 firms, the set of owner-managers at survey date did not include at least one of the founders of the firm. For these firms information relating to the human capital characteristics of the founders was checked through interviews with firms' personnel so as to be sure that it did not relate to current owner-managers.

[6] χ^2 tests show that there are no statistically significant differences between the distributions of the sample firms across industries and regions and the corresponding distribution of the population of 1974 of RITA firms from which the sample was obtained ($\chi^2(4)= 3.3$ and $\chi^2(3)= 4.8$, respectively).

ITALIAN NTBFS' ACCESS TO ITALIAN NATIONAL DIRECT SUPPORT SCHEMES

Distribution of subsidized NTBFs by industry, age and area

The composition of sample firms by sector and geographic area of operation is presented in Table 2. In particular, column 1 refers to the entire sample of NTBFs, while columns 2 and 3 report the number and the percentage of subsidized enterprises per sector and per geographic area. Overall, 123 firms (24% of the sample) have benefited from public direct subsidies granted by the Italian national authorities. Firms operating in Internet and TLC services are the least subsidized ones (17%), while for the other ICT services, there is a substantial percentage of subsidized firms within the software house (27%) and the multimedia contents (28%) sectors. Almost a third (30%) of the firms operating in the ICT manufacturing industry (computers, electronic components, telecommunication equipment, electronic instruments) have gained access to national public direct aid; the percentage is slightly less for the biotechnology, pharmaceutics and advanced materials sector (27%) and noticeably lower for the automation and robotics sectors (21%).

Table 2
Distribution of subsidized Italian NTBFs, by industry and geographic area

Industry	Firms	Subsidized firms	
	N.	N.	%
Internet & TLC services	154	26	17%
Software	151	41	27%
Multimedia contents	25	7	28%
ICT manufacturing	113	34	30%
Biotechnology & Pharmaceutics	22	6	27%
Automation & Robotics	42	9	21%
Area	Firms	Subsidized firms	
	N.	N.	%
North-West	240	52	22%
North-East	115	20	17%
Centre	89	18	20%
South	63	33	52%
Total	**507**	**123**	**24%**

Note: North-West includes the following regions: Liguria, Lombardia, Piemonte, Val d'Aosta. North-East includes the following regions: Emilia Romagna, Friuli Venezia Giulia, Trentino Alto Adige, Veneto. Centre includes the following regions: Lazio, Marche, Toscana, Umbria. South includes the following regions: Abruzzo, Basilicata, Calabria, Campania, Molise, Puglia, Calabria, Sardegna, Sicilia.

As concerns distribution by geographic area, Table 2 clearly shows that if most of NTBFs are located in the North and the Centre of Italy (240 in the North-West, 115 in the North-East, 89 in the Centre, representing 88% of the overall sample), their access to direct policy measures is fairly low (22% of the firms in the North-West are subsidized, 17% in the North-East, 20% in the Centre) compared to southern enterprises (12% of the sample) that present a greater sensitivity to access subsidies (52% of them have received a direct subsidy by the national authorities). The reason for this difference traces back to the high number of policy schemes that give exclusive or preferential access to firms located in this area of the country.

Table 3 illustrates the distribution of subsidized firms by year of application. Only 10% of subsidized NTBFs managed to access public direct funds during the year of their creation, while 26% were in an age ranging from between 1 and 3 years old, and almost two thirds (64%) of the supported NTBFs benefited from public aid when they were four years old or even older at the time of application. Again this picture raises a significant concern, since the seed and start-up phases are the most critical moments in a firm's life, and this holds true especially in high-tech sectors where young firms are more likely to suffer from important imperfections in capital markets and consequently there is more need and room for public intervention.

Table 3
Distribution of subsidized Italian NTBFs based on year of application

Firms' age at the time of application	Subsidized firms	
	N.	%
0	12	10%
1-3	32	26%
Over 4	79	64%
Total	**123**	**100%**

NTBFs' access to national direct support measures by typology of scheme

Table 4 illustrates the number and the percentage of subsidized NTBFs by typology of scheme, following the taxonomy provided by Table 1. As concerns the main objective, there are no significant differences in the number of subsidized NTBFs between those that got support for R&D investments (44%) and those being supported for undertaking general-purpose investments (42%). The percentage of subsidized NTBFs that have accessed policy schemes devoted to sustain the South of Italy is lower (31%), but extremely high if compared to the small number of NTBFs located in this area of the country. If subsidized NTBFs show a greater sensitivity to accessing discretionary rather than automatic subsidies (63% against 47%), there is an equal tendency to take into consideration the main instrument used to grant the subsidy: 65% and 67% of the subsidized NTBFs have benefited from a fiscal benefit or incentive, or a financial contribution, respectively. Lastly, most subsidized NTBFs have received direct financial aid from a policy scheme directed towards all firms (78%), while only 31% of the supported firms have benefited from subsidies that targeted new entrepreneurship or small- and medium-sized enterprises.

Table 4
Italian NTBFs' access to national direct support measures by typology of scheme

Typology	Subsidized firms	
	N.	%
Main Objective		
R&D Investments	54	44%
General-purpose Investments	52	42%
Depressed Areas	38	31%
Evaluation Method		
Automatic	58	47%
Discretionary	77	63%
Main Instrument		
Fiscal Benefits or Incentives	80	65%
Financial Contributions	77	63%
Main Target Group		
All Firms	96	78%
New Entrepreneurship or SMEs	38	31%
Total	**123**	**100%**

THE ECONOMETRIC ANALYSIS

NTBFs' access to national direct support measures

In order to highlight to what extent the characteristics of NBTFs with high-growth prospects NTBFs are also predictors of NTBFs' access to national direct support schemes, we have conducted a probit model, where the dependent variable indicates whether firms have received public financial assistance granted by the national authority (1) or not (0) during their lives. Explanatory variables capture those *founder-*, *firm-specific* and *external* characteristics that the empirical literature on firms' performance has shown to significantly influence NTBFs' growth (see Section 2). This econometric analysis will enable us to evaluate whether the overall process of allocating national public funds to Italian young high-tech firms has been efficient or not.

Explanatory variables

Independent variables of econometric models are described in Table 5. They can be subdivided into three groups. Starting from *founder-specific* characteristics, we consider the education attainments of founders; in particular as concerns graduate and post-graduate education, and we distinguish between economic and managerial studies (*Ecoeduc*) and technical and scientific studies (*Techeduc*)[7].

[7] More precisely, *Ecoeduc* measures the years spent for the attainment of degrees in economics, management, and political sciences, while *Techeduc* reflects years spent for obtaining degrees in engineering, chemistry, physics, geology, mathematics, biology, medicine, pharmaceutics, and computer science. In order to properly judge the effective

Table 5
Determinants of Italian NTBFs' access to national direct support schemes

Variable	Description
Ecoeduc	Average number of years of economic and/or managerial education of founders at graduate and post-graduate level
Techeduc	Average number of years of scientific and/or technical education of founders at graduate and post-graduate level
Specworkexp	Average number of years of work experience of founders in the same sector of the start-up before firm's foundation
Otherworkexp	Average number of years of work experience of founders in other sectors than the one of the start-up before firm's foundation
DManager	One for firms with one ore more founders with a prior management position in a company with more than 100 employees
DEntrepreneur	One for firms with one or more founders with previous self-employment experience
NFounders	Number of founders
DIncubated	One for firms located in a technology incubator at time of foundation
DMother company	One for firms that at time of foundation received some kind of aid by a "mother" company
DAcademic	One for firms with at least one individual within the founding team with previous work experience as a researcher in academia or in a public research centre
DLimited liability	One for firms with a limited liability corporate status
Age	Number of years since firm's foundation
Locdevelop	Value of the index measuring regional infrastructures in 1989 (mean value among Italian regions=100; source: Centro Studi Confindustria 1991)[a]
North-East	One for firms located in the North-East of Italy
Centre	One for firms located in the Centre of Italy
South	One for firms located in the South of Italy

Note: [a]data source is Confindustria Centro Studi (1991).

Secondly, we distinguish between founders' years of work experience in the same sector of the new firm (*Specworkexp*) and years of work experience in other sectors *(Otherworkexp)*. As stated earlier, it is widely documented in the empirical literature that it is the *specific* rather than the *generic* component of founders' work experience that is found to positively influence NTBFs' performances. The *leadership experience* possessed by founders is indicated by

level of competencies of founders, we consider the minimum length of time necessary to attain a certain degree. In order to attain an Italian graduate degree in economics, management, political sciences and most scientific degrees four years of studies are required, while five years is the minimum time for a degree in engineering and chemistry. Master and Ph.D. programmes require one and three additional years respectively, independently of the specific field.

DManager and *DEntrepreneur*. The former is a dummy variable which equals 1 if within the founding team there are one or more individuals who had a managerial position in a medium or large company (*i.e.* number of employees greater than 100) prior to the establishment of the new firm. *DEntrepreneur* equals 1 if one or more founders had prior self-employment experience. Lastly, *NFounders* represents the number of NTBFs' founders.

As concerns *firm-specific* characteristics, *DMother company* is a dummy variable that equals 1 if a new-born firm received at start-up time valuable tangible and/or intangible resources from a "mother" company (e.g. complementary technologies, access to distribution channels, after-sale services, support to entry into international markets, financing). *DAcademic* equals 1 if within the founding team there are one or more individuals with previous work experience as a researcher in academia or in a public research institution. The dummy variable *DIncubated* identifies firms located in technology incubators, while *DLimited liability* assumes 1 for firms with a corporate limited liability status.

Table 6
Descriptive statistics of explanatory variables of the econometric models

Variable	Mean	S.D.	Min	Max
Ecoeduc	0.313	0.897	0.000	5.000
Techeduc	1.815	2.228	0.000	8.000
Specworkexp	3.969	6.101	0.000	35.500
Otherworkexp	7.394	8.066	0.000	49.000
DManager	0.093	0.290	0.000	1.000
DEntrepreneur[a]	0.398	0.490	0.000	1.000
NFounders	2.736	1.801	1.000	21.000
DIncubated	0.083	0.276	0.000	1.000
DMother company	0.069	0.254	0.000	1.000
DAcademic	0.114	0.319	0.000	1.000
DLimited liability[a]	0.620	0.485	0.000	1.000
Age	9.016	5.961	0.000	23.000
Locdevelop	115.722	27.306	43.700	174.700

Note: [a]statistics on DEntrepreneur and DLimited liability are based on 437 observations.

The third group includes variables capturing the socio-economic environment in which firms are located. *Locdevelop* reflects the level of economic development in 1989 of the county where firms are located (source: Centro Studi Confindustria 1991). It is calculated as the average of the following indexes: per capita value added, share of manufacturing out of total value added, employment index, per capita bank deposits, automobile-population ratio, and consumption of electric power per capita. Three geographic dummies (*North-East*, *Centre*, *South*) identify the macro-areaion which NTBFs are located. Finally, we also introduce five industry dummies in order to take into account any possible *industry-specific* effects. Table 6 reports some descriptive statistics on explanatory variables (for the sake of synthesis we omit repetition of the statistics on geographic and industrial variables).

Results of the estimates

The results of the estimates of a probit model on NTBFs' access to national direct support measures are reported in Table 7. In particular, Model 1 includes as explanatory variables three geographic dummies (*North-East*, *Centre*, *South*) and five industry dummies[8]. In Model 2 the three geographic dummies are replaced with the variable *Locdevelop*, which is introduced in a separated model in order to avoid multiple co-linearity problems[9]. Then, Model 3 and Model 4 differ from Model 1 and Model 2, respectively, by the inclusion of *DEntrepreneur* and *DLimited liability* as explanatory variables. These variables are introduced in separate models because information on the previous entrepreneurial experiences of firms' founders and the corporate limited liability status of NTBFs were available only for a sub-sample of 437 firms. Overall, the statistical significance of the estimates is very low: values of R^2 McFadden range from 0.089 to 0.145. With very few exceptions, those *founder-* and *firm-specific* attributes that characterise young successful enterprises in high-tech sectors demonstrate a very poor explanatory power as regards Italian NTBFs' access to national public funds. As to the former characteristic, the number of founders, the level of their *specific* work experience[10] and the level of the *leadership experience* gained by them via previous managerial and entrepreneurial episodes does not exert any role in discriminating subsidized from non-subsidized NTBFs. *Ecoeduc* has a positive but statistically weak impact and only the coefficient of *Techeduc* turns out to be positive and statistically significant. However, note that Colombo and Grilli (2005b) show that economic and managerial graduate education has a greater and more statistically robust impact on Italian NTBFs' growth rate than graduate education gained in technical-scientific fields. As concerns *firm-specific* characteristics, *DIncubated* is the only variable that shows a statistically significant coefficient in all models, even if only at a 90% confidence level. The result confirms the useful role of "bridging institutions" between entrepreneurship and the public sector that technology incubators may play in Italy, as already highlighted by Colombo and Delmastro (2002). *DMother company* and *DAcademic* are never significant, while the variable

[8] The coefficients of these latter variables are overall statistically insignificant here and in all the subsequent models, and are not reported for the sake of synthesis.

[9] Apart from *Locdevelop* and the three geographic dummies (*North-East*, *Centre*, *South*) correlations among explanatory variables are generally low. The correlation matrix is available upon request from the authors.

[10] Note that we also decompose industry-specific work experience in *Techworkexp* and *Comworkexp*. The former variable measures the mean number of years of work experience of founders in the R&D, design, engineering and production departments of firms that operate in the same sector of the NTBF under consideration; the latter is the corresponding measure relating to marketing, sales and customer care functions. None of these variables present a statistically significant coefficient in the assessments of models.

DLimited liability presents a positive and statistically significant coefficient (at 95%), both in Model 3 and Model 4. The age of the firm has a positive and statistically significant impact on the probability that a NTBF has accessed national subsidies during its lifetime. Turning to variables that capture the socio-economic environment in which firms are located, as already highlighted before (see again Tables 2 and 4), southern NTBFs demonstrate a substantial advantage in accessing national public funds. More generally, national public aid flows more intensively towards young high-tech firms located in less developed areas, while the empirical literature documents that less equipped areas in terms of infrastructures and knowledge resources may depress firms' growth.

Table 7
Probit models on Italian NTBFs' access to national direct support schemes

		Model 1		Model 2		Model 3		Model 4	
a_0	Constant	-1.618***	(0.339)	-0.707*	(0.401)	-1.605***	(0.359)	-0.747*	(0.431)
a_1	Ecoeduc	0.089	(0.077)	0.102	(0.076)	0.045	(0.094)	0.073	(0.092)
a_2	Techeduc	0.107***	(0.032)	0.114***	(0.031)	0.092***	(0.036)	0.101***	(0.035)
a_3	Specworkexp	0.010	(0.037)	0.010	(0.036)	-0.003	(0.041)	0.000	(0.040)
a_4	Othworkexp	0.006	(0.013)	0.006	(0.013)	-0.004	(0.015)	-0.002	(0.015)
a_5	DManager	0.192	(0.010)	0.204	(0.009)	0.203	(0.011)	0.198	(0.011)
a_6	DEntrepre-neur	-	(0.227)	-	(0.223)	0.219	(0.244)	0.184	(0.239)
a_7	NFounders	-0.021	-	-0.005	-	-0.029	(0.156)	-0.007	(0.151)
a_8	DIncubated	0.448*	(0.237)	0.439*	(0.233)	0.452*	(0.249)	0.461*	(0.244)
a_9	DMother company	0.000	(0.281)	-0.043	(0.273)	0.028	(0.318)	-0.065	(0.311)
a_{10}	DAcademic	0.254	(0.222)	0.134	(0.215)	0.157	(0.240)	0.040	(0.235)
a_{11}	DLimited liability	-	-	-	-	0.398**	(0.163)	0.380**	(0.158)
a_{12}	Age	0.047***	(0.012)	0.045***	(0.012)	0.039***	(0.013)	0.039***	(0.013)
a_{13}	Locdevelop	-	-	-0.008***	(0.002)	-	-	-0.008***	(0.003)
a_{14}	North-East	-0.215	(0.181)	-	-	-0.183	(0.193)	-	-
a_{15}	Centre	0.026	(0.190)	-	-	-0.025	(0.211)	-	-
a_{16}	South	1.053***	(0.202)	-	-	1.077***	(0.224)	-	-
	N. of observation	507		507		437		437	
	Log-likelihood	-244.231		-255.699		-205.974		-215.670	
	R^2 McFadden	0.130		0.089		0.145		0.104	

Legend: * p < .10; ** p < .05; *** p < .01. For the sake of synthesis we omit to report estimated coefficients of industry dummies. Standard errors in parentheses.

Overall, the picture that emerges is one of an inefficient allocation of public funds granted by the national authorities to young Italian firms operating in high-tech sectors. Most of the variables that influence NTBFs' performance are found to have a very poor explanatory power as regards NTBFs' access to horizontal direct national subsidies. The risk that such policy measures have only generated *substitution effects* in high-tech sectors is thus very high.

Table 8
Multivariate and bivariate probit models on Italian NTBFs' access to national direct support measures by typology of scheme

		Model 1						Model 2			
		R&D Investments		General-purpose Investments		Depressed Areas		Automatic		Discretionary	
a_0	Constant	-2.048***	(0.525)	-2.038***	(0.576)	-2.626***	(0.766)	-1.961***	(0.560)	-2.204***	(0.452)
a_1	Ecoeduc	0.046	(0.145)	0.123	(0.103)	0.032	(0.163)	0.109	(0.101)	0.067	(0.105)
a_2	Techeduc	0.071	(0.052)	0.062	(0.043)	0.087	(0.071)	0.106**	(0.042)	0.081*	(0.046)
a_3	NFounders	-0.041	(0.078)	0.012	(0.047)	-0.020	(0.120)	-0.017	(0.046)	-0.020	(0.069)
a_4	Specworkexp	0.016	(0.023)	0.011	(0.020)	-0.003	(0.036)	0.007	(0.021)	0.017	(0.019)
a_5	Othworkexp	0.008	(0.014)	0.001	(0.015)	0.010	(0.025)	0.014	(0.014)	0.005	(0.013)
a_6	DManager	0.217	(0.357)	-0.101	(0.349)	0.504	(0.511)	-0.127	(0.336)	0.359	(0.289)
a_7	DIncubated	0.458	(0.456)	0.131	(0.338)	0.821*	(0.473)	0.186	(0.330)	0.583*	(0.341)
a_8	DMother company	-0.055	(0.657)	0.284	(0.408)	-0.664	(2.691)	0.215	(0.418)	-0.233	(0.538)
a_9	DAcademic	0.375	(0.324)	0.166	(0.229)	-0.177	(0.655)	0.325	(0.268)	-0.072	(0.360)
a_{10}	Age	0.059***	(0.022)	-0.003	(0.018)	0.075***	(0.027)	-0.017	(0.018)	0.084***	(0.018)
a_{11}	North-East	-0.204	(0.270)	-0.052	(0.279)	-0.429	(0.507)	-0.278	(0.313)	-0.156	(0.237)
a_{12}	Centre	-0.140	(0.345)	0.192	(0.268)	0.407	(0.478)	0.131	(0.244)	-0.073	(0.285)
a_{13}	South	0.072	(0.325)	0.667**	(0.259)	1.614***	(0.336)	0.681***	(0.231)	1.076***	(0.244)
	ρ (1,2)	0.133 (0.163)						0.095 (0.136)			
	ρ (1,3)	0.354* (0.188)						-			
	ρ (2,3)	-0.208 (0.263)						-			
	Log-likelihood	-396.871						-336.622			

Table 8 - continued

		Model 3				*Model 4*			
		Fiscal Benefits or Incentives		*Financial Contributions*		*All Firms*		*New Entrep. or SMEs*	
a_0	*Constant*	-1.790***	(0.403)	-2.406***	(0.505)	-1.728***	(0.415)	-2.309***	(0.573)
a_1	*Ecoeduc*	0.072	(0.098)	0.029	(0.108)	0.062	(0.090)	0.143	(0.132)
a_2	*Techeduc*	0.061	(0.037)	0.124***	(0.046)	0.069*	(0.037)	0.129**	(0.061)
a_3	*NFounders*	-0.015	(0.047)	-0.029	(0.061)	-0.054	(0.059)	0.036	(0.044)
a_4	*Specworkexp*	0.007	(0.016)	0.010	(0.019)	0.005	(0.016)	0.017	(0.028)
a_5	*Othworkexp*	0.009	(0.011)	-0.001	(0.014)	0.007	(0.011)	0.006	(0.020)
a_6	*DManager*	0.100	(0.270)	0.451	(0.289)	0.297	(0.263)	-0.221	(0.456)
a_7	*DIncubated*	0.329	(0.288)	0.629**	(0.308)	0.487	(0.304)	0.253	(0.363)
a_8	*DMother company*	0.382	(0.338)	-0.335	(0.607)	-0.276	(0.420)	0.185	(0.479)
a_9	*DAcademic*	0.299	(0.247)	0.201	(0.303)	0.125	0.267)	0.367	(0.320)
a_{10}	*Age*	0.020	(0.014)	0.074***	(0.017)	0.053***	(0.014)	0.021	(0.022)
a_{11}	*North-East*	-0.240	(0.252)	-0.054	(0.257)	-0.176	(0.224)	-0.202	(0.350)
a_{12}	*Centre*	-0.002	(0.226)	0.136	(0.272)	0.075	(0.216)	0.123	(0.381)
a_{13}	*South*	0.581**	(0.226)	1.252***	(0.233)	0.989***	(0.221)	0.748**	(0.298)
	$\rho(1,2)$	0.550*** (0.091)				0.032 (0.170)			
	$\rho(1,3)$	-				-			
	$\rho(2,3)$	-				-			
	Log-likelihood	-362.383				-330.980			

Note: * p < .10; ** p < .05; *** p < .01. For the sake of synthesis we omit the estimated coefficients of industry dummies. Standard errors in parentheses. Number of observations is 507.

Determinants of NTBFs' access to different typologies of national direct support measures

As highlighted in Section 3, NTBFs have benefited from a large number of public direct support measures with different characteristics. Thus, the following analysis aims at investigating whether inefficiencies are associated with particular typologies of policy schemes or, on the contrary, embrace the whole set of public support measures. To investigate in-depth the issue we used the taxonomy of Table 1 and ran a series of multivariate and bivariate probit models, where policy schemes are differentiated according to the following dimensions: main objective (R&D investments, General-purpose investments, Depressed areas) in Model 1, evaluation method (Automatic, Discretionary) in Model 2, main instrument (Fiscal benefits or incentives, Financial contributions) in Model 3, and main target group (All firms, New entrepreneurship or SMEs) in Model 4. In these types of models, a dependent variable takes 1 if a NTBF accessed a public subsidy with one of the above-mentioned characteristics. The analysis is run including the three geographic dummies *North-East*, *Centre*, *South* in the set of explanatory variables[11]. The results of the estimates (see Table 8) confirm the low explanatory power of *founder-* and *firm-specific* variables in all models[12]. In other words, inefficiencies are associated with all typologies of horizontal direct schemes and may not be imputed to only particular policy measures. Also, the multivariate and bivariate probit analyses enable us to gain some further insights on NTBFs' access to public direct aid.

First of all, *DIncubated* turns out to have a positive and statistically significant impact on NTBFs' access to national direct support schemes only if these latter are devoted to the support of depressed areas (Model 1), if they present a discretionary evaluation method (Model 2), and if they grant subsidies as financial contributions (Model 3). This result highlights the fact that location in a technology incubator, especially in areas characterized by a weak innovation system, may ease the information flow from legislators to entrepreneurs and allow firms to gain a first mover advantage over other

[11] Results are very similar when we substitute the three geographic dummies with the variable *Locdevelop*. They are available upon request from the authors.

[12] As concerns the *founder-specific* variables, only *Techeduc* turns out to be positive and significant in some of the equations of the models. In particular, graduate and post-graduate education in technical and scientific studies: a) do not discriminate NTBFs' access to public subsidies in terms of the main objective of policy schemes (Model 1); b) has a positive and statically significant impact on the probability of NTBFs' access to both automatic and discretionary measures (Model 2); c) has a positive and statistically significant impact only on the probability of NTBFs' access to financial contributions while not exerting any impact in the case of subsidies granted as fiscal benefits or incentives (Model 3); d) has a positive and statically significant impact on the probability of NTBFs' access to measures that target both all firms and new entrepreneurship or small-medium sizedenterprises (Model 4). Note that Wald tests accept the null hypothesis that both *founder-* and *firm-specific* variables have a negligible impact in all models of Table 8. Results are available upon request from the authors.

applicants[13]. Furthermore, technology incubators may offer consultancy services to incubated firms that help them in projecting and filling out application forms, especially as concerns discretionary subsidies, where such a process may be difficult and very time-consuming.

The coefficient of *Age* is found to be statistically insignificant in the equations regarding generic-purpose investments (Model 1), automatic subsidies (Model 2), fiscal benefits or incentives (Model 3), and new entrepreneurship or SMEs (Model 4). This implies that the hazard rate for such schemes (the instantaneous probability that a NTBF gains access to the subsidy if it has not done so before) decreases over time[14]. In other words, young high-tech start-ups present a comparatively greater sensitivity than "old" NTBFs to accessing horizontal direct national support schemes with the above-mentioned characteristics. Apart from the results of Model 4 that seem trivial, some new insights on the nature of inefficiencies may be drawn from the other findings. In particular, on an *a priori* ground, fiscal benefits or incentives are less suited than financial contributions to support very young firms, since start-ups are very likely to register negative operating margins in the short-run. But it is also true that this typology of schemes may represent the only viable way for Italian NTBFs to access direct public support measures. In fact, fiscal benefits or incentives are more often associated with automatic subsidies, while several discretionary schemes grant support as a financial contribution (see again Table 1). First, automatic rather than discretionary subsidies call for a lower level of effort on the part of firms in terms of requirements to fulfill and documentation to present; second and more importantly, automatic subsidies make firms certain to get the subsidy provided that the schemes' requirements are fulfilled. Therefore, all else being equal, firms' opportunity costs in applying for automatic subsidies granting fiscal benefits or incentives may be much lower than those corresponding to the application for discretionary support measures providing financial contributions. As a matter of fact, the youngest and smallest NTBFs may find these latter costs unaffordable, and consequently they may face severe limitations and conspicuous barriers in accessing this latter form of subsidies. Furthermore, note also that even in the case of the conspicuous participation of very young firms in the selection procedure, policy makers could use their discretionary power to favor mature and solid firms, which are usually characterized by a low probability of failure, in order to avoid negative judgments and criticism by public constituents as regards the choices they made (Lerner 1999).

Location in the South of Italy has a positive impact on NTBFs' access to all types of policy schemes, excepting those that aim at supporting firms' R&D

---

[13] For example, law no. 140/97, one of the most important discretionary schemes, includes the chronological order of applications as one of the criteria for selecting beneficiaries.

[14] If the yearly probability for a NTBF of receiving a public subsidy is positive and constant over time, we should in fact observe a coefficient of *Age* positive, statistically different from null.

activities. This result brings us to an important policy consideration: if the support to economically depressed areas and to firms' R&D investments may be both important objectives for any national industrial policy, they do not have necessarily have be pursued through the implementation of the same policy measures. In fact, although some of the measures that support R&D investments provide some incentives to southern enterprises to apply (law. no. 46/82 and law no. 317/91), southern NTBFs do not present any comparative advantage in accessing R&D subsidies with respect to NTBFs located in other areas. Allegedly, this confirms that firm-level subsidies may be of scarce relevance for NTBFs located in depressed areas if these measures are not inserted into a complex technology policy that also aims at reducing the infrastructure gap suffered in disadvantaged areas[15].

CONCLUDING REMARKS

The paper has investigated whether direct horizontal policies with general objectives are effective in supporting NTBFs, or conversely whether this typology of enterprises needs more specific and customised policy schemes (i.e a vertical technology policy approach). Young high-tech firms deserve particular attention, given their strategic role in terms of overall economic growth and the "near universal recognition of the presence of market failures in the provision of finance for new technology-based firms" (Storey and Tether 1998, p. 1049. See also Carpenter and Petersen 2002). An effective public intervention should primarily target those firms that are potentially successful and are characterised by high-growth prospects. If this is the case, the risk of generating a *substitution* effect is alleviated. This latter term describes a situation where inefficient firms manage to stay on the market longer than they would have done in a "normal" scenario. Accordingly, a subsidy may create an artificial incumbency and prevent entry onto the market of potentially efficient, but non-subsidized, firms (Santarelli and Vivarelli 2002). The second prerequisite for a successful policy measure is to target those firms with high-growth prospects that suffer from market imperfections and need public aid in order to realize their potential. If

---

[15] Note also that the correlation parameter ρ between the error terms of the two equations relative to R&D and depressed areas is positive and statistically significant at conventional confidence levels. This implies that unobserved heterogeneity (and hence unrelated to the determinants of NTBFs' performance) spurs access to both types of policy schemes. The only other ρ coefficient found to be significant is the one relative to Model 3 that highlights the presence of a positive interdependence between NTBFs' access to fiscal benefits or incentives and NTBFs' probability to benefit from financial contributions.

We estimated all models of Table 8, including the variables *DEntrepreneur* and *DLimited liability* as regressors. Estimates are based on 437 observations. The results are very similar to those presented here (available upon request to the authors) and, as concerns the two above-mentioned variables: *DEntrepreneur* has an insignificant impact in all models, while *DLimited liability* presents a positive and statistically significant coefficient only in the R&D Investments (Model 1), the Discretionary (Model 2), the Fiscal Benefits or Incentives (Model 3) and the All Firms (Model 4) equations.

this principle is observed, the policy measure has a very limited risk of generating a *deadweight* effect. For the purpose of the paper, the Italian experience is very interesting. In fact, in Italy, there has never been any direct support scheme exclusively targeted to young high-tech firms; furthermore, there has been no systematic indirect support to this typology of firms. So, the questions arise: to what extent are NTBFs able to get funds from nontargeted schemes? In other words, is public money efficiently allocated towards young high-tech firms through direct horizontal schemes?

This paper addresses these issues. Firstly, it provides a taxonomy of the public direct policy measures granted by the national authorities to Italian NTBFs. Support schemes are classified according to the following dimensions: main objective, evaluation method, main instrument used to grant the subsidy, and main target group. Then, it analyses a sample of 507 NTBFs, extracted from the RITA database, developed at Politecnico di Milano. Sample firms were established in 1980 or later, were independent at founding time, and have remained so as of 1/1/2004 (i.e. they were not controlled by another business organization, even though other organizations may hold minority shareholdings in the new firms). In the following paragraphs we outline the main results of our analysis.

Firstly, despite the nontargeted nature of policy schemes, the number of NTBFs directly supported by the national authority is conspicuous (24%).

Secondly, while most NTBFs are located in the North of the country, southern enterprises are much more likely to gain access to national public support measures. This is a consequence of the large number of laws explicitly designed to provide support for this area. Furthermore, almost two thirds (64%) of subsidized young high-tech firms benefited from public aid when they were four years old or even older at the time of application. This is particularly worthy of the attention of policy makers, since high-tech firms are likely to suffer from market imperfections more heavily when they are young and lack a track record.

Finally, the econometric analysis has highlighted that, with very few exceptions, none of the *founder-* and *firm-specific* characteristics associated with successful young high-tech firms are determinants of firms' access to direct national support schemes. This points to the high risk that horizontal policy measures only result in *substitution effects* in NTBF sectors. Furthermore, the econometric analysis has shown that inefficiencies are associated with all types of general-purpose horizontal schemes, irrespective of the laws' various objectives, evaluation methods, instruments, and target groups.

To sum up, our findings support the view that direct horizontal subsidies have been quite inefficient in supporting Italian NTBFs. As a corollary, policy makers would better turn to vertical measures that explicitly take into account the specific characteristics of high-tech start-ups. In particular, in this field governmental officials are very likely to face severe adverse selection and moral hazard problems in selecting and monitoring the beneficiaries of public support among high-tech young enterprises. Accordingly, indirect support schemes that delegate this function to specialized institutions are likely to be more effective.

In the authors' view, two types of indirect measures are worthy of attention. First, in countries like Italy that lag behind in high-tech activities, policy makers should encourage the development of effective technology incubators. In fact, in these countries there are substantial market failures as regards the provision of essential inputs to young high-tech enterprises (e.g. real estate, technical and other business services, in addition to financing). Technology incubators reduce the transaction costs connected with the purchase of these inputs, and in addition in the process of screening candidates for placement in the incubator, they signal their quality. In so doing they perform an important bridging function between new high-tech firms and the financial community, which is a very valuable service in countries where the national innovation system is less advanced (Colombo and Delmastro 2002). Secondly, in countries with a bank-based financial system there is a limited supply of venture capital to new high-tech enterprises, especially in the seed and start-up stages. Hence the development of an efficient venture capital industry should be a key target of technology policy. Viewed in this perspective, public co-financing of private equity investments would be helpful. Nonetheless, one of the reasons that deter the provision of venture capital to small young high-tech firms lies in the high fixed costs of screening investment proposals and monitoring the operations of the firms that are financed. Therefore, measures aimed at reducing the managerial and administrative costs of venture capitalists would also be important, so as to make investments in these firms more attractive. In this respect, the recent guidelines of the European industrial policy (Commission of the European Communities 2002) open new space for the implementation of more specific and targeted policy measures than what has been allowed in the recent past (see Sterlacchini 2004 for a discussion of this issue).

BIBLIOGRAPHY

Almus, M., (2002), "What Characterizes a Fast Growing Firm?", *Applied Economics*, Vol. 34, pp. 1497-1508.

Almus, M., E. A., Nerlinger, (1999), "Growth of New Technology-based Firms: Which Factors Matter?", *Small Business Economics*, Vol. 13, pp. 141-154.

Asquith, P., D. W. Mullins, (1986), "Equity Issues and Offering Dilution", *Journal of Financial Economics*, Vol. 15, pp. 61-89.

Audretsch, D., D. Dohse, (2004), "The Impact of Location on Firm Growth", *CEPR Discussion Paper no. 4332.*

Bates, T., (1990), "Entrepreneur Human Capital Inputs and Small Business Longevity", *Review of Economics and Statistics*, Vol. 72, pp. 551-559.

Becker, G. S., (1975), *Human Capital*, New York: National Bureau of Economic Research.

Blanchflower, D., A. Oswald, (1998), "What Makes an Entrepreneur", *Journal of Labor Economics*, Vol. 16, pp. 26-60.

Black, J., D. de Meza, D. Jeffreys, (1996), "House Pricing, the Supply of Collateral and the Enterprise Economy", *Economic Journal*, 106, pp. 60-75.

Breschi, S., F. Lissoni, (2001), "Knowledge Spillovers and Local Innovation Systems: a Critical Survey", *Industrial and Corporate Change*, Vol. 10, pp. 975-1005.

Brüderl, J., P. Preisendörfer, (2000), "Fast Growing Businesses: Empirical Evidence from a German Study", *International Journal of Sociology*, Vol. 30, pp. 45-70.

Brüderl, J., P. Preisendörfer, R. Ziegler, R, (1992), "Survival Chances of Newly Founded Business Organizations", *American Sociological Review*, Vol. 72, pp. 227-242.

Capello, R., (1999), "Spatial Transfer of Knowledge in High-Technology Milieux: Learning Vs. Collective Learning Processes", *Regional Studies*, Vol. 33, pp. 353-365.

Carpenter, R. E, B. C. Petersen, (2002), "Capital Market Imperfections, High-Tech Investment, and New Equity Financing", *Economic Journal*, Vol. 112, pp. F54-F72.

Chandler, G. N., E. Jansen, (1992), "The Founder's Self-Assessed Competence and Venture Performance, *Journal of Business Venturing*, Vol. 7, pp. 223-236.

Chiesa V., A. Piccaluga, (2000), "Exploitation and Diffusion of Public Research: The Case of Academic Spin-Off Companies in Italy", *R&D Management*, 30 (4), pp. 329-339.

Colombo, M. G., M. Delmastro, (2002), "How Effective are Technology Incubators? Evidence from Italy", *Research Policy*, Vol. 31, pp. 1103-1122.

Colombo, M. G., M. Delmastro, L. Grilli, (2004), "Entrepreneurs' Human Capital and The Start-Up Size of New Technology-Based Firms", *International Journal of Industrial Organization*, Vol. 22, pp. 1183-1211.

Colombo, M. G., L. Grilli, (2005a), "L'accesso al credito bancario da parte delle start-up tecnologiche italiane", *Rivista di Politica Economica,* 95 (3), pp. 165-218.

Colombo, M. G., L. Grilli, (2005b), "Founders' Human Capital and the Growth of New Technology-Based Firms: A Competence-Based View", *Research Policy*, 34 (6), pp. 795-816.

Colombo, M. G., E. Piva, E., (2005), "Academic Start-Ups and New Technology-Based Firms: A Matched Pair Comparison", 5th Triple Helix Conference, 18-21 may 2005, Turin.

Commission of the European Communities, (2002), *Industrial Policy in an Enlarged Europe*, Bruxelles, Communication from the Commission, COM (2002) 714 – 11.12.2002.

Confindustria Centro Studi, (1991), *Indicatori economici provinciali*, (Collana Industria e Territorio, SIPI, Roma).

Cooper, A. C., (1985), "The Role of Incubator Organizations in the Founding of Growth-Oriented Firms", *Journal of Business Venturing*, vol. 1, pp. 75-86.

Cooper, A. C., Bruno, A. V., (1977), "Success Among High-Technology Firms", *Business Horizons*, Vol. 20, pp. 16-22.

Dahlstrand, A. L., (1997), "Growth and Inventiveness in Technology-Based Spin-Off Firms", *Research Policy*, Vol. 26, pp. 331-344.

Dunkelberg, W. G., A. C. Cooper, C. Woo, & W. J. Dennis, (1987), "New Firm Growth and Performance", [in:] N. C. Churchill, J. A. Hornaday, B. A. Kirchhoff, C. J. Krasner and K. H. Vesper (eds.), *Frontiers of Entrepreneurship Research*, Babson College, Boston, Mass.

Eisenhardt, K. M., C. B. Schoonhoven, (1990), "Organizational Growth: Linking Founding Team, Strategy, Environment And Growth Among U.S. Semiconductor Ventures, 1978-1988", *Administrative Science Quarterly*, Vol. 35, pp. 504-529.

European Commission Joint Research Centre, (2002), *RTD Evaluation Toolbox*, Socio-Economic Evaluation of Public RTD Policies.

Evans, D., L. Leighton, (1989), "Some Empirical Aspects of Entrepreneurship", *American Economic Review*, 79, pp. 519- 535.

Evans, D., B. Jovanovic, (1989), "An Estimated Model of Entrepreneurial Choice Under Liquidity Constraints", *Journal of Political Economy*, Vol. 97, pp. 808- 827.

Feeser, H. R., G. E. Willard, (1990), "Founding Strategy and Performance: a Comparison of High and Low Growth High Tech Firms", *Strategic Management Journal*, Vol. 11, pp. 87-98.

Feldmann, M. P., (1999), "The New Economics of Innovation, Spillovers and Agglomeration: A Review of Empirical Studies", *Economics of Innovation and New Technology*, Vol. 8, pp. 5-26.

Griliches, Z., (1992), "The Search for R&D Spillovers", *Scandinavian Journal of Economics*, Vol. 94 (suppl.), pp. S29-S47.

Grilli, L., (2005), "Internet Start-Ups Access to the Bank Loan Market: Evidence From Italy", *Applied Economics*, Vol. 37, pp. 293-305.

Hamilton, O., D. Shapiro, A. Vining, "The Growth Patterns of Canadian High-Tech Firms", *International Journal of Technology Management*, Vol. 24, pp. 458-472.

Holtz-Eakin, D., (2000), "Public Policy Toward Entrepreneurship", *Small Business Economics*, Vol. 15, pp. 283-291.

Holtz-Eakin, D., D. Joulfaian, H. S. Rosen, (1994a), "Entrepreneurial Decisions and Liquidity Constraints", *Rand Journal of Economics*, 25, pp. 334-347.

Holtz-Eakin, D., D. Joulfaian, H. S. Rosen, (1994b), "Sticking It Out: Entrepreneurial Survival and Liquidity Constraints", *Journal of Political Economy*, 102, pp. 53-75.

Jaffe, A. B., (1996), "Economic Analysis for Research Spillovers: Implications for the Advance Technology Program", Wasington D.C.: U.S. Department of Commerce, National Institute of Standards and Technology, Advanced Technology Program.

Lee, I., S. Lochhead, J. Ritter, Q. Zhao, (1996), "The Costs of Raising Capital", *Journal of Financial Research*, Vol. 19, pp. 59-74.

Lerner, J., (1999), "The Government as Venture Capitalist: the Long-Run Impact of the SBIR Program", *Journal of Business*, Vol. 72, pp. 285-318.

Lerner, J., (2002), "When Bureaucrats Meet Entrepreneurs: the Design of Effective 'Public Venture Capital' Programmes", *Economic Journal*, Vol. 112, pp. F73-84.

Lerner, J., C. Kegler, (2001), "Evaluating the Small Business Innovation Research Program: a Literature Review" [in:] Charles Wessner (eds.), *The Small Business Innovation Research Program (SBIR): an Assessment of the Department of Defense Fast Track Initiative*, Washington D.C., U.S., National Academy Press, pp. 397-426.

Maskell, P., (2001), "Towards a Knowledge-Based Theory of the Geographical Cluster", *Industrial and Corporate Change*, Vol. 10, pp. 921-943.

Meyer, B. D., (1990), "Why Are There So Few Black Entrepreneurs?", NBER Working Paper n. 3537, Cambridge, Mass.

Myers, S., N. Majluf, (1984), "Corporate Financing and Investment Decisions When Firms Have Information That Investors Do Not Have", *Journal of Financial Economics*, Vol. 13, pp. 197-221.

Nerlinger, E., (1998), "Standorte und entwicklung junger innovativer unternehmen: empirische ergebnisse fuer West-Deutschland" 27, Baden-Baden.

Preisendörfer, P., T. Voss, (1990), "Organizational Mortality of Small Firms: the Effects of Entrepreneurial Age and Human Capital" *Organizational Studies*, Vol. 11, pp. 107-129.

Roberts, E. B., (1992), "The Success of High-Technology Firms- Early Technological and Marketing Influences", *Interfaces*, Vol. 22, pp. 3-12.

Santarelli, E., M. Vivarelli, (2002), "Is Subsidizing Entry an Optimal Policy?", *Industrial and Corporate Change*, 11, pp. 39-52.

Siegel, D., C. Wessner, M. Binks, A. Lockett, (2003), "Policies Promoting Innovation in Small Firms: Evidence from the U.S. and U.K.", *Small Business Economics*, Vol. 20, pp. 121-127.

Siegel, R., E. Siegel, I. C. Macmillan, (1993), "Characteristics Distinguishing High Growth Ventures", *Journal of Business Venturing*, vol. 8, pp. 169-180.

Sterlacchini, A., (2004), "Ricerca ed alta tecnologia in Italia: le basi per un rilancio" *L'Industria*, Vol. 25, pp. 675-705.

Sternberg, R., (1990), "The Impact of Innovation Centers on Small Technology-Based Firms: the Example of the Federal Republic of Germany", *Small Business Economics*, Vol. 2, pp. 105-118.

Storey, D. J., (1994), *Understanding the Small Business Sector*, Thomson Learning, London.

Storey, D. J., B. S. Tether, (1998), "Public Policy Measures to Support New Technology-Based Firms in the European Union", *Research Policy*, Vol. 26, pp. 1037-1057.

Teece, D., (1986), "Profiting From Technological Innovation: Implications for Integration, Collaboration, Licensing and Public Policy", *Research Policy*, vol. 15, pp. 285-305.

Teach, R. D., F. A. Tarpley, R. G. Schwartz, (1986), "Software Venture Teams" [in:] Ronstadt, R., J. Hornaday, R. Peterson K. H. Vesper, *Frontiers in Entrepreneurial Research*, Babson Center for Entrepreneurial Studies, Wellesley, MA, U.S., pp. 546-562.

Westhead, P., (1997), "R&D "Inputs" and "Outputs" of Technology-Based Firms Located On and Off Science Parks", *R&D Management*, Vol. 21, pp. 109-123.

Westhead, P., M. Cowling, (1995), "Employment Change in Independent Owner-Managed High Technology Firms in Great Britain", *Small Business Economics*, vol. 7, pp. 111-140.

Westhead, P., D. J. Storey, (1994), "An Assessment of Firms Located On- and Off-Science Parks in the UK", HMSO, London.

Westhead, P., D. J. Storey, (1997), "Financial Constraints on the Growth of High Technology Small Firms in the United Kingdom", *Applied Financial Economics*, vol. 7, pp. 197-201.

6

Leveraging Corporate and Country Brands – How Emerging Countries Can Find Their Way in the Global Economy. The Case of 'Brand For Poland' Project

Tomasz Domański
Marta Hereźniak

INTRODUCTION

The idea that countries can be treated as brands and that there exists numerous parallels between the mode of running a huge multinational corporation and a state is a fairly new idea in most countries. Also, in recent years one can observe an increased interest in the concept of place and nation branding among various academics representing different disciplines (management, political science, antropology) throughout the world (Kotler, Gertner, 2002; Gilmore, 2002; Papadopulous, 2004; van Ham, 2001;Walvis, 2003; Olins, 2002; Dzenovska, 2005). It should come as no surprise that this is a completely new marketing concept in Poland, inasmuch as Poland is characterized by a very young (15 years) free market economy and an even shorter period (5-10 years) of the adoption by Polish companies and public institutions of the rules of modern strategic management. Polish enterprises, regions and the country as a whole are currently undergoing an accelerated marketing course, based on the critical analysis of long-term activities and the experiences of western companies and places. The benefit to Poland is that it may therefore adopt and implement the best practices, pre-tested on foreign markets by western companies, while applying some cutting-edge techniques at the same time.

It is also worth underlining that the pace of transfer of marketing innovations, including those from brand management, is enormously rapid and requires openness to new concepts and ideas. Hence, when analyzing innovations in this area, it should be borne in mind that they are predominantly based on the creative adaptation and creative imitation of pre-tested patterns. Although the notion of innovation is often relative, they tend to result in a brand new quality, which should be treated as innovation in the context of a country where such quality was previously unknown.

From the European perspective, marketing innovations boil down to creating and communicating new values and meanings in a modern, attractive way that contributes to creating a new and coherent country image, as well as the image of its companies and regions.

This chapter examines the idea that the building up of a unique country brand (i.e. a country's attractive image and good reputation) strengthened by powerful corporate brands can vitally contribute to creating and sustaining the competitive advantage of emerging countries and facilitate their transition from the position of a supplier of commodities and unbranded goods to the position of a brand-owner, offering world-class products and services that consumers around the world are willing to pay extra for. The idea is analyzed from the perspective of the "Brand for Poland" project, in which the authors participated. This first fully professional and holistic country branding initiative in Poland was undertaken in 2002/2003 by The Polish Chamber of Commerce (PCC) and The Institute of Polish Brand (IPB) in cooperation with an international team of consultants led by Wally Olins, a world-class specialist in corporate and place branding. The authors present some of the results of this project and its implications for Poland's branding strategy. The subsequent section of the paper explores issues related to the creation of a positive *country-of-origin-effect* (COE) for the country's corporate brands. The authors posit that building a country brand must also involve the preparation of a system that strengthens, stimulates and highlights the potential 'stars' of the country's economy, i.e. the strongest corporate brands, which are capable of being internationally competitive. This position will be illustrated by examples of both countries and companies which have already adopted this approach. Finally, some recommendations will be made regarding future activities that must be undertaken in order to achieve important goals implicit in country branding.

A COUNTRY BRAND FOR POLAND - NEW CHALLENGES

Poland's membership in the European Union presents Polish companies and regions with a number of economic challenges to be faced in the nearest future.

These challenges relate to both *hard* (e.g. technological innovation, quality standards) and *soft* factors of competitiveness. The latter include creating and sustaining strong brands that would increase consumer trust and create consumer preference and loyalty towards products of Polish origin. Meeting the above challenges is connected with benefiting from the positive country-of-origin effect, that is, creating a position in which Polish companies, products, services, places, culture and human resources are recognized, generate positive associations, and represent a unique value to both foreign and domestic consumers. Moreover, strong corporate and product brands are one of the important determinants of a country's image (Phau, Pendergast, 2000) and reputation abroad and therefore are often treated as valuable national assets[1]. For Poland as a new member of the EU, attractive image and good reputation, in other words, a strong country brand, plays major role in building the country's economic, political, and social position.

It is argued in the branding literature (see: Anholt, 2003) that the creation of a unique and coherently communicated country brand, supported by product and corporate brands, creates an opportunity for the emerging countries join the developed ones. It is stressed that these are the intangible assets (including the brand) that are currently the largest source of added value, generating substantial profits for national economies. From this perspective, building a competitive advantage is understood as moving from the position of a producer and supplier of cheap unbranded goods to the position of brand-owner.

As Simon Anholt puts it (Anholt, 2003, p. 9): (the role of emerging counties in the global economy boils down to) *selling unbranded goods to richer nations at low margins. Companies in the richer nations then add large amounts of margin to the goods by finishing, packaging, branding, and retailing them to the end user. The margins on this kind of transaction have been compressed even further in recent decades as globalization has advanced, making life as a "supplier nation" an increasingly unattractive proposition.*

Aware of all the above challenges, the PCC and the IPB initiated the "Brand for Poland" project, which in fact started the public debate about communicating true, unique and attractive country identity to stakeholders around the world. However, long before the project was launched and before the country branding issues were brought up, there were certain steps undertaken to raise the awareness and profile of corporate brands in Poland – an awareness that was somewhat asleep during the communist years.

The first undertaking was the *Brand for Brands* program, initiated in 1995 by the Polish government and the PCC. The idea behind it was that corporate and product brands are key players in building and sustaining a country's competitive advantage and that Polish companies - already offering excellent products - need differentiation, trust and consumer loyalty in order to successfully operate on foreign and domestic markets. The program was therefore designed to make Polish companies realize that brands and branding should find their place at the very heart of companies' business activities. *Brand*

[1] For a detailed dicussion of this problem see: Turkheim, Piaton, 2003

for Brands was later included in the government's Industrial Development Program. Also in 1995 the Institute of Polish Brand was created under the auspices of the PCC, designed to become an intellectual center of the *Brand for Brands* program and, over the longer perspective, also a think-tank for developing and coordinating activities based on the idea of branding Poland. Work on the National Branding Program commenced in 2000, and the official inauguration took place in 2002 in Wieliczka Salt mine (probably the oldest functioning enterprise in the world). To make the initiative more complete, the concept of an Academy of Brands was developed (as the second stage of the Brand for Brands program), the aim of which was to select a group of Polish corporate brands as future partners in the national branding program and also as potential flagships of the Polish economy. A book series, *The Library of the Academy of Brands,* was created in order to publish books related to corporate and place branding by renowned authors from all over the world. These efforts were accompanied by three major image campaigns for Poland published in Time Magazine (2001, 2003), composed of three parts: *Economy under its own flag, Poland: discover and savour, Poland the heart of Europe.*

The year 2003 was a breakthrough for the Polish National Branding Program. It was the year during which the basic line of thinking about Poland as a brand was formed and clear standards were set for future activities related to national branding strategy. In July 2003 The 1st. Conference 'Brand for Poland' took place in the Polish parliament, initiating the public debate over this issue. One of the results of the conference was the government's approval of the *Economic Promotion of Poland until 2005 Program,* prepared by the Ministry of Economy, Labor and Social Policy, with the stated objective: *the development of Promotion of the Polish Brand program.*

By the end of 2003, the PCC commissioned an international team of consultants led by Wally Olins, one of the most experienced corporate and place brand consultants in the world, professor of Oxford University and the co-author or consultant of branding strategies for Spain, Portugal, the United Kingdom and many cities and regions. The primary goal of this cooperation was the articulation of the-so-called *core idea* (*central message*) of Brand Poland, which (in corporate branding terms) can be compared to a brand essence. We will return to this issue later in the article.

The Polish National Branding Program is not an isolated initiative, but rather situates itself within a strong worldwide trend towards building and sustaining strong country brands, based on the four areas of competitiveness mentioned previously. Countries such as Spain, Portugal, Norway, New Zealand, Australia, South Korea, South Africa, Estonia, Ireland, United Kingdom and Germany, to name just a few, have already started to acknowledge and appreciate the value of a strong country brand and have been using marketing and branding techniques to introduce their products, brands, human resources, destinations, culture, ideas and value systems to stakeholders around the world. If executed with organizational and intellectual commitment, as has been the case so far, the Polish National Branding Program may become one of the most innovative and coherent in the world. Many countries undertake separate promotional activities within the areas of tourism, FDI or exports, neglecting the fact that, just as in the case of corporate brands, a country brand requires coordination and coherence at both the marketing and institutional levels.

BRAND FOR POLAND – ADOPTING THE NEW COUNTRY BRANDING PHILOSOPHY

The breakthrough year of 2003 created the need to adopt modern ways of thinking about the issue of nation branding. It had to be made clear that future activities could not be a mere repetition of national promotion policies practiced in previous years, and that the foundations for a system of managing Polish image and reputation had to be established. Based on the analysis of case studies and the examination of recent place branding literature, the following principles, summarising Poland's new country branding philosophy, were devloped:

- *Strategic horizon*: country branding should be treated as a long term process (lasting 20-25 years); therefore activities, institutions and ideas that result from it should be universal and maintain their relevance and usefulness over the long-term perspective
- *Holistic approach (Olins, 1999)*: country branding should not be narrowed down to tourism promotion. A properly designed country branding program should take into account four major areas in which a country brand manifests itself: FDI, tourism, brand export, and public diplomacy[2] Operational and conceptual coordination is necessary among the institutions representimg the four 'vectors' of a country brand
- *Stakeholder involvement*: a country branding program is not an isolated governmental activity executed at the central level, but rather a multi-layered enterprise which should engage different stakeholder groups. These include, among others: companies, public organizations, opinion molders, creative and artistic circles, the media, academia, cities and regions. All these groups can participate in the process at the organizational, financial, intellectual, or conceptual levels.
- *Brand identity as a central issue (Rainisto, 2003)*: a country branding programme should be based upon an identity concept that is familiar to and accepted by major stakeholders. The concept serves as a driving force behind all the communication activities of the country, can be used across different 'products' and markets, and allows for the coordination of messages on different levels.
- *Branding is both inward and outward*: successful country branding can only take place when the society as a whole supports the idea and when the people are eager to internalize it, so that the country branding process gradually becomes a kind of social movement, and social groups and individuals start 'living the brand' (Ind, 2004).

As was stated earlier in this paper, the country branding process also requires the adoption of a systematic apporach to building and managing a country's image and reputation. Such a system allows for coordination,

[2] The concept of Public Diplomacy is thorougly examined in: Leonard, Stead, Smewing. 2002.

coherence and the effective allocation of managerial, intellectual, and financial resources.

The Polish system is based on four major pillars:

- *Institutional coordination of 4 vectors*: in the governmental document identified earlier, the institutions responsible for the major areas of building the Polish Brand were defined. These included: The Ministry of the Economy, Labor and Social Policy (brand export), Ministry of Foreign Affairs (public diplomacy), Polish Information and Foreign Investment Agency (FDI), Polish Tourism Organization (tourism) and also the PCC/IPB as a coordinating body and know-how base (see Fig. 1)

Figure 1
Institutional cooperation at the governmental level

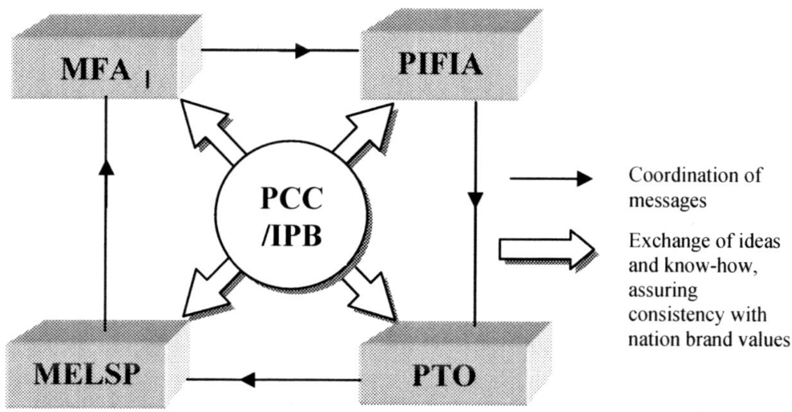

MELSP = Ministry of the Economy, Labor and Social Policy
MFA = Ministry of Foreign Affairs
PIFIA = Polish Information and Foreign Investment Agency
PTO = Polish Tourism Organization
PCC = Polish Chamber of Commerce
IPB = Institute of Polish Brand

Source*: own design*

- *Raising the profile of country branding*: as the idea of country branding is still new in Poland it needs to be popularized among the general public and the molders of public opinion. In Poland this is done through the organization of major conferences and seminars, lectures and presentations at the universities throughout Poland, media releases etc.
- *Building partnerships with companies*: as was stated earlier, product and corporate brands are among the major vectors of a country brand. In Poland the aforementioned Academy of Brands provides a platform for building partnerships between governments and corporate stakeholders.
- *Providing intellectual support*: the Institute of Polish Brand provides intellectual support to raise knowledge and awareness about the nature, methods and principles

of country and place branding. The idea is to stimulate groups and individuals to learn more about place branding, so that in the future they become experts in this field and contribute to the process of country branding. The Library of the Academy of Brands carries out consulting activities for regions, cities and companies, bringing world-class experts to Poland.

The branding principles, based on the modern way of thinking about a country's image and reputation, together with the systematic approach adopted from the beginning of the project, significantly increase its chances to become one of the most innovative and coherent programs in the world. However, the question that remains is: *to what extent will the political process influence the course of the project?*. Given the present political situation, it may be assumed that politics will decrease rather than increase the dynamics of the country branding process in Poland.

Alongside institutional coordination and coherence, account must be taken of the marketing level at which messages need to be coordinated so that they formulate a coherent, true and unique country identity which can be subsequently communicated to various audiences. In order to achieve this sort of consistency it is necessary to develop and articulate the core idea of a country brand. The core idea serves as a brand essence (Aaker, Joahimstaller 2002), and therefore should be true, unique, difficult to imitate, and able to accommodate the country's long term ambitions and aspirations. Just as brand essence, the core idea serves as an inspiration for further marketing communication and brand building activities and is designed to last for decades. The core idea must not be confused with a slogan, as the latter (unlike the idea) can be changed dependent on the situation, the market, or the product. Furthermore, the idea needs to be tailored in such a way so as to be relevant and to represent a certain value for the major areas in which a country brand presents itself, that is: brand export, FDI, public diplomacy and tourism. The idea may serve those companies which wish to emphasize their country of origin in their marketing strategies on foreign markets, diplomats, artistic and cultural institutions, creative agencies, regions, cities and every other group and organization that wishes to join the national branding program. In Spain, for instance, the national brand was based around the idea of *passion*, as an expression of a very unique way of living, working and creating, characteristic of Spanish people (Olins, 2004). The core idea of Norway is *peaceful nature*[3], expressed by this country as a tourist destination, a place to invest in, a producer of goods and an actor on the international scene. According to its core idea, South Africa is *alive with possibility* (www.imc.org.za), and this statement is indeed widely used throughout numerous initiatives undertaken within the South African branding program.

[3] The case of Norway is presented in: Leonard, 2003.

DIAGNOSIS AND THE CORE IDEA OF THE POLISH BRAND

The development of the core idea for Poland (Saffron Consultants, 2004) lasted almost a year and comprised:

- research into the image of Poland, both at home and abroad,
- around 200 in-depth interviews with foreign and domestic opinion leaders (face to face and telephone interviews)
- analysis of the branding strategies of different countries (construction of case studies),
- observations held by the members of the Saffron consultants (travels throughout Poland, collecting photographic material, observations of individual and group behavior,
- secondary data analysis (press articles, research reports, publications on Polish history and culture, promotional material, audits etc.)
- meetings, consultations and panel discussions with the representatives of companies, media, culture, academia, tourism, politics, and young people.

Initially, the team of consultants put forward three alternative versions of the core idea for Poland, which became the subject of discussions and consultations with opinion leaders from the representatives of the above mentioned groups. The final result of these consultations was the idea of *creative tension*, comprising elements of previously proposed ideas and apparently capturing the essence of Polishness. It holds a number of semantic implications and gives rise to numerous options for future application. As is stressed by the authors of the idea, *creative tension* is a state of mind dominated by constructive agitation and imagination. The processes happening in the atmosphere of creative tension are very often tumultuous, full of contradictions and even conflicts; always, however, yielding positive and outstanding results. This state is characteristic of creative and individualistic personalities, full of energy and nonconformity, and certainly not among the easiest persons to live with. This kind of person is never satisfied with the present state of affairs, but eventually contributes to progress and constructive solutions in the area s/he operates in.

This diagnosis of the Polish character was followed by the identification of brand attributes, indispensable to create and communicate messages relevant for various audiences. Initially the list included 19 attributes, seven of which were subsequently considered core attributes. These were the following: *creativity, adaptability, vitality, resilience, individualism, boisterousness,*[4] *tension*. The remaining attributes included: *imagination, flair, talent, charm, intellectualism, pragmatism, energy, passion, belief, ambition, moodiness*[5], *abrasiveness*[6]; and

[4] A tendency to make loud and sometimes uncomfortable clamour, usually but not always, positive.

[5] A tendency to show contradictory feelings and emotions, apparently without any particular cause

[6] Being harsh and straightforward with one another.

are regarded as complementary and applicable whenever any act of brand communication is designed and executed. Both the idea as such and the brand attributes have certain implications for the key vectors of a national brand, i.e. tourism, FDI, brand export and public diplomacy. The implications are presented in the table below.

Table 1
Creative Tension – implications for country brand vectors

	FDI	**Tourism**	**Public Diplomacy**	**Brand Export**
Thrust	emphasizes energy and growth, which present a fine backdrop and foundation for entrepreneurialism, value creation and investment returns.	implies that there are interesting and unusual things to see and do in Poland, that it will be an experience.	conveys that while Poland will not 'go-along-to-get-along,' it can be expected to play an active and usually constructive role in the world.	suggests that Polish passion, practicality, energy and ambition lead to Polish effectiveness and quality, not only in products and services, but also in people.
Attributes to emphasize	energy, ambition, talent, pragmatism, flair and maybe resilience	creativity, imagination, charm, vitality, individualism, boisterousness and even, perhaps, moodiness	creativity, adaptability, flair, pragmatism, passion and belief	flair, creativity, imagination, pragmatism, passion, energy, ambition
Attributes to diminish	moodiness, boisterousness and tension, individualism	none	boisterousness, moodiness, abrasiveness, and individualism	none

Source: Base on: Defining the Polish brand: Core idea, brand attributes and brand benefits, Saffron Consultants, materials of The Institute of Polish Brand.

The idea of creative tension was very well received by the majority of the groups and individuals it was consulted with (political, cultural, media and business people, tourism, FDI and export specialists, marketing practitioners and academics). It was officially presented in the Polish parliament on 6 December 2004 during the *2nd Conference: A Brand for Poland; The Core Idea* in front of over 500 participants from different backgrounds.

This core idea of creative tension was mostly liked by the discussion groups, who thought it "epitomize[d] Poles' most common national trait, 'paradoxically.'" It seemed to take account of what was termed the essentially "contrasting nature" of Poland. Others critics deemed it advantageous for its "wide scope of themes" that offer "something for everyone" and accurately encompass "these contrasts which comprise Polish passion" (Saffron Consultants).

The articulation of the core idea and the brand attributes was a central task in the Brand for Poland project. As was previously mentioned, the project is the first stage of Poland's National Branding Program. The change of both the government and the presidency in Poland has significantly delayed the process. However, as this article is being written the situation is heading towards the realization of stage two of the program. Stage two comprises the visualization of the core idea, the creation and coordination of marketing messages in key vectors of national brand, designing the brandbook, auditing and enhancing institutional cooperation, and creating the ideas for the-so-called brand centric projects and advertising campaigns.

The Polish National Branding Program requires substantial effort and numerous ideas need to be conceived and implemented. Overcoming the various barriers, including institutional, financial and mental ones[7], constitutes a great challenge both for public institutions and private enterprises, which are key actors in the nation branding process.

One of the biggest issues, involving both types of actors, is the identification of and provision of necessary support to those Polish corporate and product brands with the potential to become internationally competitive in the foreseeable future, and which can become flagships of the Polish economy and strengthen the image of the country's brand. Achieving this requires in turn expertise in the building of private-public partnerships at the operational, conceptual, and financial levels. For countries like Poland – taking their first steps in the nation branding process – it appears necessary to analyze and benchmark the activities of other countries that have already developed and implemented methods and processes to strengthen their country brand image through flagship corporate and product brands of national origin. This problem is discussed in the following section.

CORPORATE AND PRODUCT BRANDS IN NATIONAL BRANDING STRATEGIES – IMPLICATIONS FOR POLISH ENTERPRISES AND PUBLIC INSTITUTIONS

The place of origin is one of the important criteria in the consumers' decision making process. Substantial literature on the *country-of-origin-effect* demonstrates that product origin can constitute an important differentiator on the global market[8], and also that it can respond to consumers' growing preference for individualization, originality, and authenticity as well as their treatment of consumption not only as a commercial process, but also as a manifestation of their views, values and identity.

---

[7] The issue of various types of barriers is discussed in: Quelch, Jocz, 2005, pp. 229-237.

[8] See for instance: Jaffe, Nebenzahl, 2001; Papadopulous, Heslop, 2002; Baker, 2002, pp. 157-168.

When considering the possibilities of creating a positive country-of-origin effect for a country's corporate and product brand and therefore strengthening a country's overall reputation, two alternative strategies should be considered. The first comprises benefiting from the so-called *halo effect* (Han, 1989, pp. 222-229), which means focusing on the improvement of a country's overall image and then attempting to associate country brand values with certain product categories or industries and in this way building its international position. New Zealand provides a good example of country-of-origin based marketing (Beverland, Lindgreen 2002, pp. 147-167). The core idea of the New Zealand's country brand - 100% Pure New Zealand (Morgan, Pritchard, 2002, pp. 335-354) - was "transmitted" to certain product and corporate brands that fitted the country's core message. *Linded Leaves Pure New Zealand Bodycare, Evolu.Botanical Skincare of New Zealand, Living Nature, Nature Bee Pure New Zealand* or *Untouched World New Zealand* consititute examples of brands that support the idea of the New Zealand country brand, building their own equity at the same time. Links to their corporate web sites can be found at the official tourism site of the country. Another interesting set of initiatives undertaken in New Zealand is the consolidation of small entrepreneurs within a single branch under a New Zealand – related umbrella brands: *Merino New Zealand, The Wine Institute of New Zealand, The New Zealand Dairy Board* etc., using the values and imagery of the national brand. Numerous programs were developed based on cooperation with the government to enable these consolidated industries to gain a competitive position on the international market. These included marketing programs, facilitating contracts with new business partners, lobbying, quality programs, etc. Together with the governmental ministries educational campaigns were run concerning, for instance, New Zealand's food products and methods of their preparation, emphasizing how their unique qualities could be preserved and the cultural contexts in which they are consumed (e.g. workshops for chefs of exclusive restaurants operating in foreign markets).

The second strategy is based on the so called *summary effect* (Gnoth, 2002), which is based on the idea that a country's image and reputation is formed on the basis of a consumer's cumulated experience with the country's products. This strategy requires therefore the creation and promotion of at least a few high profile corporate or product brands on the international market, even without direct references to their country of origin.

It must be noted that in the case of countries like Poland the two strategies should be regarded as complementary. It seems reasonable to select brands/ branches that could benefit from country brand associations on the one hand, and stimulate the development and international competitiveness of the potential stars of the Polish economy on the other (even if they have little in common with the country brand identity at first glance). The need for simultaneous application of the two strategies derives from the fact that changing a negative or false country image is a long term process, requiring commitment from both the private and public sectors. If the first strategy is to be applied in a particular country, then it must be realized that the possibilities of application of country of origin in companies' marketing strategies is dependant upon many diverse determinants, such as: a country's overall image, characteristics of a branch, product category, a target market profile, an enterprise's strategic goals, and the willingness of managers to implement such a strategy. From this perspective, a survey of a current country image appears particularly important, due to the fact that such an image determines the type, scale and diversity of country of origin-based marketing activities. The relationship between types of country images (CI) and marketing activities is described in Table 2.

Table 2
Image complexity and marketing implications

Lacking or vague CI	Simple CI	Multifaceted CI
Country image characteristics – few general associations or wrong associations – no country equity[9]	Country image characteristics – some generic associations about the country – country equity built on one (strong) or a few country-related intangible assets	Country image characteristics – many associations and extensive knowledge about a country – high country equity built on many assets
Marketing implications – national programmes to establish some knowledge about the country – use halo effect to link country image to product attributes	Marketing implications – repeating a message will strengthen a simple country image – systematically varying the message creates multiplex associations	Marketing implications – select appropriate associations for different product categories – country image extensions across product categories

Source: adapted from Kleppe at al., 2002.

Research (Kolarska-Bobińska (ed.), 2003) shows that the current image of Poland, although evolving, still belongs rather to the first category. The associations that prevail include: former communist, middle-sized, Eastern-European, land of the Pope and Solidarity, backward agriculture, environmental pollution, vodka and sausages. But Poland has become a modern, free-market economy with Europe's largest population of university students, excellent companies and products, attractive destinations and many innovative ideas.

Given the existing perception gap, it is necessary to realize that a program designed to build and sustain the position of Polish companies and brands needs to be not only innovative, but also systematic, and that it should be linked with a national branding program, i.e. corporate/ product brands need to share some values/attributes with the country brand. To be successful such activities require the participation of both the public (government and its institutions, local

[9] That portion of consumers' affect toward a brand or product that is derived purely from the product's associations with a particular country

authorities) and the private (companies) sector. Additionally, the companies willing to use the country-of-origin in their marketing strategies should cooperate and consolidate within their branches, which should giver them a better competitive position.

With the application of the second strategy, country brand managers should rather bear in mind the sort of country image they *want to create* by building and sustaining strong and internationally competitive corporate/product brands. For instance, if a country wishes to position itself as innovative, open and creative, then the country brand managers should be looking for corporate brands reinforcing the desired image. The role of the government institution is therefore to initiate cooperation among a country's top corporations and to define the criteria for creating a flagship brand.

As a good example of this strategy we can examine the initiatives undertaken by the South Korean government, which in 2002 selected 300 brand name products, internationally competitive in terms of technology and design, and established a program to develop them as leading Korean brands. The government's plan to build strong corporate brands included the launch of a Brand Academy aimed at training some 500 brand management and design specialists annually. Another initiative involved the creation of industrial design renovation centers in every major city to help SMEs with the quality of their product design (Anholt, 2003, p. 32).

Another example of a coherent strategy is Spain's Forum of Renowned Spanish Brands[10], which was created with the participation of three institutions: Institutio Espanol de Commercio Exterior, the Association of Renowned Spanish Brands (comprising 60 Spanish companies) and the Ministry of Science and Technology. The aim of this public-private partnership is to strengthen the reputation of Made in Spain brands all over the world and the reputation of Spain as a whole.

Another instructive case is the Taiwanese government's initiative to build up the position of Taiwanese brands and to eliminate unfavorable associations therewith, such as being regarded as cheap imitations of Japanese products. Three campaign were launched: *Very well made in Taiwan, Excellence, made in Taiwan, Taiwan: your source for innovalue,* and in addition the Ministry of Economy cooperated closely with Acer on building national and corporate brands, which turned out to be fruitful for both parties[11].

Corporate and product brands, being one of the most important carriers of national identity, should be given extra attention within each country's branding strategy. Therefore the Polish system of building and managing a country brand, described earlier in this paper, should be complemented with some of the methods for supporting a country's corporate brands used in the countries mentioned above.

Given the above examples and the circumstances in which the Polish National Branding Program is progressing, both the Polish government and

---

[10] http://www.nytimes.com/global/spain/seven.html

[11] For details see: Amine, Chao, pp. 187-204.

companies have to bear in mind that there are particular sets of innovative activities that need to be undertaken by the private and the public sectors in order to manage the country-of-origin effect for the benefit of both companies and the country itself.

The major challenges on the part of **companies** wishing to implement place of origin strategies are to:

- adjust their marketing communication systems (visual identity, advertising and PR campaigns, information and educational actions to raise awareness of the brands and products) so that they reflect, even indirectly, their place of origin;
- reevaluate product, price and distribution policies to assure the best possible fit to the requirements and standards of foreign markets;
- consolidate (in the case of smaller companies of the same branch) under an umbrella brand, using at least some attributes consonant with those of the country brand.

The public sphere has also an important role to play in the process of establishing Polish companies and brands abroad. In particular this includes:

- identification of Poland's flagship brands/ products/ branches;
- assuring support for those companies that plan to use the country image in their strategies on the international market (for instance through the transfer of know how);
- setting clear standards and criteria that the companies representing Poland abroad will have to meet to be able to use some elements of country brand identity in their strategies;
- internal promotion of Polish brands and companies through special certificates and prizes;
- designing and managing a national branding program that would gain acceptance and generate companies' willingness to participate in improving Poland's image and reputation.

CONCLUSION AND INDICATIONS FOR FURTHER RESEARCH

The strategies for creating and sustaining Polish national, regional and company brands should be considered as a long-term process, the initial results of which will be observable only in 5-10 years. The impact and the scope of these results will depend on the coherence of different elements of the designed strategy, as well as upon the harmonious cooperation of various parties, integrated around a common purpose.

It must be kept in mind that, although every country/place has its own unique characteristics, the general mechanisms and barriers involved in the place branding process are very similar. Therefore the branding principles delineated for Poland, as well as the foundations established for the country branding system, may be applicable to other countries as well. The same is true of barriers such as: being dependent on the political process, financial instability, difficulties with building private-public partnerships, problems with understanding the idea of place branding by decision-makers; all of which tend to occur in many countries, cities and regions.

While there is a growing interest in branding places both in terms of theory and in practice, in terms of research the idea is still in its infancy. As such, it should become the focal point for research worldwide. There are at least several issues that need to be addressed in future studies: the model of creating a brand for a particular type of place; the methods of creating and managing private-public networks of cooperation as a part of a place branding process; and a place brand measurement methodology[12] that would allow for identification of the position and the value of a place brand. The results of such research would significantly contribute to further development of branding programs for countries like Poland, which are still at the beginning of the process and are searching for different way to make it more effective.

BIBLIOGRAPHY

Aaker, D. A., E. Joahimstaller, (2002), *Brand Leadership*, Simon&Schuster UK Ltd., London.

Amine, L. S., M. C. H. Chao, (2005), "Managing Country Image to Long-term Advantage: The Case of Taiwan and Acer", *Place Branding*, vol. 1, no 2, pp. 187-204.

Anholt, S., (2003), *Brand New Justice. The Upside of Global Branding*, Butterworth-Heinemann.

Baker, M. J., L. Ballington, (2002), "Country of Origin as a Source of Competitive Advantage", *Journal of Strategic Marketing*, no 10, pp. 157-168.

Beverland, M., A. Lindgreen, (2002), Using Country Origin in Strategy: The Importane of Context and Strategic Action, *Journal of Brand Management*, Vol. 10, no. 2, pp. 147-167.

Dzenovska, D., (2005), "Remaking the Ration of Latvia: Anthropological Perspectives on Nation Branding", *Place Branding Quarterly*, Vol. 1, 2, pp. 173-186..

Gilmore, F. (2002), "A country – Can it be Repositioned? Spain- the Success Story of Country Branding", *Journal of Brand Management*, Vol. 9, no. 4-5, April, pp. 281-293..

Gnoth, J., (2002), "Leveraging Export Brands Through a Tourism Destination Brand", *Journal of Brand Management*, Vol. 9, no. 4-5, April, pp. 262-280..

Ham van, P., (2001), "The Rise of the Brand-State The Postmodern Politics of Image and Reputation", *Foreign Affairs*, September/October, pp. 2-10.

Han, C. M., (1989), "Country Image: Halo or Summary Construct?", *Journal of Marketing Research*, Vol. 26, May, pp. 222–229.

Ind, N., (2004), *Living the Brand*, Kogan Page, London.

---

[12] So far there have been attmpts to create such a methodology, with two particularly interesting concepts: Anholt-GMI Nation Brand Index, see: http://www.nationbrandindex.com/ and brand valuation methodology that adopts a financial perspective on country brand measurement, see: Sinclair, 2004.

Jaffe, E. D., I. D. Nebenzahl, (2001), *National Image and Competitive Advantage. The Theory and Practice of Country of Origin Effect*, Copenhagen Business School Press, Copenhagen.

Kleppe, I. A., N. M. Iversen, I. G. Stensaker, (2002), “Country Images in Marketing Strategies: Conceptual Issues and Empirical Asian Illustration”, *Journal of Brand Management*, Vol. 10, no. 1, September, pp. 61-74.

Kolarska-Bobińska, L., (ed.), (2003), *Obraz Polski i Polaków w Europie*, Instytut Spraw Publicznych, Warszawa.

Kotler, Ph., D. Gertner, (2002), “Country as Brand, Product and Beyond: A Place Marketing and Brand Management Perspective”, *Journal of Brand Management*”, Vol. 9, no. 4-5, April, pp. 249-261.

Leonard, M., A. Small, (2003), *Executive Summary: Norwegian Public Diplomacy*, June.

Leonard, M., C. Stead, C. Smewing, (2002), *Public Diplomacy*, Foreign Policy Centre, London.

Morgan, N., A. Pritchard, R. Piggott, (2002), “New Zealand, 100% pure. The Creation of a Powerful Niche Destination Brand”, *The Journal of Brand Management*, Vol. 9, No. 4–5, April, pp. 335-354.

Olins, W., (2002), “Branding the Nation – the Historical Context”, *Journal of Brand Management*, Vol. 9, no. 4-5, April, pp. 241-248.

Olins, W., (2004), *O marce*, The Institute of Polish Brand, Warszawa.

Olins, W., (1999), *Trading Identities – Why Companies and Countries are Taking on each other’s Roles*, Foreign Policy Centre, UK.

Papdopulous, N., (2004), “Place Branding: Evolution, Meaning and Implications”, *Place Branding Quarterly*, Vol. 1, no. 1, November, pp. 36-49.

Papadopulous, N., L. Heslop, (2002), “Country Equity and Country Branding: Problems and Prospects”, *Journal of Brand Management*, Vol. 10, no 1, September, pp. 294-314.

Phau, I., G. Pendergast, (2000), “Conceptualizing the Country of Origin of Brand”, *Journal of Marketing Communications*”, no 6.

Quelch, J., K. Jocz, (2005), “Positioning the Nation-State”, *Place Branding*”, no 3, Vol. 1, pp 229-237.

Rainisto, S., (2003), *Success Factors of Place Marketing: a Study of Place Marketing Practices in Northern Europe and the United States*, Helsinki University of Tchnology, Institute of Strategy and International Business, Doctoral Dissertation 2003/4, Espoo 2003, http://lib.tkk.fi/Diss/2003/isbn9512266849/

Saffron Brand Consultants, (2004), *Brand for Poland. Advancing Poland’s National Identity*, The Institute of Polish Brand, Warsaw.

Saffron Consultants, (2004), *Defining the Polish Brand: Core Idea, Brand Attributes and Brand Benefits*, Warsaw.

Sinclair, R., (2004), “A Brand Valuation Methodology for Nations”, *Place Branding Quarterly*, Vol. 1, no. 1, November, pp. 74-79.

Turkheim, S., J.-M. Piaton, (2003), *Home Sweet Marque. La Marque, Embleme d’une Nation?*, Edition d’Organisation, Paris.

Walvis, T., (2003), "Building Brand Locations", *Corporate Reputation Review*, Vol. 5, No 4, pp. 358-366.

http://www.imc.org.za

http://www.nytimes.com/global/spain/seven.html

http://www.nationbrandindex.com/

7

Comparing Innovation Strategies - Wind Power Policy 1970-2000 in the Netherlands and Denmark

Linda M. Kamp

INTRODUCTION

Several Western countries started to develop renewable energy in the 1970s. Their motivations were the oil crisis and the Club of Rome report, which warned of imminent shortages of traditional energy sources like oil and gas. The renewable energy source that people had the highest expectations for was wind energy. Two of the countries that were involved in the development of wind energy were the Netherlands and Denmark. Both governments gave active support to this development. Furthermore, both countries have a comparable wind regime. However, the results of the development of wind energy in each country have been very different. In the year 2000, Denmark had a flourishing wind turbine industry, which produced wind turbines for the world market. Furthermore, at the end of the year 2000 the cumulative installed capacity of wind turbines in Denmark was 2,340 MW and wind turbines produced 15% of the country's electricity demand. In the Netherlands, the situation was far less rosy. Although 10 to 15 wind turbine manufacturers were active on the Dutch market at the beginning of the 1980s, in the year 2000 only one remained. Furthermore, at the end of the year 2000 only 442 MW of wind turbines had been installed in the Netherlands, while the target for the year 2000 had been 2,000 MW.

What are the reasons for this difference in performance? Often, location siting problems in the Netherlands are mentioned as the main reason why the Netherlands is lagging behind in the realisation of wind turbine capacity. However, our research shows that the Netherlands were already lagging behind in the 1980's, when the siting problems were not yet prominent. Therefore, we looked for other reasons to explain the differences in performance.

Research aim and methodology

The goal of this research is to investigate the reasons why the Netherlands were far less successful in developing wind power than the Danes in the period 1970-2000. We look for the answers to this question by comparing the innovation strategies that were followed by the governments of both countries in order to set up this new technology.

The research methodology used is that of comparative case study research.[1] In order to achieve validity in a case study, it is important to use multiple sources of information. Therefore, we used a wide array of data sources: interviews, scientific and technical articles, conference proceedings, press releases, policy documents, technical reports, statistics, articles in popular magazines on renewable energy, and promotional material from manufacturers. The persons we interviewed were from various backgrounds: researchers, manufacturers, wind turbine owners, etc. For validity purposes we used the triangulation method: testing and cross-checking the information gathered from various sources. We tested actors' statements against statements by other actors and against written material. Furthermore, we used the method of 'participatory observation' by way of a 3 month stay at the Dutch ECN[2] Research Centre and a 3 month stay at the Danish Copenhagen Business School.

Organization of this chapter

In this chapter we will present the results of our research. In the following section, we present our theoretical approach by explaining the concept of an 'innovation system' and the different strategies that can be followed to set up an innovation system. In section 3, we describe the Dutch case study; and in section 4 we go into the Danish case study. Both case studies will be described in two parts: one part about the large-scale wind turbine innovation subsystem and one part about the small-scale wind turbine innovation subsystem. We explain why we made this distinction in sections 3 and 4. In section 5, we draw some conclusions, go into generalized aspects, and put forward some ideas for further research.

---

[1] Most of the case study research was done as part of a PhD research, published in Kamp (2002).

[2] ECN means in Dutch: Energie Onderzoekscentrum Nederland, and in English: Energy Research Centre Netherlands.

THEORY: THE INNOVATION SYSTEM AND INNOVATION STRATEGIES

The innovation system

The concept 'innovation system' was developed at the end of the 1980s and the beginning of the 1990s by Freeman (1987), Freeman and Lundvall (1988), Lundvall (1988, 1992) and Nelson (1993, 1994). It starts from the idea that innovations are often developed within systems comprised of actors and organizations. Companies, governments, universities, banks, consumers, and other organizations all contribute in a different and interactive way to the innovation process. These actors and organizations, the relationships between them, and the institutions influencing them, together form the innovation system (Carlsson et al., 2002).

Because 'innovation system' is such a broad concept, authors can define it differently and stress the element(s) they consider the most important. However, there is a set of characteristics upon which all researchers agree. In a study of a specific innovation system, these characteristics can be used as guidelines to build the theoretical framework. Lundvall summarizes them as follows (Lundvall, 1992):

- The central focus is on technological innovation, but organizational and institutional change are considered to be important as well.
- Innovation systems in various countries are claimed to be different, and it is important to study these differences.
- The viewpoint is holistic; in other words, many determinants and their relationships are included in the analysis.
- A historical perspective is used. Innovation is seen as an evolutionary and path dependent process. Therefore, innovation can be understood best when its historical development is taken into consideration.
- Because innovation is path dependent and open ended, it is not possible to define an optimal innovation system. Since the system keeps changing, it is possible that at one moment one system is better suited for stimulating certain technological developments, whereas later on another system performs better.
- Innovation is regarded as an interactive process. Firms do not innovate in isolation, but in interaction with other actors. Innovation is influenced not only by the structures and the actors in the system, but also by the interaction between them.
- The importance of learning, and especially of interactive learning, is stressed. The accumulation of knowledge and skills is considered to be crucial. The focus is on the interactivity between the structures and the actors in the system, and on the learning processes which take place between them.
- Innovation systems consist of organizations and institutions on the one hand, and interacting actors on the other hand. Therefore, a structural view is combined with an actor-oriented view.

In short, an innovation system consists of four main components:

- the technological innovation, in this case wind turbines
- the actors involved
- the organizations and institutions involved
- the learning processes between the actors

Developing a completely new technology implies developing a whole new innovation system. Apart from the technological innovation that needs to be initiated, developed and fostered, actors need to be involved, organizations and institutions need to be set up, and learning processes between the actors need to be facilitated. These learning processes are of special importance, as they 'glue the components of the innovation system together', get the actors involved and make them develop a shared vision regarding the future of the technology developed. Learning processes help the network of actors involved become tighter.

Innovation strategies

Setting up an innovation system can be done in various ways, using different strategies. For instance, a top-down strategy or a bottom-up strategy can be used. A top-down strategy entails an active involvement from the initiating actor, in this case the national government, in setting up the innovation system, while a bottom-up strategy leaves much more to the initiatives of the actors involved. Other examples of innovation strategies are science-push innovation strategies and demand-pull innovation strategies. Science-push strategies entail a large emphasis on the development of new knowledge, which is believed to result in the development of new technology. Such strategies are characterised by large expenditures on R&D. Demand-pull innovation strategies, on the other hand, start with the market. These strategies involve the stimulation of market formation. Demand from such a market is believed to stimulate the innovation process. Demand-pull innovation strategies are characterised by market or investment subsidies and other measures designed to stimulate the development of a market.

In the remainder of this paper we will compare the cases of setting up a wind power innovation system in two countries: the Netherlands and Denmark. We will investigate to what extent the innovation systems and the innovation strategies differed in these two countries.

WIND TURBINE DEVELOPMENT IN THE NETHERLANDS

The large-scale wind turbine innovation subsystem

The Dutch NWO programme, the National Research Programme on Wind Energy, was started in 1976. Within this programme, subsidies were provided for R&D into the potential of wind energy in the Netherlands and into wind turbine building. The goal of this programme was to develop a significant wind turbine capacity in the Netherlands, consisting of a large number of large wind turbines (Pelser, 1981; BEOP, 1981). As a result of this research programme, two innovation subsystems developed: the large-scale wind turbine innovation subsystem, and the small-scale wind turbine innovation subsystem.

In the Dutch large-scale wind turbine innovation subsystem the paradigm was, beginning in the 1970s, directed towards building many large wind turbines in the Netherlands. In this subsystem a large amount of theoretical knowledge on wind turbines was gained during research projects at the Delft and Eindhoven Universities of Technology and at the ECN research centre. This knowledge was based merely on aerodynamic knowledge from the aerospace industry. Design models for wind turbines were developed and more applied research was performed into a.o. structural dynamics and the aerodynamics of wind turbines. Slowly, it became clear that wind turbines had their own characteristics and that models and theories from the aerospace industry could not be used without significant adjustment. Furthermore, in the late 1970s and the early 1980s research into tipvanes was performed by the Delft University of Technology. Theoretical research had shown that small vanes on the tips of wind turbine blades could lead to a 60-70% higher energy yield (Van Holten, 1978; BEOP, 1981). Researchers at the Delft University of Technology attempted to build tipvanes that would produce this effect in practice. Disappointingly, they did not succeed. The major part of the knowledge within this subsystem was based on R&D.

The knowledge gained was applied to three wind turbine prototypes and two commercial wind turbines. The turbine prototypes that were built consisted of two vertical axis turbines, or VATs, and one horizontal-axis turbine, or HAT. Test results were to prove which turbine type was the best in terms of energy yield and efficiency (Pelser, 1981)[3]. In 1981, a HAT-turbine, the HAT-25, was built by Stork, Fokker, Holec and Rademakers. It had a capacity of 300 kW and a rotor diameter of 25 metres. As with the VAT prototypes, the main goal of the HAT-25 project was to obtain measurement results and operational experience (Sens, 1981). The prototype was equipped with two blades and a very advanced regulation system. It could be operated with four regulating procedures (Dekker, 2000; Pelser, 1981). In this way, it was possible to test which regulating procedure functioned best. Measurement results of the turbine were satisfactory and Stork decided to develop a commercial turbine on the basis of the HAT-25 prototype. Three units of this commercial turbine, called Newecs-25, were sold to utilities in the Netherlands and Curaçao.

Within this subsystem, the actors in the manufacturing companies Fokker and Stork and the actors in the research institute and the universities of technology were completely in line with each other: they shared the goal of building a large number of large wind turbines that together would make a significant contribution to the system of national energy provision (Dekker, 2000; Van Holten, 2000; Pelser, 1981). This made co-operation and knowledge exchange between them very fruitful.

The intended turbine buyers within this subsystem were electricity production companies. The aim was to build large wind power stations which would deliver electricity to the electricity grid, analogous to other electricity production units, which were also owned by electricity production companies. However, the intended buyers were not involved in the design and

[3] Only two VAT turbines were built in the Netherlands. For more information on these turbines, see Kamp (2002) and Kamp et al. (2004).

manufacturing of the wind turbines. The design and manufacturing of the wind turbines was an ultimate science-push process: the turbines were developed by large companies and research institutes, based on scientific knowledge.

Even though they were the intended buyers, the electricity production companies were not very enthusiastic about wind energy. They were of the opinion that only a maximum capacity of 650 MWe of wind turbines could be fitted into the electricity grid, not the thousands of MWe that ECN and other research institutes envisioned. Some electricity production companies, like that of Zeeland, Schiedam and Curaçao, were willing to try operating a wind turbine. They each bought a Newecs-25 turbine produced by Stork. However, because these turbines were not tested satisfactorily, they encountered a lot of operational problems (Verbruggen, 2000). These problems did not help the electricity sector's opinion of wind energy. Stork also built a horizontal-axis turbine with a capacity of 1 MW and a rotor diameter of 45 metres, the Newecs-45 (Hensing and Overbeek, 1985). This turbine was meant as an in-between step towards a 3 MW turbine, which had been calculated to be the most cost-effective turbine (Van Holten, 2000)[4]. Only one Newecs-45 turbine was sold. Like the Newecs-25, it suffered many operational problems (Verbruggen, 2000). Because only a limited number of turbines was built, correspondingly only limited knowledge was gained.

In 1982, at the insistence of the Ministry of Economic Affairs, the SEP[5] (the Co-operation of Electricity Production Companies) became involved in the large scale wind energy subsystem. The SEP agreed to be involved in the development of a pilot wind power station, the Sexbierum wind power station. This time the SEP was very much involved in the design and manufacturing of the wind turbines. Therefore, in this project the subsystem was complete. The turbines of the wind power station were produced by Holec. The design and building of the wind turbines, however, entailed a number of problems, resulting in a large delay in the project and further dampening the enthusiasm about wind energy in the electricity sector (Hutting, 2000; Toussaint, 2000; Verbruggen, 2000).

Because of the many problems, the large financial risks, and the small home market, the large companies in the large-scale wind turbine innovation subsystem, Fokker, Stork and Holec, stopped producing wind turbines in the mid-1980s. After the mid-1980s, the aim of the Dutch wind energy policy makers was to make the knowledge developed in the large-scale wind turbine innovation subsystem applicable to small turbine manufacturers. In this way, in the eyes of the policy makers, the goal of developing a significant wind turbine capacity in the Netherlands could still be reached (NEOM, 1986).

---

[4] A 3 MW turbine was never built in the Netherlands. Only a pre-design study was performed (Kuijs, 1983). Because of the problems with the Newecs-25 and the Newecs-45, the risk of building a 3 MW turbine was considered too high (IEA, 1987).

[5] SEP means in Dutch: Samenwerkende Electriciteits Productiebedrijven and in English: Co-operation Electricity Producing Companies.

The small-scale wind turbine innovation subsystem

In the period 1976-1980, about ten small companies in the Netherlands started to manufacture wind turbines. They became interested in wind turbines because R&D subsidies into wind energy and wind turbines had been made available by the National Research Programme on Wind Energy. These small companies all had different manufacturing histories, like making steel constructions or polyester yachts and manufacturing farming equipment (Stam, 2000; Dutch manufacturers, a.n.).

In the small-scale wind turbine innovation subsystem, the knowledge developed was, in contrast with the large-scale wind energy innovation subsystem, based on competition between the actors instead of R&D. By way of trial-and-error, the first small wind turbines were built. These turbines were gradually improved and up scaled. Because the turbines were sold in the vicinity of the manufacturing companies, problems were observed and solved quickly in interaction with the users, enabling the manufacturer to learn from these problems (Boersma, 2001).

In the beginning, the turbine manufacturers encountered many difficulties in building reliable wind turbines. Therefore, ECN set up a test field in 1981. On this test field, the turbines were tested and the manufacturers received indications about what needed to be improved in their turbines (Stam et al., 1983). Because of the danger of distortion of competition, ECN was not allowed to give specific indications on how to improve the wind turbines (Stam, 2000). In addition, the turbine manufacturers received no help from each other at all: they considered each other as competitors and were not willing to share any knowledge on how they built their wind turbines (Stam, 2000).

Another problem that the Dutch wind turbine manufacturers encountered was the small size of the domestic market. The Dutch market was and remained small, because in the Netherlands no investment subsidies were available for wind turbine buyers. Therefore, payback times for wind turbines were long (Werkgroep Duurzaam-energieplan, 1984). Furthermore, wind turbine owners received only small buyback tariffs for the electricity they delivered to the grid. These two factors made the purchase of wind turbines not very attractive financially (Langenbach, 2000; Blok, 2000). The main turbine buyers were renewable energy advocates and farmers (CEA, 1993).

Gradually, the wind turbines became better and larger. However, this process went more slowly in the Netherlands then in Denmark. As a result, the Dutch manufacturers were unable to compete with the Danes on the large Californian market.[6] This factor, together with the small size of the Dutch market, caused financial problems for the manufacturers in the mid-1980s.

From the mid-1980s wind energy policy started to become actively involved in the activities of the small turbine builders. Because the wind turbine producers in the large-scale wind turbine innovation subsystem had ceased their

---

[6] This market appeared in the early 1980s because large investment subsidies were made available for wind turbine buyers in California.

activities, the small turbine builders were to be responsible for the production of efficient wind turbines that could produce a significant part of the Dutch electricity supply. Therefore, from the mid-1980s on, the research institutes and universities of technologies could only receive R&D subsidies if they made their research results applicable to the small turbine builders (NEOM, 1986). Furthermore, investment subsidies were introduced. This increased the Dutch home market, because utilities started to show an interest in buying wind turbines (IEA, 1987).

From then on, an active attempt was made to incorporate the results of R&D in the design and manufacturing process of the small wind turbine builders. Researchers from research institutes and Stork worked together with small wind turbine builders in improving and scaling up their wind turbines. However, this co-operation was sometimes difficult, since the paradigms and the approaches were completely different. This severely limited knowledge exchange. The researchers were academic trained science-based thinkers, whereas the manufacturers were 'builders' (Boersma, 2001; Verbruggen, 2000).

One manufacturer, Lagerweij, had a different approach. He did not accept direct interference of researchers into his design process, but he did use knowledge obtained by R&D by way of personal contacts in Delft and the picking up of their ideas. This resulted in gradual improvements in his small 75 kW / 80 kW turbines, for which he used a.o. ideas on flexible components developed at the Delft University of Technology (Van Holten, 2000; Boersma, 2001).

The drive towards fast upscaling and the problems involved with incorporated advanced concepts and components in their wind turbines, combined with the small Dutch home market[7] and the competition from the Danes, who offered better products, resulted in severe difficulties for the Dutch manufacturers in the 1990s. In the year 2000, only one Dutch turbine builder, Lagerweij, remained on the market.

Conclusions

Let us now examine the focus points of this chapter. What were the characteristics of the wind turbine innovation system, in terms of the actors involved and the learning processes, and what were the characteristics of the innovation strategy in the Netherlands? To answer this question, we will distinguish between the large-scale wind turbine innovation subsystem and the small-scale wind turbine innovation subsystem.

In the development of large wind turbines, the focus was on the rapid development of a large number of large wind turbines based on advanced scientific concepts. Large R&D subsidies were available. This was clearly a science-push strategy. The main actors involved were the Dutch universities of technology, the research centre ECN, and the large turbine manufacturing

---

[7] Beginning in the 1990s, the small size of the Dutch market was aggravated by the siting problems.

companies Fokker and Stork. The users of the turbines were utilities. Interactive learning between the companies and the researchers went very well, because they shared the same paradigm of research-driven technology development. However, hardly any interactive learning with the turbine users, the utilities, took place. This was because, firstly, the market was very small (only four turbines sold), and secondly, because the utilities had a very negative opinion regarding wind energy. They preferred large-scale power plants fired by gas or nuclear energy and had little trust in wind turbines. After many operational problems occurred, they became ever more reluctant to co-operate in knowledge exchange. This resulted in the development of scientific articles and models, but not in the development of wind turbines that worked well in practice. It turned out that the fast up-scaling of wind turbines was more difficult than foreseen. The science-push approach was clearly not sufficient for the development of wind turbines.

In the development of small wind turbines, the focus was on the gradual up-scaling of small wind turbines, based on learning-by-doing and trial-and-error. Also here, the main innovation strategy was the science-push approach. The market was not stimulated at all until 1986 and the only subsidies available were R&D subsidies. Furthermore, after 1986 an explicit attempt was made to incorporate the insights of science into the wind turbine innovation process. The main actors were small manufacturing companies and the owners of the wind turbines. Also ECN was involved via the test field. Interactive learning occurred between turbine manufacturers and owners, which resulted in rapid problem solving and the gradual improvement and up scaling of the wind turbines. Also interactive learning took place with ECN on the test field. However, this remained limited to the transfer of general knowledge. The interactive learning process between researchers and turbine developers after 1986 went very badly. Because of their very different backgrounds and the different mind models and concepts they used, these different actors did not succeed in transforming scientific knowledge into well developed wind turbines.

As regards the top-down versus bottom-up innovation strategies, clearly a top-down strategy was followed in the Netherlands. The government and the Ministry of Economic Affairs steered the developments to a great extent; they were even heavily involved in the technical choices made by the companies. This active steering did not yield great results.

WIND TURBINE DEVELOPMENT IN DENMARK

The large-scale wind turbine innovation subsystem

In Denmark, as in the Netherlands, a development programme for wind energy was set up in the late 1970s. The main objective was to determine under what circumstances and to what degree wind energy could make a contribution to the Danish electricity supply systems (IEA, 1985). The programme was called the Wind Power Programme. Within the programme, the research centre Risø and the Technical University of Denmark were to develop the knowledge needed to build large wind turbines. It was envisaged that large wind turbine parks owned and operated by utilities would be built by a consortium of large Danish firms. The first measurement programme that was carried out was on the Gedser turbine. This wind turbine had been built in the 1950s by the Danish technician Johannes Juul and had proved workable. It had a capacity of 200 kW, a horizontal axis and three blades (Karnøe, 1991). The Danes did not have an aerospace industry, so they could not use knowledge gained in that sector.[8]

In 1977, it was decided to build two 630 kW turbines on the basis of the specifications of the Gedser turbine, one with a pitch control system as that in the Dutch HAT-25 prototype, and one with a stall control system, as the Gedser turbine had. These two turbines were called the Nibe turbines (Karnøe, 1991). No Danish company was interested in building the Nibe turbines. Therefore, the turbines were procured on a multi-contract basis. Other actors involved were Risø, the Technical University of Denmark and the SEAS utility, which partly financed the wind turbines (Van Est, 1999). Thus unlike in the Netherlands, in Denmark the turbine owners were involved from the very beginning. Like with the Dutch large wind turbines, there were many problems with the Nibe turbines, e.g. fatigue problems in the blades and problems with the gear box (IEA, 1985).

In the early 1980s, eight more large wind turbines were built. All these turbines had a pitch control system, because the Nibe turbine with pitch control functioned better than that with stall control (IEA, 1985). Here also the utilities were involved from the beginning. All but one of these eight turbines were built by the company Danish Wind Technology. This company was established in 1981 by the Danish Ministry of Energy and the SEAS utility (Van Est, 1999).

All wind turbines suffered problems with the blades and the gearboxes (IEA, 1990; Heymann, 1998). Building wind turbines proved to be more expensive and risky than expected. Furthermore, no large Danish company appeared to be interested in building large turbines. In the early 1990s, the Danish state sold its shares in Danish Wind Technology (Karnøe, 1991). By that time, the Danish small-scale wind turbine innovation subsystem had demonstrated its ability to manufacture reliable well-working wind turbines that were far cheaper than the wind turbines developed by the large-scale wind turbine innovation subsystem.

In the Danish large-scale wind turbine innovation subsystem, as in the Dutch, the emphasis was on R&D by way of research and measurement programmes. Knowledge exchange went better in the Danish large-scale wind turbine innovation subsystem than in the Dutch, because in Denmark the turbine buyers were involved from the beginning and shared the framework goals of the research institutes and the turbine producers: building large turbines that could make a significant contribution to the Danish energy supply.

---

[8] As explained in section 4, this in fact turned out to be an advantage for the Danes. Wind turbines were found to have their own characteristics, and therefore models and theories from the aerospace industry could not be used without significant adjustment.

The small-scale wind turbine innovation subsystem

In Denmark, as in the Netherlands, a small-scale wind turbine innovation subsystem was developed beginning in the late 1970s, and was relatively independent of the wind energy R&D programme set up by the Danish state. The first wind turbine producers in this subsystem were adherents of the grassroots movement and small entrepreneurs. These actors rediscovered the Gedser wind turbine and started developing wind turbines based on this example.

The adherents of the grassroots movement were attracted to the idea of small locally-owned and locally-governed power production units, instead of large power production units that were centrally-owned and centrally-governed by the utilities. Furthermore, they saw renewable energy as an absolutely essential substitute for environment-polluting fossil fuels and for the nuclear power plants that were planned by the Danish utilities (Jørgensen and Karnøe, 1995). A famous turbine developed by left-wing oriented people was the Tvind turbine. It was developed by teachers and students of the Tvind school with the help of people from different educational backgrounds. The design, blade profile and calculations were performed with the help of engineers (Karnøe, 1991).

By 1978, about ten small wind turbine companies had developed. Many of these had previously manufactured agricultural equipment. Their knowledge was based on the manufacturing of machines, and they learned slowly, by way of trial-and-error, how to manufacture and improve wind turbines (Karnøe and Garud, 2001). They obtained their knowledge from previous wind turbines, like the Gedser and the Tvind turbine, from their own trial-and-error experience in the design and production of wind turbines, and from the turbine users, either individually or collectively during so-called Wind Meetings.[9] The turbine companies' design philosophy was to build wind turbines that worked reliably and safely (Karnøe, 1995). They were under pressure to improve their turbines, especially because the performance of their turbines was made public in the magazine Naturlig Energi. This magazine was set up by the Danish Windmill Owners Association, and its pages described the performance of the several types of turbines. Because they were organised, the users created a strong selection environment for the first Danish turbine builders (Karnøe and Garud, 2001; Heymann, 1998).

All kinds of problems with a.o. rotor speeds, gear boxes, burned out generators, and broken yaw systems were handled. Design and construction were based upon trial-and-error and simple rules of thumb (Karnøe, 1995). The manufacturers were used to this way of working and they refrained from taking risks. Gradually, practical and hands-on knowledge about the poorly understood technology accumulated. On the basis of this knowledge, the design rules were gradually improved. Design and development problems stemmed from turbine failures or from construction problems. The failures were often solved by

[9] During the Wind Meetings, knowledge and experience on wind turbines were shared between wind turbine manufacturers, owners, and researchers.

making the turbines more solid, or, in other words, by 'throwing metal on the problem'. This method increased the lifetimes of the Danish wind turbines by limiting aerodynamic loads and preventing dynamic vibrations (Karnøe, 1995).

In 1978, a wind energy department was created at the Risø research centre. Because the research centre only received financing for three years, their strategy was to be of immediate service to the wind turbine manufacturers. If the manufacturers could be convinced of the usefulness of the research Centre, it could in the future get its financing through orders from the manufacturers themselves (Dannemand Andersen, 1993). Therefore, the goal of the members of the research Centre was not to develop the technically best wind turbine, but to develop a wind turbine industry. In this way, a tight network between wind turbine producers, owners and the Risø research centre developed within the small-scale wind turbine innovation subsystem. Because most turbine producers chose to follow the technology guideposts established by the Gedser turbine, they produced the same turbine type, i.e. a three-bladed stall-regulated wind turbine (Karnøe, 1991). This made the exchange of knowledge very efficient. Therefore, knowledge exchange went very well within the subsystem.

Another favourable circumstance was the size of the Danish home market. Already in 1979, investment subsidies were introduced (Van Est, 1999). This made buying a wind turbine far more attractive than in the Netherlands. The relatively large home market gave the Danish turbine manufacturers the opportunity to produce a relatively large number of wind turbines and learn by doing during the process. Furthermore, the relatively large user group, which had organised itself into the Windmill Owners Association, was able to act as a strong party during negotiations on buy-back tariffs with the utilities.

In the early 1980s, the size of the Danish home market decreased. However, at the same time a very large market arose in California, because large investment subsidies for wind turbine buyers were introduced there (Van Est, 1999). Because the Danes produced relatively good wind turbines and were able to prove this with the statistics published in the Naturlig Energi Magazine, they were able to capture a large part of the Californian market. In 1985, they sold 2,000 wind turbines to California (Karnøe and Garud, 2001). This greatly enhanced knowledge development. However, the demand in California was different from that in Denmark. In California, the buyers wanted larger and more cost-effective wind turbines (Van Est, 1999). This forced the Danes to speed up technology development and to sell turbines that had not yet been thoroughly tested (Karnøe, 1991). This resulted in severe technical problems for the Danish manufacturers.

In 1986 the Californian investment subsidies expired. Exports declined and came to a halt in 1988 (Gipe, 1996). However, the Danish market had started to grow again beginning in 1985. In that year, the utilities had signed a 100-MW agreement, which meant that they had to install 100 MW of wind turbines within the next five years (Van Est, 1999). This enabled the Danish turbine manufacturers to make a new start. The Wind Turbine Guarantee Company was set up to guarantee the long-term financing of large export projects (Van Est, 1999). One of the conditions that the manufacturers had to meet in order to qualify for the guarantees was that their turbines had to be approved according to a new, more rigid approval system (Hvidtfelt Nielsen, 2001). The manufacturers were required to set forth their knowledge in a more formalised way (Dannemand Andersen, 1993). Furthermore, they had to up scale and improve their turbines further, because the utilities' demand was for relatively large and cost-effective wind turbines. Turbine design therefore gradually changed from a trial-and-error process to a more R&D based and formalised process (Dannemand Andersen, 1993). With the help of Risø, the Danish manufacturers succeeded in meeting the utilities' demand and building up a strong position on the world market.

Conclusions

Let us now return once again to the focus points of this chapter. What were the characteristics of the wind turbine innovation system, in terms of the actors involved and the learning processes, and what were the characteristics of the innovation strategy in the Netherlands? To answer this question, we will distinguish between the large-scale wind turbine innovation subsystem and the small-scale wind turbine innovation subsystem.

In the development of large wind turbines, as in the Netherlands, the focus was on large, science-based technology development: a clear science-push strategy. This subsystem was furthermore characterised by a top-down innovation strategy because of the large governmental involvement. The network of actors involved was more complete, which resulted in more knowledge exchange between turbine users and producers than in the Netherlands. However, the large costs and technical problems forced this development to come to a halt in the mid-1980s.

However, the development of small wind turbines was very successful in Denmark. The innovation system was complete and functioning well. Because of market subsidies, a large group of users developed. Therefore, many turbines could be produced and a lot of knowledge was transferred between the turbine producers and users. Furthermore, the paradigm at the Risø research centre was more hands-on than science-based, especially in the 1980s. This made knowledge transfer between Risø and the other actors very effective. Gradually, this tight network of actors succeeded in building up a strong knowledge base and industry, one of the leading wind industries in the world. In terms of innovation strategy, we see that in this subsystem a bottom-up strategy was much more prevalent. Much more was left to the initiatives of the actors themselves: companies, turbine owners and researchers. As to science-push versus market-pull, we see a gradual shift over time. Until the 1990s, the focus was clearly on market-pull: stimulation of market development, and little attention was paid to the development of science and input of science into the innovation process. However, this gradually changed, because over time the wind turbines became larger and larger and more and more complex machines. Therefore the input of science became more and more important. Co-operation between turbine producing companies and scientific researchers developed and professionalised over time.

CONCLUSIONS AND DISCUSSION

Conclusion

The goal of this research was to investigate the reasons why the Netherlands were far less successful in developing wind power than the Danes in the period 1970-2000. We looked for answers to this question by comparing the innovation strategies that were followed by the governments of both countries in order to set up this new technology. Which conclusions can be drawn from this historic case study for the future, especially for policies on renewable energy sources? In Table 1 below our main conclusions on innovation strategies and innovation systems are summarised.

Table 1
Main conclusions on the innovation strategies and innovation systems in the four wind turbine innovation subsystems

	Top-down of bottom-up strategy	Science-push or demand-pull strategy	Completeness of the innovation system	Main learning processes
Large-scale wind turbine innovation subsystem in the Netherlands	Top-down	Science-push	Intended buyers not involved	R&D-based learning
Small-scale wind turbine innovation subsystem in the Netherlands	Top-down	Mainly science-push	Small group of buyers; researchers marginally involved	Learning-by-interacting between companies and turbine owners
Large-scale wind turbine innovation subsystem in Denmark	Top-down	Science push	Intended buyers only marginally involved	R&D-based learning
Small-scale wind turbine innovation subsystem in Denmark	Bottom-up	Mainly demand-pull, later on more and more science-push	Complete and well functioning	Learning-by-interacting between all actors involved

Based on the case materials presented we can draw some very distinct conclusions about innovation strategies and innovation systems. As to top-down versus bottom-up, we see that in the only case where the bottom-up strategy was followed it was successful in developing a wind turbine industry with a strong position on the world market. An explanation for this is that a bottom-up strategy will cause more actors to become involved and make them less passive about awaiting governmental actions. This will stimulate knowledge exchange between them.

As to science-push versus demand-pull, we see that in the only successful subsystem a mainly demand-pull strategy was used, especially in the beginning. The explanations for this are:

- Because wind turbines are very complex machines and need to function in rapidly changing wind conditions, building up large experience in their use 'in the field' is very important. Therefore, it is important to build up a large market at an early stage.
- This large market can, if it is well organised, have a large input in the innovation process, which will result in better and more market-adjusted products.

On the subject of innovation systems, we can conclude that a complete and well-functioning innovation system is a prerequisite for successful innovation. This means that all major actors (technology producers, technology owners (or envisioned owners, researchers and policy makers) have to be actively involved from an early stage. This will facilitate learning-by-interacting between all the actors involved, which is another prerequisite for successful innovation.

Furthermore, based on the case materials examined we can draw some other conclusions:

- We can conclude that building up a strong home market is very important. Thus, setting up policies to stimulate the development of renewable energy sources should at least include the stimulation of market development. The best way to do this is subsidize per kWh produced. This will stimulate technology developers to develop cost-effective technology. The existence of a strong home market is important for two reasons. Firstly, it will enable technology producers to sell more. Secondly, it will promote knowledge exchange between owners and producers. This knowledge exchange will be more successful of the owners are organised, as in the case of the Danish small wind turbines.
- We see that often technology is developed without taking into consideration the wishes of (future) owners. This was e.g. the case in the development of the Dutch and Danish large wind turbines. These turbines were developed based on scientific theories and models, without looking at who the owners were going to be and what they wanted. In the Netherlands, the envisioned owners of the large turbines were the utilities, who were not enthusiastic about wind energy at all. The technology was more or less forced upon them. This hindered knowledge exchange to a great extent. Another example can be found in the development of photovoltaic solar panels, which is also very much science-based. The wishes of actors, such as designers of buildings and future owners, are not taken into consideration at all. Changing this would result, in our view, in a larger market.
- The case of the large wind turbine development shows that developing large machines in a constantly changing environment (different wind speeds) is difficult and risky. Therefore, thorough testing and the taking of small technological steps is important. The way in which the Dutch manufacturing company Stork underestimated the jump from a 300 kW to a 1 MW wind turbine explains a lot of the problems with the 1 MW turbine. Therefore, in our view, designing off-shore wind turbines with a lot of new science-based concepts for deep water is very risky. However, this is precisely what the Dutch development programme on off-shore wind energy is striving for. In our view, taking smaller steps from shallow to deep

water and from proven technology to new concepts would significantly reduce the risks. And since wind energy is not popular in the eyes of the general public, because of the so-called Not In My BackYard problem, it is important that the Dutch development programme succeeds in developing good, reliable turbines and does not turn into a failure.

- The case of the Danish development of small wind turbine shows that the building up of a strong network and involving several kinds of actors is important. Because in Denmark the wind turbines were often owned by a whole village or large parts of a village, the siting problems were less severe than in the Netherlands. The villagers shared both the advantages and the disadvantages of the wind turbines. The same is true for siting policy. It is important to involve policy makers at all levels, both local, regional and national. In the Netherlands, local policy makers were not involved in the wind turbine siting programme of 1991, which resulted in a lot of opposition from the municipal councils.
- Setting up a long-term consistent policy is important. Since, as it appears from the cases, the development of wind turbines is expensive and risky, it is important that financers can be sure of that the subsidising schemes will remain in place in the coming years. Recently, we saw that both in Denmark and in the Netherlands subsidising schemes were halted. This poses a great threat to the further development of wind turbines.

Discussion

To what extent are the results of this research applicable to other cases? In our view, the general conclusions we draw on setting up a wind turbine innovation system in the Netherlands and Denmark are applicable to setting up such an innovation system in other countries as well. We even think that our conclusions, except for the third one, which is specifically about wind turbines, are applicable to other kinds of innovation systems as well. Not only those concerning other energy technologies, but also those concerning other kinds of technologies.

This would be an interesting starting point for further research: applying this theoretical research method to other innovation systems, e.g. the innovation system for solar cells or some other non-energy technology and comparing the set-ups of such a system in two countries.

BIBLIOGRAPHY

Andersen, E. S. and B. A. Lundvall, (1988), "Small National Systems of Innovation Facing Technological Revolutions – an Analytical Framework', [in:] C. Freeman and B. A. Lundvall (Eds.) *Small countries facing the technology revolution*, London: Pinter Publishers.

BEOP, (1981), *Perspectieven voor windenergie, Nationaal Onderzoekprogramma voor Windenergie in Nederland 1976-1981, resultaten en aanbevelingen*, Petten.

Blok, K., (2000), Personal Communication, Professor at Department of Science, Technology and Society at Utrecht University.

Carlsson, B. and S. Jacobsson, (1997), "Diversity Creation and Technological Systems: a Technology Policy Perspective", [in:] C. Edquist, *Systems of Innovation – Technologies, Institutions and Organizations*, London: Pinter Publishers.

Carlsson, B., S. Jacobsson, M. Holmén, and A. Rickne, (2002), "Innovation Systems: Analytical and Methodological Issues", *Research Policy*, Vol. 31, pp. 233-245.

CEA, (1993), *CEA Database on the Wind Turbines in the Netherlands*, Rotterdam.

Dannemand Andersen, P., (1993), *En analyse af den teknologiske innovation i Dansk vindmølleindustri – Herunder prøvestationen for vindmøllers dobbeltrolle som forskningsinstitution og godkendende myndighed*, Copenhagen: Samfundslitte-ratur.

Edquist, C. (ed.), (1997), *Systems of Innovation – Technologies, Institutions and Organizations*, London: Pinter Publishers.

Freeman, C. and B. A. Lundvall (eds.), (1988), *Small Countries Facing the Technology Revolution*, London: Pinter.

Garud, R., (1997), "On the Distinction Between Know-How, Know-Why, and Know-What", *Advances in Strategic Management*, Vol. 14, pp. 81-101.

Grin, J. and H. van de Graaf, (1996), "Technology Assessment as Learning", *Science, Technology & Human Values*, Vol. 21, pp. 72-99.

Hedberg, B., (1981), "How Organizations Learn and Unlearn", [in:] P. C. Nystrom and W. H. Starbuck (Eds.), *Handbook of Organizational Design*, New York: Oxford University Press.

Hensing, P. C. and H. H. Overbeek, (1985), "Nederlands grootste turbine operationeel in Wieringermeer", *Energiespectrum*, December 1985, pp. 254-260.

Heymann, M., (1998), "Signs of Hubris; The Shaping of Wind Technology Styles in Germany, Denmark and the United States, 1940-1990", *Technology and Culture*, Vol. 39.

Hutting, H., (2000), Personal Communication, KEMA, Arnhem.

Hvidtfelt Nielsen, K., (2001), *Tilting at windmills*, Academic Thesis, Århus University.

IEA, (1985), *Wind Energy Annual Report 1984*, Paris.

IEA, (1987), *Wind Energy Annual Report 1986*, Paris.

IEA, (1990), *Wind Energy Annual Report 1989*, Paris.

IEA, (2000), Experience Curves for Energy Technology Policy, Paris.

Jørgensen, U., and P. Karnøe, (1995), "The Danish Wind-Turbine Story: Technical Solutions to Political Visions?", [in:] A. Rip, T. Misa and J. Schot, (Eds.), *Managing Technology in Society; The Approach of Constructive Technology Assessment*, London: Pinter.

Kamp, L. M., R. E. H. M. Smits, & C. D. Andriesse, (2004), Notions on Learning Applied to Wind Turbine Development in the Netherlands and Denmark, *Energy Policy 32 (14)*, 1625-1637.

Kamp, L. M., (2002), *Learning in Wind Turbine Development: A Comparison Between The Netherlands and Denmark*, PhD thesis, Utrecht University.

Karnøe, P., (1991), *Dansk vindmølleindustri: En overraskende international succes; Om innovationer, industriudvikling og teknologipolitik*, Copenhagen: Samfundslitteratur.

Langenbach, J., (2000), E-mail Communication, Wind Turbine Owner and Wind Energy Advocate.

Lundvall, B. A., (1988), "Innovations as an Integrative Process – From User-Producer Interaction to the National System of Innovation", [in:] G. Dosi et al. (eds.), *Technical Change and Economic Theory*, London: Pinter Publishers.

Lundvall, B. A., (1992), *National Systems of Innovation – Towards a Theory of Innovation and Interactive Learning*, London: Pinter Publishers.

Neij, L., (1999),"Cost Dynamics of Wind Power", *Energy*, Vl. 24, pp. 375-389.

Nelson, R. R., (1993), *National Innovation Systems – A Comparative Analysis*, New York: Oxford University Press.

NEOM, (1986), *Integraal Programma Windenergie (IPW) 1986-1990*, Utrecht.

Nooteboom, B., (2001), *Learning and Innovation in Organizations and Economies*, New York: Oxford University Press.

Pelser, J., (1981), "Evaluatie Resultaten Nationaal Onderzoekprogramma Windenergie", Energiespectrum, June 1981.

Rosenberg, N., (1982), *Inside the Black Box: Technology and Economics*, Cambridge: Cambridge University Press

Sahal, D., (1981), *Patterns of Innovation*, Reading MA: Addison Wesley.

Sens, P. F., (1981), "Het Nationaal Onderzoekprogramma Windenergie", *Energiespectrum*, July/August 1981, pp. 172-177.

Stam, W. J., H. J. M. Beurskens and J. B. Dragt, (1983), "Het ECN windturbinetestveld: algemene ervaringen en werkwijze", *Energiespectrum*, October/November 1983, pp. 274-281.

Van Est, R., (1999), *Winds of Change; A Comparative Study of the Politics of Wind Energy Innovation in California and Denmark*, Utrecht: International Books.

Werkgroep Duurzaam Energieplan, (1984), *Teruglevertarieven voor windelektriciteit*, Delft.

Stam, W. J., H. J. M. Beurskens and J. B. Dragt, (1983), "Het ECN windturbinetestveld: algemene ervaringen en werkwijze", *Energiespectrum*, October/November 1983, pp. 274-281.

8

Knowledge and Technology Transfer from Academia to Business – Polish Perspectives

Dariusz M. Trzmielak

INTRODUCTION

At the heart of each knowledge transfer from academia to business is a relationship between an organization with a base in knowledge (universities, R&D institution) and a business venture. This transfer of knowledge enables both private and public sector research organizations to apply their invention to a commercial project, which is crucial to the development of new products or services.

The expected benefits of implementing the Polish offset programs (as a result of the agreement between Lockheed Martin Corporation and Polish Government for purchase of F16 jet fighters) approach include an increasing ability to incubate new ventures and the opportunity to identify pre-qualified strategic partners quickly. These new ventures should be able to take on a leading role in major Polish and international scientific ventures, and to capitalize on emerging opportunities through early and improved access to leading-edge technology.

Technology transfer is one of the most important issues affecting the technological environment. Based on their knowledge of the key components of the technological environment and its relationship to commercialization processes, scientists and inventors have become painfully aware of the fact that many new concepts for products are no longer be valid when it comes to implementing them in the market. In other words, in the area of commercialization it is necessary to expend considerable effort aimed at gaining a better understanding of the factors that impact on the implementation of technology and the adoption of new products.

OBJECTIVES

The purpose of the chapter is to provide scientists and inventors with comprehensive information which will aid in their understanding of the process of knowledge transfer from academia to business and form a basis for technology transfer from R&D departments or university units to industry. Programs focusing on high technology sectors, stimulated by governments, international organizations, or private firms, are currently the source of significant knowledge transfer components. Unfortunately, many Polish representatives of academia and the business community rely mostly on donations and programs sponsored by governments or public institutions and stumble around in the dark when it comes to the actual technology transfer process. Their lack of knowledge about the commercialization process prevents them from implementing product concepts in the market. The university offset program (an Offset Commitment Agreement between University of Łódź and State University of Texas at Austin signed on 24th of July 2003 as a part of the agreement between Lockheed Martin Corporation and Polish Government) in Poland is designed to address the problems mentioned above, based on the assumption that the biggest obstacle to technology transfer among organizations and companies is the lack of a broad market vision.

This text is divided into three parts. The first part of the article outlines the essential aspects of the innovation and commercialization processes. The second part focuses on educating professionals and providing them with specialized skills in the management of knowledge, aimed above all at accelerating the transfer of technology-based ideas into products and services in the marketplace. The third part examines how to support entrepreneurs by providing human resources, funding, technologies and company databases, and assisting in the establishment of partnership networks at the early and crucial stages of product development.

The author also provides examples for individuals who want to become successful entrepreneurs, technology commercialization leaders, and professionals, i.e. those who:

- are interested in the creation of new products, processes, companies, or economic branches
- work in national and international companies
- are involved in technology transfer at a university, R&D laboratory, incubator, or science park
- work in governmental and nongovernmental institutions and organizations involved in financing scientific research

- are interested in general management of the technology commercialization processes
- want to learn how to start and operate high technology ventures, either as new start-ups or spin-offs or within an existing organization.

INNOVATIONS, KNOWLEDGE AND TECHNOLOGY TRANSFER

Innovations become widely known and adopted in three dimensions: in the social system, in the business system, and in the research and development process. Innovation adoption in a social system is dependent on the actions of individual customers. Consumers and habitual end-users will make the ultimate decision as to what is new and interesting and useful, and what is not. The way in which innovations are accepted is often spontaneous and not capable of being controlled.

Exchanges of innovation between companies within the framework of an economic system spreads innovation in general in the business system. Innovations move from the place of creation to production and development companies. The spread of technological achievements and entrepreneurship most often comes about through business activities, thus the more they flourish throughout a country the more conducive is the atmosphere for innovation.

Innovation as creative and dynamic (Lang, 1992) has a major impact on research and development processes, and its creativity and dynamics are decisive in the commercialization of new products or services. They create ideas and new products that can then be tested by the R&D institutions (vertical knowledge transfer) (Korcz, 2003). The inventions and new solutions discovered then take shape and materialize (in the case of new products) and move on to the assessment stage. In order to understand a market and be able to assess the market potential of new technologies, it is necessary to attain a better understanding of the commercialization process by learning, practicing, and experimenting. Commercialization of new technologies and new products follows a positive review at this stage.

Innovations can move both within a single country or from one country to the other. Partnerships between institutions extend the range of innovations and create more and better opportunities for progress or advancement. International networking enables collaboration between universities and industries on strategic, high-level research projects. This creates a communication network between and within organizations, industries, and business systems.

Many organization external and internal factors impact on innovations. Above all the sector specialization of regions and trends in regional policy shape the cooperation between the main players on the market: business, academia, and government. Salas et al., (2003) argue that the local availability of an appropriate labor force and reasonable labor costs are main regional advantages. We believe that these opportunities attract first of all large companies, which in turn by cooperation with SMEs construct value added partnering. SMEs and micro ventures are attractive partners in knowledge and technology transfers because their competitiveness is less dependent on labor costs than large regional or global companies.

Growing attention is being devoted to the importance of knowledge for economic development internationally. There is still however a considerable gap between the interest shown in the role of knowledge and an analytical understanding of the issue. B. A. Lundvall, defines knowledge on the taxonomies theory developed by Aristotele. This definition refers to universal theory "know-why, practice "know-how" and "local wisdom" (Lundvall, 2005).

Knowledge is the accumulation of information in the human brain, yielding independent and distinct feelings. As a consequence, there is a lower risk in the decision process (Mynarski, 2004). Knowledge also consists of competencies, traditionally considered as immanent in the practical activities of businesses as they carry out actions such as running their business or managing new ventures. The great inventors and important authors of new technology should also possess an experience-based intuition that enables them to recognize and understand a complex situation quickly (tacit knowledge). Lundvall also identified organizational and individual competencies (Lundvall, 2004). Organizational competencies are inherent in the organization's routines and problem-solving methods. What kind of intellectual resources does the company have? What kind of information is gathered? Which activities and experiences will lead to success? In contrast to organizational competencies, individual competencies can also be identified. These are the abilities and skills which take shape in the learning process, peculiar to each individual. All kinds of knowledge can be transformed into information that could easily be transferred from academia to business and vice versa. Knowledge, understood as the ability and skill to carry out an action, is usually attained through practice and experience.

The type of knowledge frequently employed is that within the domain of the universities. The question thus arises: if knowledge is universally recognized to create national wealth, how will it be transformed into business; and if not, how does spread from world institutions? The concept underlining the Polish university offset program has several components designed specifically to enhance the transformation of academic and governmental knowledge and technology to business.

Successful knowledge transfer depends on the perception of academic programs by business and business's desire to obtain good results. Both partners (universities / research institutes and commercial entities) must be knowledgeable and willing to cooperate. From the commercial perspective, incubated firms and flourishing companies seek to expand or enhance their technology or new technical solutions, and wish to receive management knowledge. From the academic side, universities wish to exploit the developments of their faculty or staff in order to become more market oriented (Mohr, 2001). This is indispensable today, in light of the competition between Poland and other member states of the European Union, as well as other emerging countries. The mechanism of knowledge transfer from academia to business is also a road map for technology transfer. Technology transfer leads to the development of new products and services, while knowledge transfer enables the process to function efficiently. One of the major tasks involved in knowledge transfer is the use of higher-order organizing principles such as rules, directives, and routines.

Gomory indicates the importance of the "ladder process", the step by step practice of generating new scientific knowledge and then converting that knowledge into new products (Gomory, 1994). Functional items are created as the culmination of cumulative scientific research, Thus the process moves towards increasing practicality. Commercial products emerge out of scientists' labs, and they can succeed by providing them with a significant business opportunity.

Technology transfer is connected with technological innovation (Rogers et al., 1994) and dependent upon any aspect of company technological activities (Myers, 1999). The technology commercialization process usually means the transfer of technological innovation from R&D institutes, universities and companies to potential clients, i.e. firms that have at least the "know why" and sources to compete via innovation. The key to the commercialization of new technology is transfer. The results and benefits of commercialization are the development of new products or services, and their implementation on the market. Technology transfer is the part of commercialization process, opening up communication channels via:

- spin-offs ,
- licenses,
- publications,
- seminars and conferences,
- educational and training programs,
- cooperation networks.

The ability to transfer technology depends on the foundation of knowledge in universities and R&D institutes, but above all on intelligence generation, intelligence dissemination, and response implementation (Rupik, 2004). These are the determinants of the degree to which scientists and authors of new ideas can participate in commercialization process. There are, moreover, two things needed to bring technology to the market. One is the proper analysis of market information. The other concerns knowledge of the market processes that permit new technology to be absorbed. Jolly adds that not only do market processes need to be understood, but that the problems which arise in the process need to be addressed (Jolly, 1997). Finally, technology becomes transferred because its users incorporate know-how and know-why and see added value in the long term. If the transferee focuses only on the immediate benefits and technical aspects, then the transfer of technology might appear inappropriate and the potential new products and services would not reach the market. Parallel knowledge transfer allows for the creation of future product added value for potential end users.

University and business transfers

The process for transferring knowledge and technology from academia to business can best be understood by first taking a quick look at the commercialization process. The representational nature of commercialization often makes an impact on the development of new technologies in universities and on there implementation by private companies in business. The business context traditionally provides that transfers go to the business branch (business to business; academia to business). It is important to notice, in context of the Polish example, that primary transfers went to academia (academia to academia- A to A). As we might suspect, these transfers of A to A have come about primarily because of the lower level of development of the Polish high-tech sector. Although inventions depend on the creativity of scientists, they undoubtedly are shaped by political dimensions. The components of the commercialization process are designed to find market forces and decrease to a minimum the risk that valuable new products and services will not be developed. However the Polish market, in addition to its need for financial sources, needs knowledge on how to use the existing Polish intellectual capital to create national wealth. In this environment, market barriers must be reduced by government or nongovernmental organizations. Polish universities and R&D institutes are now encouraged to compete freely in high level quality areas. This means that they have to learn quickly from those organizations which have practices and know how built up over many years.

The beginning of the new century (indeed millennium) has become the time for an innovation strategy boom in many Polish organizations. Polish universities have set a goal to establish incubators, innovation centers, new science and technology parks, and technology transfer offices. Their activities should have a strong impact on technology transfer from academia to business. The new organizations are key components in Polish innovation policy, encouraging scientists, inventors and authors of new ideas to participate in the creation of Polish national wealth. Unfortunately this part of the innovation infrastructure has had little impact on the process of technology commercialization. This situation may be compared to that of an engine without fuel. The great problems are a lack of venture capital and best practices, leading to a lack of funding. There are a lot of interesting new technologies at Polish universities, but only a few university inventors want to be entrepreneurs. These who wish to engage in business need the tools for market potential assessment (Benson, Philpott, 1990). One of the milestones in alleviating this problem is a change in the university management mentality. The inventors have to understand the commercialization process and see their own role in the creation of spin-offs.

The technology transfer process from academia to business goes through four stages.

First is the research stage. A technology transfer office or innovation center gathers information concerning new scientific research achievements. Then the university technology manager provides IP services – if the invention is a novelty the transfer technology office applies for patent protection. Protection of intellectual property is the main goal in this stage.

The *second* stage consists of technology and market assessment. This step involves the development of hypothetical technology profiles with differing combinations of those attributes and values which have market utility.

The *third* phase of technology transfers refers to business models for developing technology. By discussing a variety of forms of business alliances and cooperation which fit the expected market activities, it is possible to cut down the risk involved in market operations.

The *fourth* stage focuses on finance and production sources for future operations. Venture capital investments are sought and new ventures endeavor to improve the implementation process. The main aim is to be ready to launch the new product or services.

STARTING A NEW VENTURE ON TECHNOLOGY-BASED IDEAS

Technology commercialization is also persuasion process, involving striking a balance between the groups of stakeholders. It is understood that scientists often have little knowledge about the business environment, marketing, and technology launching; similarly stakeholders often have little knowledge about inventions and their benefits. Both technology players should be the targets of persuasion. The inventors have to know what kind of forces are essential to the commercialization process, and the stakeholders have to know what is worthy of their money and what kinds of factors can make technology a success story (e.g. human, capital, knowledge, or information). Market uncertainty forces the potential technology authors to study the nature of various customer mentalities and the buying processes that enable them to be effective on the marketplace (Potwora, Duczmal, 2001). Commercialization of technology is the converting of an idea into tools and machinery improve production. Jolly (Jolly, 1997) adds " to cause something having only a potential income-producing value to be sold". Ideas include people knowledge, so the transfer of knowledge makes technology possible and effective. Thus, in order to share technology among customers it is essential for laboratories, industry, R&D institutes, local government, and universities to use knowledge as a means of increasing their competitiveness. In dealing herein with transfers from academia to business, the same questions arise: if the transfers go from universities to business, what is the market? What kind of problems would be solved? What kind of firms are ready to obtain knowledge? Using human motivation language (Maddoch, 2000) each company representative adopts to his or her work environment by imitating what the more successful companies do. There is a considerable gulf that exists in the market between those who want to be innovative and those who actually take steps to be innovative.

By examining the Polish market data that let's us show why innovation gives new perspectives, we can learn a lot about the transfer of knowledge. One thing that we learn is that innovativeness' potential depends on the transfer of know-how and know-why from big foreign companies to their subsidiaries in

Poland, as well as on transfers to small and medium sized firms unknown on the global market. Another thing we learn is that the innovation in Poland is mostly driven by the adoption of technology abroad through imports. A third thing we discover is that the quality of basic science is very high, but there is a low level of interest in the application of science. In 2002 Poland spent 0.59% of their total revenue on R&D (Innowacje, 2004), which is far below the OECD average. The Polish economy suffers from its very low level of R&D financing and from weak links between universities and business. The public R&D sector needs to be adjusted to the requirements of business. That is why programs focusing on knowledge and technology transfer are crucial and they undoubtedly will produce new Polish high-tech sector.

The findings show that the commercialization of new technology is correlated with macro trends (Żołnierski, 2005), but the most significant examples of commercialization took place in the ICT sector. Another limitation is that Polish entrepreneurs have focused their attention on the hard material factors of innovation such as machines, computers, and appliances. While this improves the innovation potential over the short term, rapid productivity growth begins to decline if the 'soft' factors, which stimulate an increase of effectiveness, are ignored. The penetration and imitation strategies have to be changed into new products (the product innovations are the one of the main strategies of SMEs and Micro Firms, figure 1) or new market development. This is the fundamental issue for new project management in the high-tech sector. The role of innovation centers, technology incubators, technology transfer offices, and technology and science parks has become essential because these are the places where a relatively easy technology transfer to SMEs can occur. The soft factors are central to the management of innovation. The perspectives of SMEs, where the transfers of know how and know why have to be financed by owners or government organizations, are primarily designed to exchange ideas, broaden knowledge and vision, and stimulate new thinking. By providing SMEs with training, education, and network access, they are encouraged to use their own human resources. The knowledge transfer from academia to business encounters barriers, but the afore-mentioned services help accelerate changes in organization strategy and managers' mentality and bring about the opportunity to attain world standards.

The micro sector faces very similar problems. The numbers of businesses cooperating with academia is low. Few of them have patents or an IP protection policy. Access to licenses to use new technologies is much lower. This suggests that size matters in the transfer of knowledge (Żołnierski, 2005). It is evident that the role of government and non-governmental organizations in helping innovative companies to generate added value for the market is significant.

Figure 1
SMEs and Micro Firms innovation strategies on Polish market

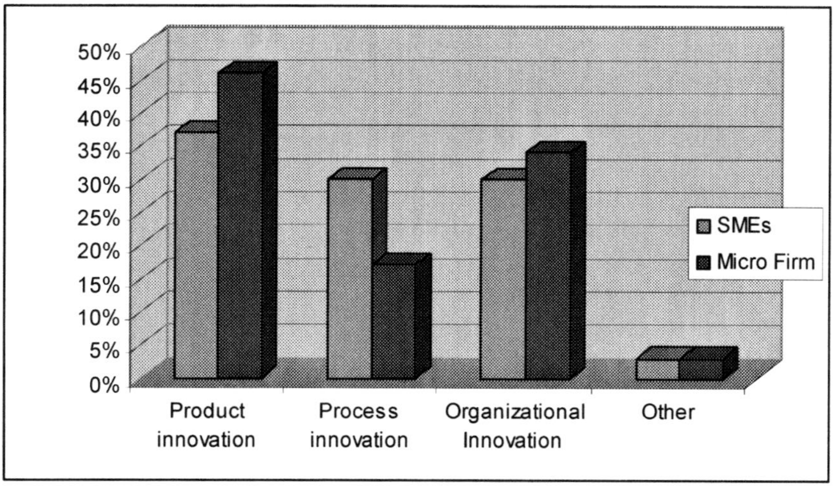

Source: A. Zołnierski, 2005.

Inventors' proposals and the new ideas of company representatives cannot be the concern of the technical community only. The biggest threat to the survival of innovative concepts is the lack of a broad market vision. Innovators are often unable to see the broader picture of market demand and the forces that shape the commercialization process. In order to avoid an improper market definition, which would restrict the discovery of market opportunities, new ventures need knowledge and need to know how to open the windows of opportunity in order to access a broad array of customers' needs. This kind of knowledge provides a better understanding of potential substitutes and competitive threats. The people who develop new products and try to implement them in the market need to link the world of technology with the world of business. They are used to speaking of the needs of the technology community, but they need to share its language with people who are involved in business strategies, marketing, and financing. Both groups need to acknowledge that new technology is futile unless it creates business value (added value for the customers) (Clark, 1994).

In Poland, where there are not many success stories of new high-tech ventures and where promising sectors of the Polish economy (biotechnology, nanotechnology, clean energy) are actively seeking partners, business experts, and promising market scenarios, the transfer of knowledge from high level universities and R&D institutes to the business world is crucial to the growth and acceleration of the country's development.

Hirsch and Peters (2002) suggest that "one of the most difficult problems in the new venture creation process is obtaining financing". On the basic of our experience in Polish high-tech and SMEs sector we can add that a second need has to be addressed: the need for appropriate tools to use knowledge to start a new venture. During the growth stage of a venture the entrepreneurs need to consider some important management skills such as team building, marketing, and business plan creation. These skills often are critical to the success of a venture. Generally, new Polish ventures do not have ready access to appropriate human resources because Polish universities and R&D institutes have been not market oriented. Each year the Polish government reduces the amount of its subsidies. This forces the universities to adopt a new science policy. The inventions, teaching, trainings and advisory services are being adapted and are creating new goals focusing on the European and global markets.

UNIVERSITY PROGRAMS' SUPPORT FOR ENTREPRENEURS

Central East European countries began to rebuild their economies by utilizing the most modern technologies. The Lodz Program and the indirect Offset Program is focused on turning SMEs toward the global market. The transfer of IC2 Institute University of Texas know-how results in more rapid change of SMEs involved in the programs. They have been implementing high quality technologies at lower prices, which has come to be seen as one of the crucial benefits of the programs.

It is important to point out that the Polish national and Lodz governments expected a higher number of emerging companies in the high tech sector, but the new panacea did not produce the expected results. While the Polish and Lodz high tech sectors slowly and inevitably began to realize new programs, matching them with American companies' programs proved more difficult. The first year of cooperation brought about only a few framework agreements signed between Polish and American potential partners. The source of this weak result probably lies in differing mentality, methods of organization, and in the lack of financial services. The way new ventures are organized to utilize technology, and the methods used to manage and motivate managers, are different. New technology firms require an investment and therefore have no opportunity to cover the cost of knowledge and technology transfer. The added value given by Offset Program has not been seen.

Opportunities for technology commercialization vary from firm to firm. Technological opportunities arise from productivity and the prospects for the purchase of new technologies. Purchase opportunities arise either from expansion of the demand for the technology, or from the replacement of a current technological base (Gauvin, Sinha, 1993). These two factors are decisive in assessing the market potential of a new technology. Scientists and entrepreneurs who work on the commercialization of new technology have to take the market into account. Otherwise they often are not able to use the ladder process and evaluate technology opportunities in a practical way. Incremental innovations mean nothing if there is no need and demand for them.

When we look at the Polish university, we find that its structure is not new technology oriented. We can agree with the opinion that "the market plan for academic positions, particularly in comprehensive universities, places a high value on publication" (Neel, Snyder, 1991). The publications, and educational and scientific achievements, are important to the academic community. The Offset Program has had a great impact on Lodz and on the Polish university community, but goal which should be attained is the improvement of service activities to obtain added value for potential owners of spin offs, inventors, and faculty members. The professionalism of university staff has been isolated and insulated from the threats and opportunities facing the country. The pressure of changes in the European marketplace is supposed to rebuild the infrastructure of Polish university sector. The difficulties involved in the reconstruction of the business high tech sector arise from the fact that universities do not face the same set of market pressures.

Launched in May 2004, the one-year program is taught in Polish and in English by staff both from the University of Łódź and the University of Texas at Austin, as well as entrepreneurs, industry experts, and technology leaders from Polish and international institutions.

The main programs are: MSSTC (Master of Science in Science Technology Commercialization), ATAC (Accelerated Technology Assessment and Commercialization), Technology Incubator and M&M (Manning and Matching).

The objective of the MSSTC Program is to educate professionals with specialized skills to manage knowledge and, above all, to accelerate the transfer of technology-based ideas into products and services in the marketplace. The program prepares individuals to become successful entrepreneurs and technology commercialization leaders in their organizations and institutions.

The MSSTC Program is specially designed for professionals who:

- are interested in the creation of new products, processes, companies, or economic branches,
- work in national and international companies,
- are engaged in technology transfer at a university, R&D laboratory, incubator, or science park,
- work in governmental and nongovernmental institutions and organizations fostering the financing of scientific research,
- are interested in general management of technology commercialization processes,
- want to learn how to start and operate high technology ventures, either as new start-ups, spin-offs, or within an existing organization.

The ATAC (Accelerated Technology Assessment and Commercialization) Program is targeted at research teams and entrepreneurs involved in high technologies and all those who are interested in discovering the potential commercialization of innovative services and products. In the first edition of the program 16 of the most promising technologies were selected and then evaluated by ATAC workshop participants. During the workshops the participants were first introduced to the Quicklook method of technology assessment, then evaluated the selected technologies commercial viability, and then they presented preliminary conclusions in the form of presentations and/or reports.

The authors of the most promising technologies will have the opportunity to receive assistance in their commercialization and incubation from the Innovation Center and the High Technology Accelerator. Both institutions will help them to establish contacts with Polish, American, and European partners.

The incubator program targets the communities of scientists, inventors, and innovators, as well as start-up companies. In order to reach the market innovative technologies or products which are result of research work require resources beyond the possibilities of academic institutions. One of the options is setting up companies which stem from higher education units (so called spin-off and spin-out companies), or implementing scientific ideas in other ways. The same goes for start-ups and innovators who have already started or plan to launch a business to commercialize their ideas.

The incubator program has been created to support:

- new high-tech enterprises, which are being developed by scientists, innovators, and academic and R&D institutions
- start-up high-tech companies
- individuals who develop high-tech technology, products or services in order to bring them to the market

Scientists, innovators, financing specialists, and marketing and management experts, assisted by the highly educated incubator staff, are able to build a team to run a new venture firm which implements technologies and ideas "under construction'. As all members of the group are specialists in their field, everyone can expand their competency and capabilities, forming a professional team.

US and Polish/EU companies benefit from this offset program by having the opportunity to quickly identify pre-qualified strategic partners in the European Union. Possible business scenarios include:

- License Agreements
- Joint Ventures
- Service Contracts
- Joint Development Projects
- Strategic Alliances
- Outsourcing Agreements
- OEM Manufacturing
- Localization and Marketing Agreements

Companies which take part in M&M (Mining & Matching) benefit from the Program by establishing and maintaining business relations between Polish and American companies, spurring the transfer of novel technological solutions from the USA, to the USA, and from joint enterprises on Polish, European and American markets.

The entrepreneurs or scientists who work in the areas of technology development and implementation on the market need the following:

- assistance in assessment of the market potential of an advanced technology or high-tech product or service while it is being developed by a company, as well as during the technology commercialization process
- access to information – market and industry sector analysis
- assistance in preparation of the enterprise business plan
- formulation of an implementation and development strategy for an innovative enterprise
- the ability to transfer novel technologies and innovative solutions to Polish companies
- to acquire financing from outside sources
- support in the management of intellectual property rights
- to seek business partners both at home and abroad
- assistance in negotiations with potential partners and investors
- help with constructing the best possible model of collaboration with a business partner
- assistance in locating businesses in Poland, based on cooperation with the Accelerator and the Polish national and local governments, as well as technology and industry parks
- promotion of the company at home and abroad

Table 1
Results of Łódź University Offset Program

Activities 2004-2006	Results
International cooperation in tech transfer	19 agreements
Innovation and technology projects assessment	300 projects
Business plan assessment	7 business plans
Technology Incubator Program	12 companies participated
International seed capital and VC investors cooperation	Establishing business relations with 3 Polish and 3 international seed capital and VC investors
Training programs in the area of technology transfer, technology commercialization	300 trainees
Masters Studies program focused on science and technology commercialization	25 graduates (R&D Institutes, Science and Technology Parks managers)

Source: Innovation Center and High-Tech Accelerator University of Texas and University of Lodz Offset Program Executive Summery, 2006.

The above-mentioned programs yield added value described as (see, Table1 for more statistics):

- acceleration of product or service development
- professional evaluation of a technology which shows its market impact
- absorption of new, competitive technologies
- optimizing the business model of the enterprise

- establishing cooperation with already-on-the-market companies (which reduces costs of entering a new market)
- promotion of the company at home and abroad
- access to financial resources (like seed or venture capital)
- assistance of specialists

CONCLUSION

University offset programs enable Polish incubated and emerging firms to get to know the value and market advantages of their new technologies, and can match them with companies in the United States. In meeting the challenges of a technology's link with customer requirements, Polish ventures can participate in three main types of programs: training and educational courses, acceleration, and network programs.

These programs seek to support those inventors and companies that have successfully combined their own research activity with business engagement. It does not oblige the beneficiaries to carry out the specific projects originally indicated by them. They are free to decide whether to continue their ongoing program or to participate in new ones. The major challenge that the Polish innovation organization may face in the near future is connection with the global economy. The change in the environment of Polish firms may force them to seek new partners based on access to markets and the scientific community.

High-tech industry has become a source of new ventures that stimulate a country's development, but it has become clear that the best ideas in the world will not be put into the commercialization process without knowledge concerning the intellectual property system, venture funds, market roles, sector standards, production procedures, end-users' motivation to purchase etc. The Polish university offset program emphasizes the added value of applying core competence in business and technology transfer, and enables participants to see the broader picture of market demand and the forces that shape the commercialization process.

BIBLIOGRAPHY

Benson S., T. Philpott, (1990), *Polemicists*, nr 2, pp. 4-5.

Clark, K. B., (1994), "What Strategy Can Do for Technology", [in:] K. B. Clark, S. C. Wheelwright (eds.), *The Product Development Challenge. Competing through Speed, Quality, and Creativity*, A Harvard Business Review Book, Boston, pp. 61-68.

Gauvin S., R. Sinha, (1993), "Innovativeness in Industrial Organizations: A Two-stage Model of Adoption", *International Journal of Research in Marketing*, no 10, p. 169.

Gomory R. E., (1994), "From the "Ladder of Science" to the Product Development Cycle", [in:] K. B. Clark, S. C. Wheelwright (eds.), *The Product Development Challenge. Competing through Speed, Quality, and Creativity*, A Harvard Business Review Book, Boston, pp. 70-76.

Hirsch R. D., M. P. Peters, (2002), *Entrepreneurship*, McGraw-Hill Irwin, Boston.

Jolly V. K., (1997), *Commercialization New Technologies. Getting from Mind to Market*, Harvard Business School Press, Boston.

Korcz K., (2003), „Międzynarodowa dyfuzja innowacji – podejście marketingowe", [in:] H. Brdulak, T. Gołebiowskiego (red.), *Wspólna Europa innowacyjność w działalności przedsiębiorstw*, Difin, Warszawa, pp. 58-71

Lang F., (1992), *Märkte und Produkte Sinnvollplanen und Steuern. Unternehmenspraxis Marketing*. Maria Lang Verlag, Endingen.

Lundvall B. A., (2005), "Innovation Policy and Knowledge Management in the Learning Economy", [in:] D. Gibson, M. Heitor, A. Ibarra-Yunez (eds), *Learning and Knowledge for the Network Society*, Purdue University Press, pp. 25-56.

Lundvall B., (2004), *Innovation Growth and Social Cohesion. The Danish Model*, Edward Elgar, Cheltenham.

Maddoch, R. C., (2000), *Motigraphics, The Analysis and Measurement of Human Motivation in Marketing*, Quorum Books, London.

Mohr J., (2001), *Marketing of High-Technology Products and Innovations*, Prentice Hall, New Jersey.

Myers, D. D., (1999), "Strategic Technology Transfer", [in:] R. C. Dorf (ed.), *The Technology Management Handbook*, CRC Press, pp. 17-1-35.

Mynarski, S., (2004), „Informacje w świecie współczesnym", [in:] B. Gregor (ed.), *Marketing - Handel – Konsument w globalnym społeczeństwie informacyjnym*, Wydawnictwo Uniwersytetu Łódzkiego, Łódź, T. 1, pp. 55-65.

Neel C. W., W. T. Snyder, (1991), "Building Responsiveness Universities: Some Challenges to Academic Leadership", [in:] M. J. Stahl, G. M. Bounds (eds.), *Competing Globally Through Customer Value. The Management of Strategic Suprasystems*, Quorum Books, New York, pp. 67-74.

„Patent i innowacja - szansa czy utrapienie firmy?" (2004), *Innowacje*, nr 8.

Potwora W., M. Duczmal, (2001), *Marketing. Strategie i struktury*, Polska Akademia Nauk, Katowice-Opole, Opole.

Rogers E. M., S. Takegami, J. Yin, (2003), "Lesson Learned About Technology Transfer", [in:] D. V Gibson, Ch. Stolp, P. Conceição, M. V. Heitor, Praeger (eds.), *System and Policies for the Global Learning Economy*, Westport, pp. 369-382, (source: D. V. Gibson, E. M. Roger, (1994), *R&D Collaboration no Trial: The Microelectronics and Computer Technology Consortium*, Harvard Business School Press, Boston).

Rupik K., (2004), „Orientacja a wiedza przedsiębiorstwa – wybrane relacje i ich uwarunkowania", [in:] B. Gregor (ed.), *Marketing - Handel – Konsument w globalnym społeczeństwie informacyjnym*, Wydawnictwo Uniwersytetu Łódzkiego, Łódź, T. 1, pp. 83-91.

Salas C. P., G. Susunagi, I. Aguilar-Barajasa, (2003), "The Role of Innovation In Regional Economic Development: Some Lessons and Experiences for Policy Making", [in:] D. V Gibson, Ch. Stolp, P. Conceição, M. V. Heitor (eds.), *System and Policies for the Global Learning Economy*, Praeger, Westport, pp. 111-136.

Simon H., (1996), *Hidden Champions: Lessons from 500 of the World's Best Unknown Companies*, Harvard Business School Press.

Żołnierski A., (2005), *Potencjał innowacyjny polskich małych średniej wielkości przedsiębiorstw. Raport z badań*, Polish Agency for Enterprise Development, Warszawa.

PART II

LEARNING AND ICT IN REGIONAL INNOVATION SYSTEMS AND CLUSTERS

9

Effective Regional Innovation Policies - Lessons for Polish Regions

Tadeusz Markowski

THE CONCEPT OF A REGIONAL INNOVATION SYSTEM IN PRACTICE

The concept of regional innovation systems (RIS) is, in its nature, linked to territorial production systems as well as to the concept of a so-called innovative environment. An innovative environment is seen as a particularly complex cultural, institutional and research-related infrastructure which facilitates the creation and implementation of innovations[1]. A regional innovation system is, on the other hand, seen as a separate interdependent system of relationships between organisations which operate within this system in order to achieve the entire system's goals, and which cannot exist without this system. This approach to the RIS refers to features which are specific for typical biological systems (Markowski 1999, Stawasz 1997).

Researchers dealing with regional socio-economic systems, including the regional innovation systems, understand that the organisational structures in many regions are far too dispersed and atomised to have developed into a complex and interdependent regional innovation system (Jewtuchowicz,

[1] These issues have been dealt with for over a dozen years by the European Research Group for Innovative Environment – GREMI (the acronym is derived from the French name of the Research Group).

Matusiak, Stawasz 2001, Todtling, Sedlacek 1997). It seems that the increasing complexity of organisations and global relationships is a natural feature of the contemporary information society. We can foresee that globalisation will, on the one hand, lead to an alienation of certain international business services from regional systems[2], while on the other hand it will create new international horizontal relationships.

Perhaps RIS is only an intellectual concept, a desired model. Perhaps any attempt at creating a real, fully-fledged innovation system is destined to fail. This remains to be seen. In order to verify this hypothesis we would need to analyse the types of innovations being generated in the atomised and underdeveloped structures of haphazard relationships, and then to compare them with the types of innovations which are specific for an already developed RIS[3]. Or perhaps the only difference lies in the rate of innovation creation, the sustainability of competitive advantages possessed by units within the system, and the appearance of organisational innovations on the territorial mezzo-structure level. However, we intuitively assume that the creation of a RIS would provide a new quality and a whole new impulse as far as the possibilities of economic development for a region are concerned.

We assume hypothetically that the creation of horizontal relationships and interdependencies, specific for a complex organisation in any given region, will prove to be beneficial and that such a system will provide a new quality in an organisational and spatial sense. Such a new organisational mega-structure is seen as a spatial and territorial organisational network which promotes additional innovations (an innovation synergy), and thus further increases the economic potential of a region and its competitive advantage. Thus, the creation of a fully-fledged RIS is a starting point for laying down new endogenic socio-economical foundations for regional development. Such a system allows the Knowledge Intensive Business Service sector (Miles 1999) to develop.

The issue of defining a region remains one of the most serious problems in research concerning the regional innovation systems. First of all, we need to determine how to define the area of an RIS. Should we do this in accordance with the spatial diffusion and absorption of innovations, or within an already existing administrative division? If we assume that the Internet and information technologies cause a growing divergence between geographical and economical availability, then treating a region defined by administrative borders as a unit constituting an innovation system loses any salient sense. However, the practice of research on innovativeness of regions shows that most researchers still concentrate on regions as defined and delimited by administrative boundaries. Having said that, it needs to be emphasised that despite a growing divergence

---

[2] For example financial centers of multinational companies, which employ a cheap local labor force while serving external companies.

[3] When referring to a developed RIS, we assume that the atomised organisational structures, which have already been in existence in the region and which belong to other (supra-regional) innovation systems, are incorporated into the regional system.

between geographical and economical availability[4], the importance of the availability factor (understood as a general economic category) for regional development is not decreasing, even if we do have access to Internet technology.

Various types of human activity will be characterised by varying spatial scopes in which innovations will be generated and diffused. Innovation is not a linear process. It is an interactive process within networks of many actors, strongly related to business cycles (Van Geenhuizen, Nijkamp, 2006). It may start from typically local scopes, which are determined by a local sub-culture, and end up in a global one. Therefore the RIS is a spatially pulsating and cyclical system. It does seem however that every RIS needs a fixed central entity which serves as a stabiliser and which constantly refreshes the system's innovation capabilities, regardless of the changes occurring in the production and service structure. In such a case attempts may be made at singling out certain composite spatial and functional systems which are the poles (or axes) of regional innovation systems, and which ought to be subject to an innovation policy. Various types of entities may perform the role of RIS poles – starting with regional capitals through to entire metropolitan areas, and finishing with smaller production-oriented territorial units (i.e. clusters). Since innovation policy concentrates on the central, key elements of an innovation system, we are going to pay less attention to its administrative boundaries and the blurred limits of its relative territorial influence.

An innovative region may not be closed within administrative borders. A RIS is not a coherent and closed system which is in equilibrium as far as the supply and demand for knowledge are concerned (Héraud 2003). Héraud emphasises that there is no universal model of a RIS. Every such region may have its own specific characteristics. There are some regions which develop mainly due to science – they may be financed by the public or the private sector or both – or some which rely on traditional sectors, resources, or even small enterprises. Every such territorial production system has its own system of interactive relationships allowing for the exchange and acquisition of knowledge. This implies that creating an innovation strategy requires that all these potential aspects and conditions, which change as the system develops, be identified and well-understood.

A well-developed RIS allows all participants (actors) of the system to mutually enhance their individual learning processes. A RIS as such provides an impulse for creation of a specific regional good – jointly acquired conscious knowledge and tacit knowledge. At the same time the fact that a RIS may be perceived as a kind of peculiar infrastructure for the learning process makes it like a club in itself.

No credible and composite research exists at the moment which would confirm the existence of fully developed and relatively autonomous regional innovation systems,. Carrying out such research would entail huge costs, including the preparation of very precise surveys and guaranteeing the

---

[4] Economical availability is primarily related to the costs of overcoming the resistance of space, and is a function of money and time.

credibility of information received. However, there is a growing number of partial research studies and case studies available in developed countries, which seem to allow for a positive verification of the concept of a RIS in Polish conditions.

CONTEMPORARY TENDENCIES IN INNOVATION POLICY

Innovation policy is changing its character because of globalisation. Traditional innovation policy concentrated mostly on supporting certain specified technologies (Schienstock, 1999). Currently, most national innovation policies are moving towards a more general approach – supporting everything that may help generate innovations. Such an approach is caused by a growing uncertainty with regards to what the driving forces of the economy are and what determines future economic success, as well as by the short-term nature of competitive advantages based on specific technologies. This does not mean that such a general policy ought to fully replace a technology-based policy. However, the experiences of highly developed countries, world leaders in terms of generated innovation streams, show that the traditional technology-based policy no longer yields the expected success. Perhaps it may be assumed that the world leaders will move towards a general innovation policy, while the countries which are lagging behind will increasingly focus on the traditional policy of supporting specific technologies. Having said that, it is necessary to emphasise that developing countries like Poland, need to implement the general innovation policy more and more extensively.

The economic sense of performing each detailed socio-economic policy by public authorities is justified by the fact that market mechanisms are imperfect. In the case of traditional (selective) technology policy, the need for intervention arises from the fact that technology transfer and restructuring of the economy do not adapt quickly enough to new demand signals from the market. Technological reorientation may thus be performed using budgetary support systems and certain regulations. A new innovation policy is horizontally oriented and directed at the business, social and cultural environment of enterprises. Its influence is far greater, as it intervenes in many sectors of the economy which are responsible for the innovative capacity of entire territorial production and service provision systems. Some examples of imperfect markets where a general innovation policy is used include an infrastructure market, investment markets, an imperfect institutional system, imperfect flow of information, underdeveloped social capital etc. Developing such a model of an innovation policy results in treating the territorial (regional) production systems as complex systems for generating innovations.

General innovation policy is oriented towards creating institutional conditions favourable for generating innovations. Schienstock (1999) points out that the general innovation policy which is being formed nowadays performs three basic tasks. Firstly, it supports organisational restructuring of enterprises. Secondly, it supports the development of business relationship networks. And thirdly, it develops organisational and institutional infrastructure. General innovation policy is carried out on all levels of public authority's outreach. Its importance is increasing for several reasons. One is that enterprises, even global ones, expect governments to support their activities in this sphere. At the same time globalisation restricts the individual countries' capabilities for carrying out separate macroeconomic policies. For this reason certain areas, like labour markets, social issues, and lately innovation policies as well, tend to be the subject of public authorities' political activity. This tendency reflects the social expectations and also the real capabilities of public authorities to undertake actions.

In European Union countries we may observe innovation policy at all levels of territorial organisation; the Commission's policy, national policies, regional policies and, to a varying extent, also local policies. Innovation policy may also be observed globally – in the form of bilateral or multilateral agreements on joint space, sea or Antarctic research, but also in the form of exchanges of scientists and students as well.

Various specific spheres of activity may be outlined in every policy. An innovation policy is a set of activities performed by the public authorities in the economy in order to influence innovative processes. Schienstock (1999, p. 435) outlined 9 areas of an innovation policy and compared them with the involvement and role played by various levels of public authorities – starting from the EU level and ending with the regional one (Table 1).

Table 1
The involvement of different policy levels in various fields of innovation policy

Policy areas	EU level	National level	Regional level
Creation of a policy vision	Important	Very important	Important
Setting up and strengthening knowledge infrastructure	Important	Very important	Important
Support knowledge transfer	Important	Important	Very important
Procurement policy	Important	Very important	Less important
Finance innovations	Important	Very important	Less important
Support for innovation networks	Important	Very important	Very important
Support for the use of knowledge intensive business services	Less important	Very important	Very important
Support for organizational change and intelligent applications of ICTs	Less important	Important	Very important
Improve human resource management	Important	Very important	Very important

Source: Schienstock (1999, p. 435).

As we see, the role of the regional level is particularly great when it comes to supporting the transfer of technologies, promoting the use of IT technologies, and improving human resource management. Important tasks performed at the regional level also include creating visions for development and implementation of the infrastructure used for creating and developing knowledge. If we take these spheres of a regional policy and compare them with the local level, it will certainly be the case that the scope of innovation policies will be the same, at least in the economically strong local units (i.e. urbanised communities and metropolitan areas). It needs to be emphasised that regional innovation policy ought to be co-governed and co-managed by the regional public authorities and the local public authorities.

Nowadays we can observe in some EU countries further decentralization in innovation policy. For example in the Netherlands the national government delegates more responsibility for the creation of a policy vision and the strengthening of knowledge infrastructure. We can say that there is a continuing process of strengthening the public innovation policies on the regional and local levels.

The experiences of many countries, including Japan, show that traditional technology policy sooner or later leads to a situation whereby enterprises become oriented only towards seeking financial and legal privileges linked with the particular technologies which are preferred by the government. Therefore in the long-term such companies lose interest in investing in their R&D base. The general outline of differences in an approach adopted by a traditional technology policy and the general supporting policy has been presented by Schienstock (1999), Table 2.

In stressing the need for implementation of a regional innovation policy, we need to bear in mind that regional and local authorities have only limited capabilities for undertaking intervention in imperfect markets. These limits are also caused also by the natural imperfection of governing institutions. This aspect may not be overlooked; otherwise we succumb to the unrealistic assumption that every intervention undertaken by public authorities would bring about the creation of a new regional innovation system or improve an already existing one. However, it may be the case that the current state and development level of institutions with territorial authority is not sufficient for undertaking an effective policy which would match the new requirements and conditions. If so then it is necessary to improve the institutional capacity of the authorities themselves.

Specialists have, for many years, been trying to find ways to overcome the imperfect nature of public authorities (Van Geenhuizen and Nijkamp, 2006; Benner, 2003; Hohn and Neuer 2006). This imperfect nature has many causes. It is a result of low intellectual and social capital, bureaucratic pathologies, the political nature of local governments, and regional and local political culture (which of course depends on the quality of the previously mentioned social capital).

Table 2
Comparison of direct technology policy and conditions-enabling policy

Aspects of policy	Traditional direct technology policy	Conditions-enabling innovation policy
theoretical model of technological policy progress	cascade model	recursive model of the innovation process, system approach
legitimating innovation policy	market failures	all areas of systematically weak performance in terms of innovation
Research orientation	support for basic research	focus on problem oriented research
control of technological progress	direct state control through public research institutes and large research projects	indirect state control through vision creation
policy integration	Technology policy as science policy	innovation policy as part of economic policy closely linked with other policies (labor market policy, education policy etc.)
companies' knowledge stock	exploit existing knowledge stock of companies more efficiently	add to firms' knowledge and capabilities by bringing a greater degree of access to external knowledge
policy focus	technological innovations	social innovations (organizations and institutions)
stages of technological development included	little bearing on the market	all innovation activities
Instruments	lower R&D costs through innovation grants, R&D subsidies, R&D tax breaks and procurement policy	catalytic type of instruments (creation of innovation networks, investment in communication infrastructure etc.)
implementation strategy	direct state intervention	use of monitoring and mediating institutes
policy type	policy as a process of decision stages (setting goals, developing programmes, implementing projects)	policy as continuous learning process
policy aim	to produce economic growth through supporting technological progress	balanced socio-economic development (sustainable development)
role of the state	sovereign economic regulator	facilitator and orchestrator of private economic interests
actors	nation state	division of tasks among regional, national, and EU levels

Source: Schienstock (1999, p. 432).

The concept of imperfect government may be extended to any governing institution, and has been developed in institutional economics. It attempts to make the approach of both neo-classical economics and the theory of welfare economics more realistic. Both these approaches see the imperfect nature of the market as a problem. However the similarities end here, as both groups offer different solutions. Neo-liberalists demand a greater expansion of the free market, claiming that this would lead to overcoming imperfections in the market. On the other hand supporters of the welfare economics theory demand greater intervention by the government, claiming that this leads to a better allocation of resources in the economy.

Both these theories assume that either it is hypothetically possible to create a perfect market, or that the government is perfect and thus capable of objectively correcting any market imperfections. However, practice shows that neither group is correct and that at least some of their assumptions are unrealistic. In practice we are always faced with an imperfect market and an imperfect government, which encompasses its entire system of governance, including territorial authorities.

It is therefore necessary to ask several key questions. How can an imperfect government overcome the imperfect nature of the market? What are the learning processes of all actors on the market? How can learning processes of the government and territorial authorities be characterised? The features of public authorities and causes of their imperfect nature are being researched and analysed in the political sciences, sociology of power, and lately also by institutional economics. Institutional economics attempts to explain the mechanisms behind the functioning of contemporary economics. The economic models have until now assumed that it would be possible to find rational ways in which the economy could operate, if only the political influences were eliminated. This assumption however is unrealistic – there can never be an economy without politics in the real world.

Regional authorities need to be equipped with certain competencies and instruments in order to be capable of steering a network. These instruments ought to be derived from the features of the economic system, but as we have seen they are not yet been fully understood. The reason for this is that the system is an imperfect one (certain elements and relationships are missing). Often one finds only some of the elements, connected by unclear and very weak relationships. Sometimes these relationships are also missing. The fact that there are different stages of development of regional innovation systems means that the need for intervention also differs. The required skills of the regional authorities, as well as the scope of governmental support, likewise differ. We must understand that the Polish system of governance is geared towards hierarchic governance rather than towards partner-like and network-based governance. We still need to learn how to govern and manage in a different way, and understand the theoretical usefulness and see the necessity for implementing such an approach. This is true not only for the economies undergoing systemic transformation, but also for economies which are considered developed market economies. Marshall (2003), who wrote about regional governance systems in

Great Britain, shows that many Scottish, Welsh or Northern Irish regions which were trying to build a federation in the 1990s were also having problems with their approach to governance. They tended to centralise the system of governance in administrative regions and make it hierarchic, thus remaining attached to the principle of ruling rather than moving towards co-governance (Marshall 2003).

So, what is "regional governance"? Fürst (2003) wrote that „regional governance" is more than just inter-community co-operation. Regional governance is characterised by, firstly, a mixture of an institutional system and a network-oriented legal system; secondly, the co-operation of sub-social systems between the state, the economy and society; and thirdly, it is based on a paradigm assuming a change in the way that regional problems are treated, by moving towards an approach which assumes a concern for regional welfare and a readiness to implement co-operative self-governance methods.

Theoretically speaking, regional self-government ought to, both legally and formally, manage a region through regional governance. In Polish practice we can observe that many actions are only simulated, while the so-called participation of partners is restricted to the personal participation of small groups of people, who are formally considered to be representatives of various institutions within the region. Such token participation often restricts the possibilities of other partners who would like to become actively involved. What is more, sometimes the acknowledged representatives lobby only in accordance with their own interests, blocking the development of other institutions and organisations and restricting the influx of external investments, thus hampering development of local competitive businesses.

Governance is defined as management based on partnership and network systems (Schwab and Kübler, 2001; Herrschel and Newman , 2002; The Reform of Metropolitan Governance, 2000). Regional innovation systems are open networks. Therefore any attempts to manage such a system and to stimulate its development need to be based on a concept of governance which assumes that all business environment institutions, local governments, and governmental authorities make up a network of partners.

Certain processes which are taking place, like the fragmentation of urban areas, the metropolisation of space in great cities or, put differently, the transformation of cities into urbanized regions, naturally span across administrative borders and demonstrate the need for management to be understood in the way presented above. In order to manage such regions it is necessary to adopt a partnership approach and create networks between many public and private organizations. Managing in this way requires new institutional forms, both formal as well as informal, which will be based on mutual trust between partners.

INNOVATIVENESS OF LOCAL AUTHORITIES AND REGIONAL INNOVATION POLICY

Increasing the innovativeness of regions and creating regional innovation strategies has become the challenge of our times. Efficient and professional local and regional self-governments are playing an increasingly important role in the act of improving the competitiveness of enterprises and regions. Local governments are becoming a significant link in any regional innovation system, taking responsibility for co-management of the entire process. Local authorities may support the development of business environment institutions, co-create a climate for pro-innovative activities, mobilise their inhabitants to implement new products or managerial processes, and create a local entrepreneurial environment. In order to exert a lasting impact on the local territorial and production system a modern and innovative administration needs to be introduced in town hall. Thus local government administration also needs to become innovative, constantly learning and up-to-date with the latest changes. Otherwise it will not be able to inspire local and regional development. These are strategic objectives with regards to improving local governmental institutions. These objectives stem from the objectives of the structural policies and regional policies implemented by countries which are interested in stimulating the competitiveness of the economy through the development of regional innovation systems.

Unless the significant role played by local authorities is emphasized, the creation and development of regional innovation systems, understood as integrated and modern systems for the management of local government organisations and the socio-economic development of territorial units, will be crippled. According to research on the contemporary globalising economy, efficient public authorities are becoming increasingly important for the organization of a competitive external business and spatial environment (Cook, 1997; Gertler and Wolfe, 2002; Hassink and Lagendijk; 2001, Morgan; 1977). The increasing role of the quality of the institutional business environment arises from the fact that it is an immobile asset. Therefore, it may often strive to attain a sustainable competitive advantage in the production and service sector, which is itself becoming increasingly mobile in space.

The quality of social capital[5] is another important aspect of the immobile environment which may determine the location of a contemporary enterprise. Improving the quality of social capital is possible only in the long-term, spanning several generations. The activities performed by the formal and informal institutions and organisations, including governments and local governmental authorities, are significant in this aspect. Adopting innovative governance methods and techniques for the management of socio-economic processes may have a long-lasting impact on the competitive advantages of entire nations. If an innovative mechanism works efficiently as far as

[5] Social capital may be defined as mutual trust and the capability for co-operation between social entities which operate within a given territory (see also J. Łobocki 2003).

governance is concerned, then the economy gains a source of a lasting competitive advantage. If on the other hand generating and diffusing innovations is beset by barriers in the system of governance, then the entire economy will be characterised by lower competitiveness.

There are numerous tools and techniques which may be used to improve the efficiency with which public authorities operate with regards to administration and managing the development of a territorial unit. Creating pro-innovative attitudes amongst politicians and local officers is one of the ways to increase the innovation absorption level, as well as to generate modern techniques and management processes in public administration. The chances of success in this sphere depend greatly on the local social capital.

SOCIAL CAPITAL AND LOCAL GOVERNMENT'S ABILITY TO IMPLEMENT AN INNOVATION POLICY

Social capital is often referred to as 'relational capital'. Its territorial dimension needs to be emphasised – it means that relationships between people and institutions depend on economic accessibility, which so far is inseparably tied to geographical accessibility. The set of relationships which make up territorial relational capital may not be precisely marked out. However, we must realise that some relationships, which may be identified as common and characteristic of a larger territorial system, may provide a unique quality of capital at the local level. This depicts the complexity caused by a kind of local synergy, which may not be imitated or may be imitated only in a limited scope in places where humans perform carry out economic activities.

Relational capital becomes increasingly important in creating external sources of competitive advantages for enterprises (Fukuyama 1999). It is the key to creating the so-called innovative environment which is vital to the success of enterprises in a globalising economy. Spatial differentiation in the quality of social capital causes spatial differences in the level of socio-economic development and also causes further polarisation of space. While relational capital is not a new economic category, its importance to the organization of socio-economic processes is increasing and becoming ever more apparent. If the paradigm presented here is deemed viable, then we may conclude that it is possible to exert an increasing influence on the local economy through the shaping of relational capital.

In the discussion concerning an innovative region we therefore need to address the issue of the conditions in which relational capital becomes an inspiration and an agent for the transmission of innovations. We also need to consider the role played by various institutions and organisations, including public authorities, in shaping relational capital in terms of creating an innovative environment. Social capital is a club-like public good. Its development partly relies on the system of institutions and organisations within a territorial organisation like an innovative system.

CLUSTERS AS AN INSTRUMENT OF REGIONAL INNOVATION POLICY

In the light of international experiences (Newlands, 2003; Hallencreutz and Lundequist, 2003; Lundequist and Power, 2002; Gancarczyk and Gancarczyk, 2003) the concept of clusters seems to be working in practice. Many practical initiatives show that despite local differences in approaches to the problem and the sometimes significant deviations from the theoretical model[6] presented by Porter (2000), this concept remains a source of the practical implementation of local and regional development strategies. It is an inspiration for contemporary pro-innovative regional policy and is closely tied with the creation of regional innovation systems.

Sweden is an interesting example of positive experiences with regards to development of clusters. Many of the Swedish experiences in this sphere were gathered and described by Hallencreutz and Lundequist (2003) as well as Lundequist and Power (2002). Thirteen analysed examples of regional clusters, which are all members of the "Klustergruppen association", are used by these authors as background for presenting four extremely differentiated cases; starting with the Småland-Blekinge cluster in Southern Sweden which is based on the aluminium industry and old agricultural traditions, through to the medical Öresund region cluster and the telecommunications and IT Karlskrona region cluster and, finally, ending with Norrbotten cluster of vehicle testing located in Northern Sweden. The last of the previously mentioned clusters is based on the peculiar features of Sweden's cold winter climate, which provides access to frozen lakes which are used by car manufacturers from all over the world to test their technologies and cars in harsh winter conditions from November to April. The detailed descriptions of these cases may be found in the cited literature. For us however the practical conclusions are of major importance.

First of all, we see that the strategies for creating such clusters differ greatly. Perhaps further analyses of an increasing number of positive cluster examples will make it possible to describe the general principles of cluster creation, as well as the principles of intervention and support by public authorities, in more detail. Nonetheless the authors state that regardless of the size and branch peculiarities of a cluster there are several strategic phases in cluster support policy. The exact sequence of activities will of course differ, as it remains dependent on the given conditions of each case.

Sometimes the support policy will include the promotion of clusters within a region as well as promotion of the region itself and its new capabilities, i.e. those gained by co-operation with and between the entrepreneurs making up the new cluster. In other instances the most important elements of the support policy may be catalysing the relationships between producers, supporting the contacts

---

[6] According to M. Porter (2000 p. 16), clusters are a geographical concentration of mutually linked specialist firms – manufacturers, service and material suppliers, sub-contractors, business environment institutions of certain types (i.e. universities, agencies, trade organizations) which co-operate and compete at the same time.

between actors, and creating and developing co-operation networks between entrepreneurs, business environment institutions, and universities and laboratories. Such activities seem particularly important in our Polish conditions. Many scientific and research units in the post-socialist economy have not yet established sufficient horizontal relationships. They perceive all their activities and development only in the context of vertical support from the budget. However, the policy of supporting innovations always needs to seek its own internal characteristics and there should be a unique logic behind the functioning of each individual cluster.

Foreign experiences show that clusters are not made from scratch. Certain prerequisites and initial conditions must be in existence before a cluster may be set up. Its success depends primarily on private entrepreneurs. The role of support and intervention performed by the public sector is relatively small, although it may sometimes be significant or even necessary for increasing the speed with which the process of cluster development is unfolding. The public sector performs its role through active promotion and reducing or eliminating various infrastructural, cultural, social and environmental barriers, including the low quality of housing, poor aesthetics of a city, use of public space etc.

Having educated local leaders and a steady support policy independent of the changing political conditions are key factors of success. The experiences also show that development of strong regions never comes about on its own accord. Certain specific localized prerequisites must be met, even though the politicians may often perceive them as unimportant. Several years ago nobody was capable of predicting that a well-developed academic centre in the field of accounting at the Faculty of Management at the University of Lodz would prove to be the decisive factor in the organization of a financial and accounting centre for Philips' units in Europe. Usually the factors become known *post factum*, when a new activity is launched and the market verifies whether the localisation is correct.

According to concept of RIS its success depends strongly on the performance of its urban system. The most important RIS are those which constitute a large agglomeration or metropolitan region. Poland is characterised by a distinct asymmetry of metropolitan areas which serve as regional generators of innovation. Some areas still remain under the dominating influence of one or several production and research sectors. The Warsaw agglomeration is by far the most diversified, and therefore balanced, innovation centre. It is followed by the agglomerations of Poznan, Wroclaw, Cracow and Lodz. Individual sectors still remain dominant in the other cities however. For example, the steel and coal industry still dominates in the Upper Silesian conurbation, while Gdansk and Szczecin are still relying heavily on the shipyard industry. Furthermore, the organisation of science and research in all Polish academic centres is dependent on the scientific and research units in Warsaw. This situation is one of the remnants of centralisation of power, whereby the capital city was placed in a dominant position over the remaining cities and concentrated both the human as well as intellectual capital. This structure still remains intact and is even strengthening its grip – the organisational and financial asymmetry with regard to the process of allocating research funds has not been changed in any fundamental way. The old system is also kept up by the fact that it allows the units which are located in Warsaw to lobby directly for an allocation of funds from the central budget.

The growing spatial polarisation of the research potential between the academic centres in Warsaw, Cracow, Wroclaw and Poznan at the expense of the remaining academic centres has become a dangerous threat insofar as the creation of equal development capabilities between regions is concerned. The regions characterised by a low scientific and research potential will surely lose out when faced with the new strategies of international corporations and companies based on knowledge – strategies, which are concentrating on identifying and using the competitive advantages of regions. The changes in the strategies for financing research which are being implemented by the huge competing international corporations confirm this thesis. These corporations are gradually withdrawing their funds for financing fundamental theoretical research, and even some applied research as well. The global competitive pressure forces them to restrict their funding only to the kinds of research which may be quickly commercialised and which are closely linked to their development strategy. Their experience demonstrates that research results which are not linked directly to commercialisation or which do not closely reflect the company's specialisation very often trickle over to their competitors. Therefore international corporations are undertaking close co-operation with universities in order to reduce the costs of their R&D activities. Philips serves as an example of a company which adopted such a strategy. Also other great TNCs which move their production to cheaper countries have begun implementing this strategy. Once their activities become established they initiate and gradually develop co-operation with the new, but often very robust, academic centres, while reducing their own R&D activities and facilities. Therefore, cities which lack strong scientific, research and didactic centres will become of marginal importance. Fortunately we can observe that in Polish big cities there is growing collaboration between universities and internationally renowned companies: *Philips Lightening* in Lodz, *Motorola and Delphi* in Krakow, *Intel* in Gdansk, *Siemens* in Warsaw.

CONCLUSIONS

The RIS concept is agreeably attractive and internally coherent enough to play a useful role in the process of managing regional development, even though it has not been fully tested and proven empirically. It is a concept which allows for the construction of regional innovative development strategies in a logical and theoretically transparent way. It fits in well with the contemporary approach to planning local economic development, which takes into consideration the growing uncertainty of the environment. Furthermore, it is also compliant with the current paradigm of progress and development, which states that economic development is based on knowledge. Knowledge, together with networking, are becoming the most crucial of forces spurring the development of the world economy.

The success of an innovation policy in Poland will depend greatly on the level of attained knowledge, but also in no small measure on the development of social capital as well as on the innovativeness of local and regional authorities. An efficient administration in city hall is equivalent to efficient management in a territorial unit. In Poland, the frequent and deep personnel exchanges after each election, which cut the continuity of acquiring knowledge, make it difficult to achieve this goal. Therefore it seems necessary to introduce regulations which would stabilise the position of local government staff. The creation of a civil corps of local government officers seems like a good idea. Such a corps could even be restricted to certain key fields of activity which require a great degree of professionalism (like managing local and regional development, spatial planning, property management, education). What is more, the minimum standards of knowledge should be clearly defined for people desirous of holding these key posts.

These factors seem far more important in post-communist countries than in countries where democracy is historically embedded. The communist system has left deep, negative and relatively long-lasting traces in the way that these societies think and act. In such conditions the high intellectual capital and sophisticated technical skills may be hindered by a lack of trust between people or, in other words, by low social capital, which makes it much harder to build lasting relationships based on cooperation and strong networking among involved organisations. Obviously such a situation increases the costs of companies' activities. Realization of this may spur the creation of specific programmes aimed at developing social capital. Innovative local and regional authorities, supported by the European Union and the Polish government, could play a significant role in this process. This requires that some special programmes and reforms aimed at increasing the institutional capacity of regional and local governments in Poland be implemented. This national policy should be accompanied by research into the legal, organizational and financial conditions which must be fulfilled by territorial authorities in Poland in order to effectively participate in decentralized horizontal regional innovation policy.

An efficient administration in territorial authorities is equivalent to the efficient management of a territorial unit. In Poland frequent and deep personnel changes after each election, which cut the continuity of acquiring knowledge, make attainment of this goal difficult. Therefore it seems necessary to introduce regulations which would stabilise the position of territorial government staff. The creation of a civil corps of regional and local government officers seems like a good idea. Such a corps could even be restricted to certain key fields of activity which require a great degree of professionalism (like managing local and regional development, spatial planning, property management, education). What is more, the minimum standards of knowledge should be clearly defined for those persons desirous of holding these key posts.

Effective regional innovation policy will not be possible without further research in identifying networks which support and barriers to the development of interactive relations between the actors belonging to regional innovation systems. Unfortunately regional and local governments in Poland are not able to finance this research. In this case support from the Polish national government and EU co-financial support would be necessary.

When assessing the importance of the cluster concept for Polish innovation policy it may be concluded that development of production-oriented clusters within an RIS should be based on the already existing initial formations of such clusters. The role played by public authorities in developing clusters may include activities aimed at developing regional social capital apart, from the classic governmental activities aimed at creating infrastructural improvements. Such an approach makes it very likely that the adoption of continuing wise policies by regional and local authorities will allow newly developing clusters to become lasting fixtures, which will enter into strong relationships with the scientific and research centres of the largest Polish metropolitan areas.

BIBLIOGRAPHY

Benner, C., (2003), "Learning Communities in a Learning Region: the Soft Infrastructure of Cross-firm Learning Networks", *Environment and Planning A*, Vol. 35, pp 1809-1830.

Cook, P., (1997). "Institutional Reflexivity and Rise of the Region State", [in:] G. Benko and U. Stromayer (Eds) *Space and Social Theory: Interpreting Modernity and Post-Modernity*, Oxford: Blackwell.

Gertler, M, D. A. Wolfe, (eds), (2002), *Innovation and Social Learning: Institutional Adaptation in an Area of Technological Change*, Basingstoke: Palagrave Macmillian, Houndsmill.

Fukuyama, F. (1999), *Social Capital and Civil Society*, The Institute of Public Policy. George Mason University.

Fürst, D., (2003), „Steuer ung auf regionaler Ebene versus Regional Governance“, [in:] *Aktionsraum Region - Regional Governance, Bundesamt für Bauwesen und Raumordnung, Informationen zur Raumentwicklung*, Heft 8/9, p. 441 – 451.

Gancarczyk, J., M. Gancarczyk, (2003), *Regionalny model wspierania małych i średnich przedsiębiorstw* [A regional model of supporting SMEs], Doctoral dissertation, University of Lodz.

Hallencreutz, D, P. Lundequist, (2000), "Spatial Clustering and the Potential for Policy Practice: Experiences from Cluster-building Process in Sweden", *European Planning Studies*, Vol. 11, No. 5, p. 534-47.

Hassink R, A. Lagedijk A., (2001), "The Dilemmas of Interregional Institutional Learning, Environment and Planning C", *Government and Policy*, Vol. 19, pp. 65-84.

Hėraud, J. (2003), "A Regional Innovation Systems and European Research Policy: Convergence or Misunderstanding?", *European Planning Studies*, Vol. 11, No. 1, p. 42-56.

Herrschel, T. and P. Newman, (2002), "Governance of Europe's Regions", *Planning, Policy and Politics*, Routledge, London and New York.

Hohn, U. and B. Neuer, (2006), "New Urban Governance: Institutional Change and Consequences for Urban Development", *European Planning Studies*, Vol. 14, No. 3, pp. 291-298.

Jewtuchowicz, J., K. B. Matusiak, E. Stawasz, (2001), *Zewnętrzne determinanty rozwoju innowacyjnych firm* [External Determinants of Development for Innovative Enterprises], a publishing series monographs, dissertations, reports, studies; Department of Economics, University of Lodz.

Lundequist, P., D. Power, (2002), "Putting Porter into Practice? Practices of Regional Cluster Building; Evidence from Sweden", *European Planning Studies*, Vol. 10, No. 6, p. 685–704.

Łobocki, J., (2003), "Znaczenie kapitału społecznego w budowie konkurencyjności regionu" [Importance of Social Capital for Building Regional Competitiveness], [in:] A. Klasik, (ed.), *Zarządzanie rozwojem lokalnym i regionalnym w kontekście integracji europejskiej* [Managing local and regional development in the context of European integration], Biuletyn KPZK, No. 208, Warsaw.

Markowski, T., (1999), *Zarządzanie rozwojem miast* [Managing development of cities], PWN, Warsaw.

Marshall, T., (2003), "Regional Governance in the United Kingdom", [in:] *Aktionsraum Region - Regional Governance, Bundesamt für Bauwesen und Raumordnung Informationen zur Raumentwicklung*, Heft 8/9, pp. 523–535.

Morgan, K., (1997), "The Learning Region: Institutions, Innovation and Regional Renewal", *Regional Studies*, Vol. 31, pp. 491-503.

Miles, I., (1999), "Services in National Innovation Systems: From Traditional Services to Knowledge Intensive Business Services", [in:] G. Shienstock, O. Kuusi, (ed.), *Transformation Towards a Learning Economy*, Finnish National Fund for Research and Development, Sitra 213, Helsinki.

Newlands, D., (2003), "Competitions and Cooperation in Industrial Clusters; The Implication for Public Policy", *European Planning Studies*, Vol. 11, No. 5, pp. 521–532.

Porter, M., (2000), "Location, Competition, and Economic Development: Local Clusters in Global Economy", *Economic Development Quarterly*, No. 14, pp. 15–35.

Schienstock, G., (1999), "From Direct Technology Policy Towards Conditions-Enabling Innovation Policy" [in:] G. Shienstock, O. Kuusi, (ed.), *Transformation Towards a Learning Economy*, Finnish National Fund for Research and Development, Sitra 213, Helsinki, pp. 420–441.

Stawasz, E., (1997), "Przegląd podstawowych pojęć: innowacje, transfer technologii, krajowy i regionalny system innowacji, polityka innowacyjna [An Overview of Basic Terms: Innovations, Technology Transfer, Stimulation of Innovations, Choice of Expertise], [in:] T. Markowski, E. Stawasz, R. Zembaczyński, (ed.), *Instrumenty transferu technologii i pobudzania innowacji, wybór ekspertyz* [Instruments of Technology Transfer and Innovation Stimulation, Chosen Expertises], Task Force for Structural Policy in Poland, Warsaw, Publishing House Przedświt.

The Reform of Metropolitan Governance, OECD Observer - Policy Brief, (2000), www.oecd.org/publication/Pol_brief.

Todtling, F., S. Sedlacek, (1997), "Regional Economic Transformation and the Innovation System of Styria", *European Planning Studies*, No 5, pp. 43-64.

Van Geenhuizen, M., P. Nijkamp, (2006), "Learning Regions in an Evolutionary Context: Policymaking for High Technology Firms", *Int. J. Entrepreneurship and Innovation Management*, Vol. 6, No. 3, pp. 265-282.

Schwab, B., D. Kübler, D., (2001). "Metropolitan Governance and the 'Democratic Deficit', Theoretical Issues and Empirical Findings", conference paper, "Area Based Initiatives in Contemporary Urban Policy", Danish Building and Urban Research and European Urban Research Association, Copenhagen 17 -19 May 2001.

10

The Cluster Linkage – A Case Study of Policy Made by a Japanese Local Government: The Sendai-Finland Wellbeing Center Project

Gen Amano

INTRODUCTION

In this paper, the industry cluster policy developed by the Japanese government is briefly reviewed and evaluated from the point of view of a local government. The Sendai-Finland Wellbeing Center Project is discussed as a case study of cluster policy made by a local government, in the sense of cluster linkage with overseas industrial clusters.

JAPANESE CLUSTER POLICY

In reviewing the industrial cluster policy of the Japanese government, it is worthwhile discussing some pre-WWII history. Tokyo, Nagoya, Osaka and Kitakyushu were Japan's heavily industrialized areas before WWII because of the accumulation of military related industries and the steel industry. This continues to be true today. Because those areas are located along the coast line bordering the Pacific Ocean, it is called "Pacific Belt Zone" (Fig. 1). The Pacific

Belt Zone has been leading the Japanese economy. However, that concentration raised environmental issues in the 1970's. The national government also recognized that the concentration constituted, to an excessive degree, a national risk.

Figure 1
Heavily Industrial Zone

Source: the City of Sendai

The national government began to investigate the possibility of establishing ONE science park as a national technology center in Japan. This science park is called "Technopolis," which was a word coined by a Japanese bureaucrat to represent a combination of Technology and Polis, which means 'State City' in old Greek. The national government planned to establish the "Technopolis Law," however local governments exercised very strong lobbying on the national government, and the law was greatly modified in 1983, designating 26 areas for "Technopolis" (Fig. 2). As a result, the law lost its power to establish a single innovative center that might have international competitiveness. The aim of the law was shifted from the enhancement of Japan's national competitiveness to so-called "balanced development of the Japanese land." The content of the law was also shifted from generating innovating science parks to mainly attracting high-tech plants or factories to each area.

Figure 2
Technopolis Areas designated by the Technopolis Law (1983)

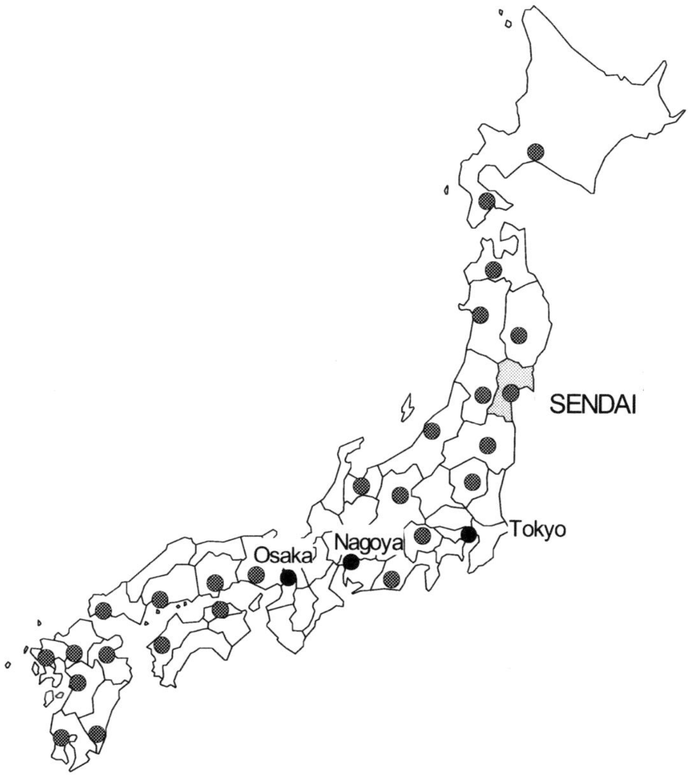

Source: the City of Sendai

In the 2000's, the national government shifted its industrial policy dramatically toward a policy of academia-industry collaboration. The Ministry of Education, Culture, Sports, Science and Technology (MEXT) started the "Knowledge Cluster" project, and in 2002 designated 12 areas (Table 1). It is a 5-year long project designed to promote technology transfer from universities to industries, and its budget is approximately $60 million for each fiscal year. The Ministry of Economy, Trade and Industry (METI) separately started its own "Industrial Cluster" project, and in 2001 designated 19 projects, spread out all over Japan (Table 2). The designated fields of industries vary from area to area (Fig. 3); however, the budget has been shifted to the regional bureau of METI. The budget is approximately $350 million for a fiscal year.

Table 1
Knowledge Clusters by MEXT (2002)

Name of the Cluster	Target Fields
Sapporo IT Carrozzeria Cluster	IT
Sendai Cyber Forest Cluster	IT
Nagano/Ueda Smart Device Cluster	Nanotech, Materials
Hamamatsu Optronics Cluster	IT, Life Sciences
Keihanna Human L^3 Cluster	Life Sciences, Environment, IT
Kyoto Nanotech Cluster	Nanotech, Materials
Saito Biomedical Cluster	Life Sciences
Kobe Translational Research Center	Life Sciences
Takamatsu Rare Sugar Biocluster	Life Sciences
Hiroshima Biocluster	Life Sciences
Fukuoka System LSI Design and Development Cluster	IT
Kitakyushu Human Technology Cluster	IT, Environment

Research into clusters such as Austin in the United States and Oulu in Finland became popular among policy makers in Japan, and these top-down approaches have been implemented by Japanese ministries. These separate policies have been strongly criticized because of their ineffectiveness.

Although the change of policy is evaluated positively in the sense of utilization of academia-industry collaboration, the policies that have created by the national government have been trying to scatter governmental investment equally to all over Japan. While this is worthy in terms of satisfying domestic inter-regional conflicts of interest, it is not an effective way to attain international competitiveness, compared to concentrated governmental investment in a limited number of clusters. Some Japanese local governments have found that this kind of equal development policy does not create those differences among regions which generate their domestic and international competitiveness.

Because of the declining birthrate and the growing proportion of elderly people in Japan, the issue of government finances issue is very serious, and there is a strong demand for a small central government. This also accelerates the movement toward decentralization of national governmental power to local governments. Apart from the shrinking policies of the central government, the local governments are trying to have unique industrial promotion policies, and some policies are aimed at gaining international competitiveness for their own regional industrial cluster. The Sendai-Finland Wellbeing Center Project, introduced in the next section, provides a good example of one such policy.

Table 2
Industrial Cluster Projects by METI (2001)

Name of the Project (Regional Bureau for METI)	Target Fields
Hokkaido Super Cluster Promotion Project (Hokkaido)	Biotechnology, IT
Industry Promotion Project for IT, Life Science and Cutting-edge Manufacturing (Tohoku)	IT, Health , Manufacturing
Industry Promotion Project for a Recycling-oriented Society (Tohoku)	Environment, Energy
Regional Industry Revitalization Project (Kanto)	Manufacturing
Fostering Bio-Ventures (Kanto)	Biotechnology
Fostering IT-Ventures (Kanto)	IT
Project to Create Manufacturing Industry in Tokai Region (Chubu)	Manufacturing
Tokai Bio Factory Project (Chubu)	Biotechnology
Project to Create Manufacturing Industry in Hokuriku Region (Chubu)	Manufacturing
Bio Five-Star Company & Tissue Engineering Project (Kansai)	Biotechnology
Active Manufacturing Industry Support Project (Kansai)	Manufacturing
Kansai Information Technology Cluster Promotion Project (Kansai)	IT
Kansai Energy & Environment Cluster Promotion Project (Kansai)	Energy
Project to Form Next Generation Key Industries in the Chugoku Region (Chugoku)	Manufacturing
Project to Form Recycling-Oriented and Environmentally-friendly Society (Chugoku)	Environment
Shikoku Techno Bridge Plan (Shikoku)	Health and welfare, Environment
Kyushu Recycle and Environmental Industry Plaza (Kyushu)	Environment
Kyushu Silicon Cluster Project (Kyushu)	Semiconductor
Okinawa Industry Promotion Project (Okinawa)	Commercial Information, Health, Environment, Processing

Figure 3
Japan's Cluster Policies in the 2000's

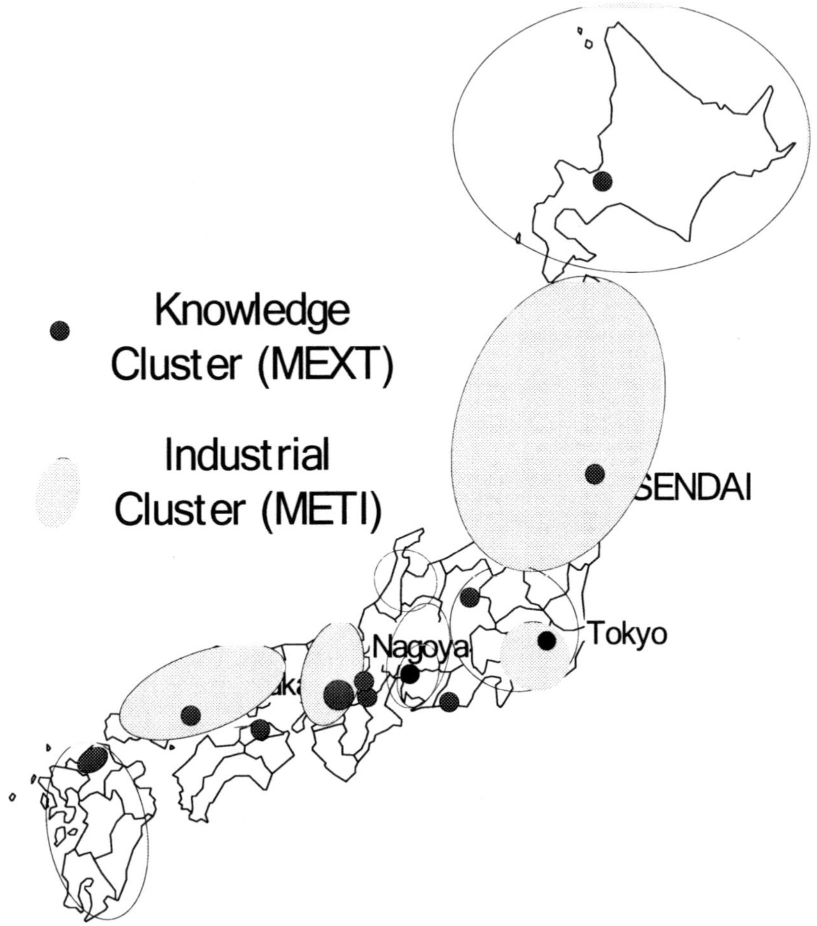

Source: the City of Sendai

THE SENDAI-FINLAND WELLBEING CENTER

Background

Sendai is de facto the capital city of the North East (Tohoku) Region of Mainland Japan. Sendai has a population of 1 million, compared to the 10 million in population of the Tohoku region. Sendai has $51 billion in Gross Regional Product (2002), which is equivalent to Hungary.

Sendai is home to many branch offices of Tokyo headquartered companies, and it has been said that it is a "Branch Office Economy". In other words, Sendai has been positioned in the economic hierarchy topped by Tokyo; however, Tokyo seems to be losing its international competitiveness. These days, more overseas enterprises locate themselves in Shanghai and Singapore than Tokyo. The City of Sendai established a New Industry Creation Plan in 2002, and it aims to be independent from the Tokyo hierarchy and to implement direct exchanges with overseas countries. The plan includes a clustering policy for the welfare equipment industry. Its target is to create assistance technology in support of independence of the elderly, taking into account the expanding market in the advanced counties.

Finland has been evaluated as the world's most competitive country, despite the fact that it is a rather small country with 5.2 million in population. The origin of its competitiveness is its engineering technology, such as mobile communication technology and machine technology. Finland is also highly respected as an advanced welfare country. Finland is aiming to create a new industry combining engineering technology and care technology. This is part of its effort to rearrange its economic structure from a Nokia-monopoly economy. The Finnish Foreign Trade Association (Finpro), National Research and Development Center for Welfare and Health (STAKES), and Finnish Institute initiated the internationalization of this effort.

The City of Sendai and Finpro started the discussion in 2000 to produce an international project to create high value added welfare equipment and services for the expanding elderly market, aimed at providing them with an independent life. The cooperation is based on the concept that cooperation among clusters enhances their competitiveness. The City of Sendai and Finpro entered into an agreement in 2003, which contains the details of the project, with 7 other participating organizations, including universities, governmental bank and governmental organizations.

With the newly established "Sendai-Finland Wellbeing Center," the City of Sendai is aiming to promote the health and welfare related industrial cluster in the Sendai region. Through collaboration with Finnish universities and companies via sales alliances, localization, and joint R&D, the City of Sendai is also seeking to internationalize its regional industries so that they can attain international competitiveness.

STRUCTURE OF THE PROJECT

The Sendai-Finland Wellbeing Center has been in operation since March 2005. It has two functions:

1. R&D Unit

   The R&D Unit functions as an incubator for private companies and universities. The companies and universities of both Sendai and Finland are conducting R&D activities for health and welfare services and equipment.

The users' voices are easily introduced into this process of R&D activities via collaboration with the Care Unit, which is located adjacently.

2. Care Unit
   This is a nursing home for the elderly and has 100 independent rooms for long-term care, 20 rooms for short stays, and day service for 15 elderly persons. The Tohoku Welfare Corporation, which is affiliated with the Tohoku Welfare University (Tohoku Fukushi University), provides advanced care services created by a fusion of Japanese and Finnish care concepts aimed at enabling the elderly to realize an independent life. This unit also provides information concerning the "needs" of the elderly, and evaluates the services and equipment developed in the R&D unit.

Figure 4
The Structure of the Sendai-Finland Wellbeing Center Project

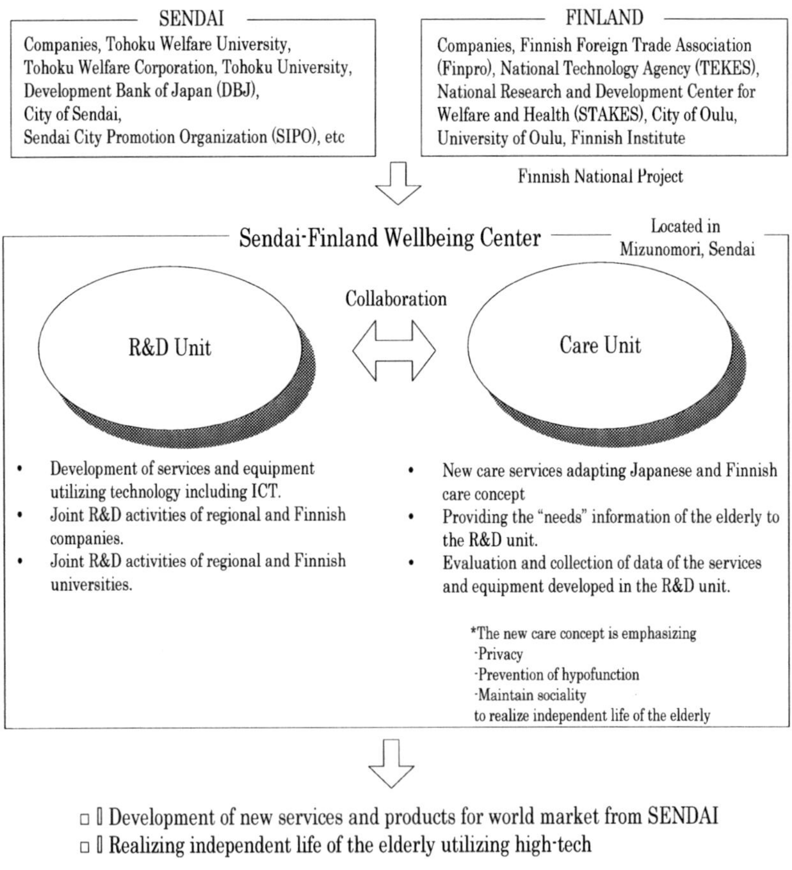

Source: The City of Sendai

Figure 5
A view of the Sendai-Finland Wellbeing Center

Source: the City of Sendai

MAIN FEATURES AND CONCERNS OF THE PROJECT

Industry Clustering by Internationalization

Promotion of industrial clusters is realized not only with domestic perspectives in mind, but also with regard to the international point of view. Owing to the internationalization of trade and industry, this is indispensable if an industry cluster is to attain world competitiveness. For example, a rational target for a start up company would be a niche market, and in this case it would be necessary, in order to enlarge its turnover, to include overseas' markets in the business plan. For this project, the participating SME's are targeting European and Asian markets, including the expanding Chinese market. Not only in terms of the scale of the markets, but also for the competitiveness of the products, internationalization is very effective to create innovation if the product is strongly based on the culture or habits of the single market.

Case 1

Company "A" is located in Sendai and is producing wheel chairs. This company "A" enhanced the lightness of the wheel chair, utilizing new materials. It did this because the Japanese market has to satisfy the care needs of persons who would like to handle and transport wheel chairs easily. Company "B" is also a wheel chair company, located in the EU. The advantage of the wheel chair of company "B" is the positioning of the barycenter. In the EU market, the barycenter position is more important for the user who sits on the wheel chair and handles it. Companies "A" and "B" enter into an alliance and are producing the light weight and well positioned barycenter wheel chair.

While regional SMEs may encounter difficulties in internationalization, by concentrating on a specific field, in this case the welfare and health care field, the resources of public support are able to be utilized efficiently and effectively.

Cluster Linkage and Co-incubation

One of the essential elements of a cluster that might be highlighted is its internal network. In addition to that, having a network to the outer world is also important. This project is based on the linkage between the Sendai cluster and the Finnish cluster. More specifically, the linkage between the City of Sendai and City of Oulu should be highlighted. The City of Oulu is one of the Finnish participants in this project and is famous as an ICT cluster. Sendai and Oulu are collaborating to promote the industry in both regions, and the two cities and their affiliated business promotion organizations entered into a "Co-incubation Agreement" in November 2005. This agreement ensures that, for example, the SMEs of the Sendai region will be supported by the Oulu cluster when they start a business in Oulu, and vice versa. The SME's of both regions will have special support when they become tenants of the incubators in both regions. Clusters are able to complement each other with their resources. For example, Japanese precise technology for electronics and Finnish human oriented simple design have been creating functional and easy to use, yet still beautiful, welfare equipment.

It is assumed that the linkage of strong overseas clusters will be more important than ever in facilitating the progress of internationalization and in the sense of branding or "Co-branding" of the cluster. It is important to utilize the existing brand power of the other cluster, as in "Cluster with simple Scandinavian product design."

Project Incentives for Industrial Clustering

There are a variety of ways to promote industrial clustering, such as the "Seeds Oriented" approach, "Monetary incentive" approach, or "Infrastructure incentives" approach. The City of Sendai had adopted "Project incentive" type of promotion for clustering. It is easier for SMEs to start their business or developmental activities in a cluster by participating in projects that have already been proposed or commenced by the cluster. If the project is also financially supported by the governmental sector, it becomes even easier for them to be located in the cluster. In this way the City of Sendai and other organizations, including Finnish governmental organizations, are attracting the widest range of welfare and health R&D projects. The City of Sendai is also preparing a policy to encourage the SMEs to remain permanent residents in the cluster even after the success of a given R&D project.

A User Oriented R&D Process

Many technology push type R&D activities have failed; therefore it is crucial for an R&D type cluster to be close as possible to the voices of end-users.

This project is promoting a "Co-design" concept, emphasizing the collaboration between users and developers or researchers. This is the reason why the Sendai-Finland Wellbeing Center consists of the nursing home and R&D facility for companies and universities. The "users" of the welfare equipment include not only the elderly people themselves, but also their families and care persons. It is very important for researchers to listen to the voices of the families and care givers because quite often they are the persons who manipulate the equipment and who pay for it. The families and care persons have a notion of the services that they would like to provide to the elderly people, as well as how to realize the services; thus they are motivated to utilize new equipment. Therefore, the R&D processes that are being carried out at the R&D Unit are is focused on knowing, in the first instance, the services that the families and care persons would like to realize. These kind of guidelines of the R&D process enable the transition from a mere research cluster to a business cluster.

Figure 6
User-oriented R&D Process

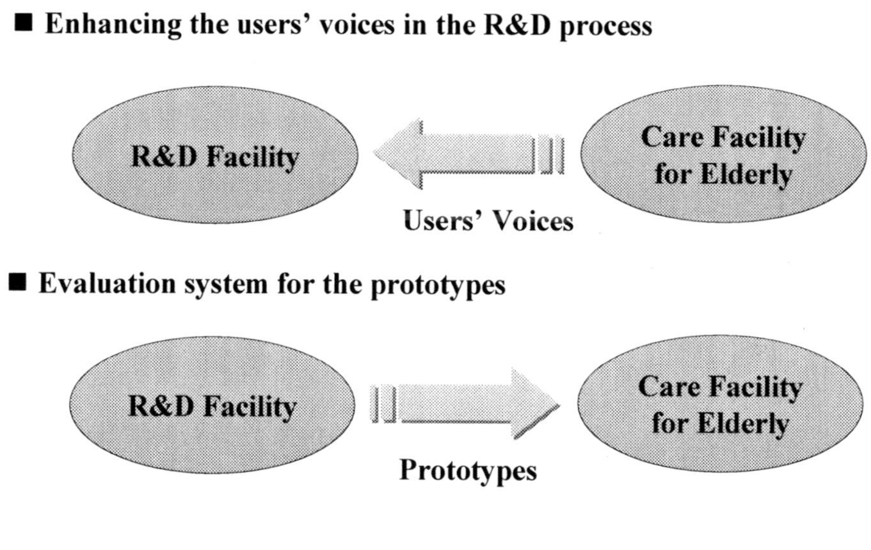

Source: The City of Sendai

Case 2

Company "C" has a business plan to provide internet contents through cable TV sets to elderly people. In order to assure more participation of the elderly, company "C" was planning to create a new type of key board as an easy to use interface. However, company "C" faces difficulties in gathering users' voices by questionnaire sheets. The observation is more effective for the product evaluation if it is for the elderly people. Sendai-Finland Wellbeing Center arranged a university professor "D", who is an expert in product design, and company "C" started a R&D project at the center with the professor. The elderly students in the personal computer class are cooperative in this R&D project, and videos are taken of each student manipulating the keyboard in order to analyze the easiness of use.

Source: the City of Sendai

PRESENT ACTIVITIES OF THE PROJECT

As tenants at the R&D Unit, six Japanese companies and universities are as well as six Finnish companies and universities are active (as of July 2006). They are supported by the business development directors dispatched by Japanese and Finnish sides. In addition to the residents of the R&D Unit, the center works as the one-stop support center for the companies located in both countries, and the business development directors have been supporting 47 business matches between 28 regional companies and 32 Finnish companies. It should be noted that not only international business matches, but also domestic business matches are important for a cluster. Thirty-three business matches have been created among 26 regional companies. In addition to the business matches, 14 R&D projects have also been started. Examples of the R&D projects are listed below:

- Sensor in bed to detect body perspiration
- Standardized ICT platform for various care equipment
- Promoting a digital community utilizing the Web through TV set.
- Digital simulation movie for the activities of daily life (ADL) as a communication tool among care persons.
- Usable remote control unit for electric appliances

The R&D Unit is an incubator, but it is also a show case for the prototypes that are developed at the R&D Unit. Therefore, the R&D Unit is open to the public and more than 6,000 potential customers visited it in FY 2005. This is also an effective way to gather users' voices from the visitors.

Case 3

Company "E" is a tenant company at the R&D Unit, and the researchers of the company have daily occasion to meet potential customers such as the elderly people, their families, and care persons who visit the R&D Unit. The researchers have been trying to collect the "problems" that the potential customers have in their every day life relating to the elderly people. This is because the solution to those "problems" may turn out to be a business opportunity.

A visitor who is a manager of a care facility has had problems with the hectic daily reports among care persons at her facility, and the researcher of the company had a frank conversation on this matter at the R&D Unit. Company "E" found this is a common problem among relatively small facilities and developed an easy entering system utilizing a PDA-like mobile phone at a reasonable price. Immediately after the release of the product, the facility manager bought one system, and other relatively small facilities are following.

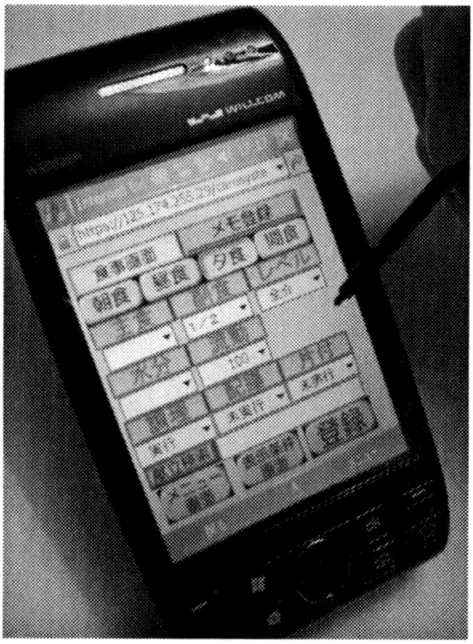

Source: the City of Sendai

THE ROLE OF THE MUNICIPAL GOVERNMENT

There have been ongoing discussions about the birth of industrial clusters; the policies that can create clusters (top-down approach); as well as policies that can NOT create clusters (bottom-up approach). Entrepreneurship has been focused on the bottom-up approach, and its ideal situation is a high-mobility society. A municipal government does not find its place in the bottom-up approach. However, there have been always contradictions when we think about regional economic development. The regions that need to be developed are often those with low-mobility. In this case, a municipal government must take the initiative and implement suitable policies. Especially in terms of internationalization or cluster linkage, there still seems to be a role for a municipal government to play. The municipal government can act as a strong mentor and coordinator to bridge the overseas clusters and regional clusters consisting of SMEs.

CONCLUSION

The Japanese national government is hardly able to implement its cluster policy to enhance the international competitiveness. The present national cluster policy is for the equal development of domestic Japan. However, some local governments are aware of the risks of being in the Tokyo topped economic hierarchy, and are trying to implement direct linkages with overseas clusters in order to maintain international competitiveness.

The Sendai-Finland Wellbeing Center Project is a challenging project designed to promote a cluster with linkage to an overseas cluster, and it leads to the concept of Co-Incubation and Co-Branding.

BIBLIOGRAPHY

The City of Sendai, publication materials.

11

What Role for Knowledge Access in Innovation by Small Furniture Manufacturers in Indonesia?

Marina van Geenhuizen
Nurul Indarti

SETTING THE SCENE

New knowledge, be-it technical or organizational, and the concomitant competition have been powerful drivers of economic growth in the world economy. It is commonplace to refer to knowledge as the primary input into economic processes and as a crucial condition for the ability of companies, communities and individuals to participate successfully in the global economy, both in developed and developing countries (Reich, 1991). One of the key activities to take advantage of knowledge by companies is knowledge management. In its most basic form, knowledge management can be seen as the explicit and deliberate building, renewal and application of knowledge to maximize a company's knowledge-related effectiveness and returns from its knowledge assets. Knowledge management is connected with different types of innovation, organizational adaptation and effective decision-making (e.g. Salleh and Goh, 2002).

However, small and medium-sized enterprises in developing economies may not be aware of knowledge management as an explicit and deliberate activity. In small artisanal firms, parents traditionally pass off craftsmanship and commercial wisdom to their children, and innovations are often undertaken on demand of local buyers (contractors) with support of knowledge exchange in the local communities. Besides, many small firms use learning-by-doing and copying from other firms in the cluster as sources of new knowledge. This local orientation and concomitant reliance on informal and tacit knowledge – often with a relatively poor content – tend to prevent SMEs to obtain the knowledge necessary for innovations (upgrading) to increase competitiveness in a global market (e.g. Beerepoot, 2005; Humphrey and Schmitz, 2000). However, it is often not known how pervasive these patterns are, due to a lack of systematic and quantitative studies.

Against this background, the current study aims to examine systematically the ways in which small firms in a clustered industry in a developing economy obtain new knowledge and how this is connected with different kinds and levels of innovation. Indonesia is taken as an example of developing countries and the furniture industry is taken as an example of traditional manufacturing in such countries facing a fierce competition in global value chains and a concomitant urgent need for upgrading of business practices to global standards.

THE JEPARA CLUSTER IN INDONESIA

The world furniture market grows significantly from year to year. In 2003, the total world furniture market amounted to US$ 71.64 billion, against US$ 53.19 billion in 1999 (www.intracen.org). Around 140 countries, including Indonesia, compete with each other in increasing market share. In 2003, the value of furniture export from Indonesia was US$ 1.57 billion (a share of 2.19% in the world market) giving a rank at place thirteen (Table 1). After the large depreciation of the RUPIIA in 1997, export competitiveness of local enterprises increased in Indonesia and the sector grew rapidly in the wake of the crisis.

Table 1
Leading furniture exporters in the world

No.	Country	Export value in 2003	
		US$ 000	% share
1	Italy	9,653,259	13.47
2	China	9,035,240	12.61
3	Germany	6,238,988	8.71
4	Canada	4,907,069	6.85
5	Poland	3,871,695	5.40
6	Mexico	3,740,772	5.22
7	USA	3,607,357	5.04
8	France	2,614,329	3.65
9	Denmark	2,398,546	3.35
10	Belgium	2,017,653	2.82
11	Spain	1,728,431	2.41
12	Malaysia	1,620,656	2.26
13	**Indonesia**	**1,569,628**	**2.19**
14	Austria	1,555,761	2.17
15	Sweden	1,472,233	2.05
	Total world export values	71,642,747	100

Source: http://www.intracen.org/

The world furniture market today is still dominated by old players, like Italy and China, but there is a tendency that the market share of Eastern European countries is increasing. For example, between 1999 and 2003, furniture export values of Poland increased by almost 50 per cent, causing a 5th place as a furniture exporter in 2003. Particularly Indonesia is facing an increased competition from China (already in 2nd place) and still small but quickly growing Vietnam. The contribution of the furniture industry to Indonesia's GDP is not large. Data, provided for a more comprehensive category, i.e. wood products, indicate a share of less than 5%. Nevertheless, the industry contributes, aside from other exporting manufacturing industries, to counterbalance the dependence on exports of crude oil.

One of the largest furniture manufacturing clusters in Indonesia is that of Jepara in the central part of the island Java. This cluster contains about 3.700 companies and exemplifies a quick growth in the years following the crisis (by 50%, between 1997 and 2002). The cluster contributes for $ 111.73 million dollar or 7.3 per cent to the national furniture export (2003), indicating the existence of various other large furniture clusters. We selected the Jepara cluster for our study because it grew quickly after the crisis and because – despite the fierce competition with China and Vietnam – it remains attracting customers, indicating a sense for adjustment among the furniture manufacturers (Loebis and Schmitz, 2005). However, most of the Jepara manufacturers are small and seem 'captured' in value chains with low price competition while relying on knowledge from local buyers. A systematic look into patterns of innovation and underpinning knowledge may clarify the direction of the adjustments and potential weaknesses or opportunities herein. The research questions addressed in this study are as follows:

(1) How do small manufacturers achieve new knowledge? To what extent are knowledge sources internal or external to the firm, and local (within the cluster) or global? What is the concomitant level of codification of the knowledge underpinning innovation?

(2) What is the relation between the use of particular knowledge sources and innovativeness, and what are – given specific subcontracting relations of the firms - the potentials for upgrading business performance and increasing competitive power to a global level?

The remaining paper is divided into five sections. First, we briefly discuss various theoretical perspectives and concepts used in the current study. This is followed by a section on the design of the empirical work. Next, we present the results concerning the ways in which small furniture manufacturers achieve new knowledge as well as their patterns of innovation. In a separate section, we examine differences in knowledge gaining between firms that are more innovative and those that are less innovative. We conclude the paper with a discussion of the results and potential ways for upgrading of business operations of low innovative manufacturers, including some brief policy recommendation and future lines of research.

KNOWLEDGE AND INNOVATION

Barney (1991, p. 101) defines resources as "all assets, capabilities, organizational processes, firm attributes, information, knowledge, etc. controlled by a firm that enables the firm to conceive of and implement strategies that improve its efficiency and effectiveness". Resources-based theory (Barney, 1991) argues that resources that are heterogeneous in an industry and perfectly immobile across firms in an industry will give sustained competitive advantages to a firm, provided that these resources are also valuable. According to Porter (1985), a valuable resource will enable (1) to reduce costs and thus enhance a firm's cost leadership strategy; or (2) to increase demand for its products due to unique characteristics, thus provide an improved differentiation position. Hence, to be in forefront of the market, a firm should always seek for valuable resources that are not ubiquitous or easy to gain.

New knowledge can be seen as a key resource in competition. Provided that it is - to a certain extent - unique, difficult to transfer and to be imitated, new knowledge gives competitive edge through various types of innovation. In this context, various authors have stressed the importance of people, not technology, as the basic "ingredient" in any innovation (Claver et al., 1998; Woodman et al., 1993) because people determine the innovation capability and the motivation for innovation (Prajogo et al., 2004). An important characteristic in this context is absorptive capacity, that enables firms to recognize the value of new external knowledge and to assimilate and apply it effectively (Cohen and Levinthal, 1990).

Strategies to increase new knowledge can be divided into those focusing on internal sources and those focusing on external sources. Employing internal sources may follow a structured path (planned) like in-house research and development and small scale experimentation, but it may also be incorporated in the way of working, like learning-by-doing. External sources encompass a wide range of sources. According to Afuah (2003), knowledge sources - as *functional* sources of innovation - can be classified into five major categories: (1) internal value-chain functions; (2) external value-added chain of suppliers, customers, and complementary innovators; (3) university, government and private laboratories; (4) competitors and related industries; and (5) other nations or regions. Among SMEs in developing economies the first two categories and concomitant actors may play a strong role, e.g. internal value-chain functions by family members and friends, and external value-adding activity, mainly by local buyers (contractors) and suppliers.

Achieving new knowledge may be rather expensive. Ogawa (1998) gives various reasons for high costs of transfer, like the nature of the knowledge itself - e.g. a high complexity and a protected status of the knowledge - and the type of channel used for transfer, like in the situation of embodied knowledge when buying new equipment. However, in most developing economies where small firms operate within the confines of subcontracting relationships, new knowledge is often provided for free by buyers. But this knowledge is also one-sided including a limited scope for subcontractors, because it aims at increasing the competitiveness of the buyers in first place (e.g. Geriffy, 1999). Achieving knowledge from buyers tends to cause a focus on product innovation, such as an improved product or new design. Particularly if the subcontracting relations give only small margins, the few organizational (managerial) and process innovations that take place tend to support survival, rather than significantly improve business performance. Aside from buyers (contractors), new knowledge originates in the local community - neighbors, family and friends in the same business - as mainly tacit knowledge. Different from clusters in the *developed* world, this knowledge is *disconnected* from global knowledge flows and tends to reproduce traditional practices (e.g. Humphrey and Schmitz, 2000, 2002). The above ideas indicate the existence of a relationship between the knowledge sources used and the types of innovations introduced in business practice (e.g. Darroch and McNaughtan, 2002; Kristiansen et al., 2005). This calls for an appropriate applied study on the basis of systematic fieldwork.

According to Neely et al. (2001) innovation within a company is a multidimensional concept. An innovation can be defined as "an idea, practice, or object that is perceived as novel by an individual or other unit of adoption" (Rogers, 1995, p. 11). An important aspect that differentiates between innovations is the degree of newness. An innovation as a product or process may be new only for the firm concerned, but at the other extreme, it may be new for the whole sector in the world (e.g. Johanneson et al., 2002). Innovation can also be seen as an activity. Wissema (forthcoming) argues that "innovation is the successful introduction of something new, successful as shown by acceptance in the market or other use". This implies that the activity that leads to an innovation is only partly a technical process, scientific or not. It is very much also a commercial process putting the new product in the market, implicating that knowledge about the market and demand for new products is crucial.

CHARACTER OF THE EMPIRICAL STUDY

This study employed an extensive fieldwork among small firms in the furniture industry in the district of Jepara. Most of the small firms, being home industry and family business - are not registered and therefore, the sampling was based on visibility of the manufacturers from the streets and on the researchers' knowledge about presence and development of furniture industries in the various villages. The sampling was stratified in such a way that the selected firms fairly represent the large segment of very small firms and the small segment of somewhat larger firms. Accordingly, most firms are independent businesses and fall in the class of less than 25 employees (80%). The remaining ones are subsidiaries and somewhat larger but still less than 100 employees. Data were collected in April and May 2005 on the basis of personal face-to-face interviews and this fieldwork produced a sample of 90 manufacturers. According to estimations by the furniture association, the sample represents about 3 per cent of the population of furniture manufacturers in Jepara.

Interviews supported by a semi-structured questionnaire were carried out by well-instructed students and staff members of the Gadja Madah School of Management (Yogyakarta). This way of working enabled to provide respondents with additional explanation concerning the more complicated questions, like on the difference between tacit and codified knowledge and the level of innovativeness (ranging from new to the firm to a break-through in the world) and to keep an eye on correct answering if necessary. Note that the answers given reflect to a large extent the self-image of the furniture manufacturers and this image may be somewhat colored by a certain optimism among them, also reported in other studies (e.g. Loebis and Schmitz, 2005).

The analysis presented in the next section is descriptive. It depicts the pattern of knowledge gaining and the pattern of innovation. The section that follows hereafter, however, presents a comparative analysis between more innovative and less innovative firms, revealing some indications of causal relations between knowledge gaining typical for clusters in developing countries and level of innovativeness of the firms concerned.

A SMALL VARIATION IN USE OF KNOWLEDGE

We measured the importance of various knowledge sources on a five-point scale, with 1 as virtually no importance and 5 as very strong importance. New knowledge among the small Jepara furniture manufacturers appears to be partly developed within the firms and partly derived from outside sources (Table 2). Learning-by-doing and buyers (customers) are considered as the most important knowledge sources, witness an average score of 4.15. In addition, in-house experimentation (development) ranks among the most important ways of achieving new knowledge (a score of 3.96). It appears that the manufacturers make less often use of sources that link them with mainly global knowledge, like consultation of specialists, research institutes, and government agencies (scores between 2.01 and 1.76). Also, the Internet is not often used as a source of knowledge (a score of 1.73), a situation that can be explained by a poor access to this source (just a little more than 10% of the manufacturers has access to the Internet). Knowledge sources that are an exception to the small use of global sources, are magazines (newspapers), exhibitions and television (radio) (scores between 2.58 and 2.42). The dominant pattern of local knowledge sources implicates that knowledge necessary to make the companies familiar with up-to-date standards and requirements to compete in global markets reaches them only to a limited extent.

The types of knowledge used fit into the previous pattern. Mainly tacit knowledge is the basis for knowledge exchange for almost 40% of the innovations. This may include knowledge exchange between the manufacturers and buyers, family members and local friends on interpretations of, for example, ways of copying new furniture design (style). Slightly more common is the use of a combination of tacit and codified knowledge (42.3%), whereas mainly codified knowledge is used in supporting a minority of the innovations (20%). An example in this context is a manual delivered with new machinery. A dominant use of local, tacit knowledge is also found in a study of artisanal furniture industry in the Philippines (Beerepoot, 2005).

Table 2
Knowledge sources, average scores on importance

	Currently	*Near future*	*Significance of t-test outcome*
Mainly local origin			
Learning-by-doing	4.15	4.47	**
Experimentation (development)	3.96	4.37	**
Buyers	4.15	4.38	*
Suppliers	3.27	3.95	**
(Other) business partners	3.66	3.95	**
Friends, neighbors, etc.	2.61	2.89	**
Religious affiliations	1.97	2.22	**
Competitors	2.72	3.23	**
Mainly global origin			
Industry association	2.01	2.58	**
Research institutions, universities	1.76	2.24	**
Consultants	1.79	2.07	**
Exhibitions	2.56	3.71	**
Government agencies	1.97	2.51	**
Magazines, newspapers	2.58	3.38	**
Television, radio	2.42	3.33	**
Internet	1.73	3.48	**

*p>0.05, **p>0.01

Source: Fieldwork

Of course, the ways of accessing new knowledge by the Jepara manufacturers are not the result of a rationally established knowledge management. These manufacturers act upon family traditions and trust within the local community. At the same time, there is an overall ambition to make a greater use of all sources of knowledge, including the ones of global origin (Table 2). The differences between future importance and current importance are significant for all knowledge sources. However, most manufacturers suffer from constraints in realizing their ambitions. As much as 94.7% of them are facing obstacles and most of these obstacles (80.2%) are financial (Table 3). We may understand from the relatively small current and also future use of research institutions (universities) and consultants that financial barriers are particularly active in accessing these knowledge sources. Among non-financial barriers, the missing capability to understand highly complex subject matter and foreign language appear to be a major set of barriers (together 10.5%). A large physical distance to knowledge sources is of minor importance.

Table 3
Obstacles to access new knowledge

Obstacle	*N*	%
Financial	199	80.2
High level of complexity of new knowledge	25	8.1
Large physical distance to knowledge source	11	4.4
Language barriers	6	2.4
Others	10	4.8

N = 251 (obstacles).

Source: Fieldwork

By focusing on the manufacturers' expectations in the near future in more detail, it becomes clear that they particularly expect an increase of using global knowledge sources (Table 2), namely the Internet, exhibitions, and through magazines and newspapers. Apparently, financial obstacles are no hindrance to access these particular sources in the near future.

A RATHER ONE-SIDED PATTERN OF INNOVATION

In this section the pattern of innovation will be examined. We measured the importance of innovations by asking the manufacturers the three most important innovations on the basis of contribution to turnover. Next, we made a distinction between product, service, process, market, logistic and organizational (managerial) innovations. The level of innovativeness could be determined by distinguishing between the context in which the innovations are new (type of newness), i.e. only new for the firm, new for the sector in the region, new for the sector in Indonesia, and new in the world.

Product innovations appear to be the most important category (Table 4). Out of 270 reported innovations, product innovation (31.9%) stands out as most significantly contributing to turnover. New product design and new types of products are the most common product innovations. The second and third most important ones are market (26.7%) and logistic (23.0%) innovations, respectively. If we focus on innovations that contribute most to turnover, the role of product innovations is even stronger. Almost half of these innovations are product innovations (47.8%). The same holds for market innovations, but not for logistics innovations. The latter innovations follow from a situation in which reserves of good quality raw material (wood) are shrinking and manufacturers are forced to search for new sources, new suppliers or an innovative organization of transport of wood over larger distances to the factories. Thus, logistics innovations contribute to the survival of the firms, not to extra income. Further, process innovations are of minor importance, whereas organizational (managerial) innovations are very rare and confined to large companies. The typical Indonesian culture in which uncertainty avoidance is high (Hofstede, 1991) contributes to an explanation of the virtual absence of the latter innovations.

Table 4
Type of important innovations a)

Type of innovations	*First important innovation*		*All three most important innovations*	
	N	*%*	*N*	*%*
Product	43	47.8	86	31.9
Service	4	4.4	28	10.4
Process	4	4.4	19	7.0
Market	25	27.8	72	26.7
Logistics	14	15.6	62	23.0
Organizational	0	0.0	3	1.1
Totals	90	100	270	100

a) Importance on the basis of contribution to turnover

Source: Fieldwork

A dominant position of product innovation as observed among the small Jepara furniture manufacturers is in line with research findings for SMEs in other developing economies, e.g. in Tanzania. Kristiansen et al. (2005) observe that most innovations among small garment and furniture industries occur in products, while process, organizational, and logistics innovations are rare.

With regard to newness of the innovations, the vast majority of innovations (81.0%) is merely new for the firm. 11.8% of the innovations are considered to be new for the sector in the region, 5.3% are considered new for the sector in Indonesia, and the remaining innovations (1.9%) are considered new in the world. The few latter innovations – to be considered as radical or a break-through – mainly include product innovations. However, the vast majority of product innovations (79.8%) is merely new for the firms, indicating that these are often incremental in nature, connected with learning-by-doing and demand from buyers (Table 5). By comparing the different types of innovations, it becomes clear that the newness of the other types of innovations, except for logistic ones, tends to be even stronger confined to the own firm than in the case of product innovations (up to 85.7%). Logistic innovations tend to be most often new outside the firm. This pattern may be explained by the fact that in-bound and out-bound logistics are by definition taking place outside the confines of small firms and their localities.

Although the interpretation of the previous patterns is hampered by a lack of standards for comparison or results of comparable research, we may conclude on the basis of newness of innovations that the level of innovativeness among SMEs in the Jepara furniture industry is relatively modest. This complies with the pattern of knowledge creation and exchange through in-house activity and local networks of mainly subcontractors in other studies. However, the situation is not that bleak. There is some differentiation in innovativeness between the companies in Jepara. A small segment of the manufacturers is able to produce radical product innovations. In de next section, we take a closer look at differences between more innovative firms and less innovative firms in terms of knowledge sources and types of knowledge used. We particularly examine the idea that a dominance of local sources and tacit knowledge is associated with low levels of innovativeness.

Table 5
Types and newness of innovations

Type of innovation	*New for the firm*		*New for the sector in the region*		*New for the sector in Indonesia*		*New in the world*	
	n	%	*n*	%	*n*	%	*n*	%
Product	67	79.8	9	10.7	4	4.8	4	4.8
Service	24	85.7	2	7.1	2	7.1	0	0
Process	16	84.2	3	15.8	0	0	0	0
Market	58	85.3	8	11.8	2	2.9	0	0
Logistic	45	73.8	9	14.8	6	9.8	1	1.6
Organizational	3	100.0	0	0.0	0	0.0	0	0.0
All innovations a)	213	81.0	31	11.8	14	5.3	5	1.9

a) non-response of 2.6%.

Source: Fieldwork

DO KNOWLEDGE SOURCES MATTER IN INNOVATIVENESS?

We carried out a comparative analysis between firms engaged in different levels of innovativeness. The level of innovativeness was determined as follows: a weight was assigned to each of the three most important innovations based on the degree of newness (i.e. new for the firm (1), new for the sector in the region (2), new for the sector in Indonesia (4), and new in the world (8). For each firm, the sum of the weighted outcomes of the three most important innovations was calculated as an overall score. Firms facing a score below the average score were considered as less innovative (65 out of 90) while those with a score above average were considered as more innovative (25). Of course, the weighting procedure is subjective in a way, but it clearly reflects a difference in valuation of innovations, i.e. between the more local, incremental innovations on one end of the spectrum and the more radical innovations, new in a national or global context, on the other end. In terms of structural characteristics, the more innovative firms tend to be often somewhat larger, older and more often a subsidiary of larger firms.

It appears that more innovative firms make a smaller use of in-house learning (experimentation) than less innovative firms and that the former firms also make less use of buyers (customers) to gain new knowledge; but only the difference concerning the latter source is significant (mean difference of 0.28) (Table 6). This outcome supports the idea that disclosing new information mainly through buyers hampers innovation among small subcontractors.

Table 6
Knowledge sources, average scores on importance

	Less innovative firms	*More innovative firms*	*Significance of t-test outcome*
Mainly local origin			
Learning by doing	4.16	4.12	
Experimentation on purpose	4.00	3.87	
Buyers (customers)	4.22	3.94	*
Suppliers	3.22	3.36	
(Other) business partners	3.61	3.76	
Friends, neighbors, etc.	2.53	2.82	*
Religious affiliations	1.86	2.26	*
Competitors	2.69	2.79	
Mainly global origin			
Industry associations	1.95	2.17	
Research institutions, universities	1.72	1.86	
Consultants	1.78	1.81	
Exhibitions	2.46	2.85	*
Government agencies	1.84	2.27	*
Magazines, newspapers	2.52	2.77	
Television, radio	2.37	2.61	
Internet	1.68	1.95	

*p>0.05, **p>0.01

Source: Fieldwork

For more innovative firms, all other mainly local and mainly global knowledge sources tend to be more important compared with less innovative firms, which indicates that the former tend to use more knowledge in supporting production processes and innovation. In particular, a relatively strong use (compared with low innovative firms) is made of exhibitions and government agencies, witness mean differences that are significant. These sources apparently give a relatively strong push towards higher levels of innovation by transferring global information.

Remarkably, more innovative firms make also a relatively strong use of local knowledge sources through friends, neighbors and religious affiliations. This seems to contradict our expectation that local knowledge in developed economies is not favorable for upgrading business according to global standards. However, more innovative firms may have developed the ability – through knowledge of global standards - to filter local knowledge and merely absorb those components that support an improved performance. In addition, if we focus on the types of dominant knowledge used, it appears that more innovative firms make a stronger use of codified and a mix of codified and tacit knowledge compared with less innovative firms (80.0%

versus 50.8%) (Table 7). This difference is significant. Thus, the idea that using mainly tacit knowledge exchange is not a favorable mechanism in clusters in a developing economy, is supported by our data on the type of knowledge underpinning innovative activity.

Table 7
Types of knowledge underlying innovations

	Codified and mixed codified /tacit	Mainly tacit	All firms
Less innovative firms	33 (50.8%)	32 (49.2%)	65 (100.0%)
More innovative firms	20 (80.0%)	5 (20.0%)	25 (100.0%)
All firms	53 (58.9%)	37 (41.1%)	90 (100.0%)

Chi-square = 6.37, $p < 0.05$

Source: Fieldwork

CONCLUDING REMARKS

The discussion on the use of particular types of knowledge by small and medium firms in developing economies and potential impacts of this use on the level of innovativeness was central in this paper. By taking the furniture industry in Jepara (Java, Indonesia) as an example, a dominance of in-house learning and knowledge gaining from buyers (contractors) could be identified, alongside a small use of sources that directly connect the firms with global knowledge. In addition, about 40% of all innovations is based on mainly tacit knowledge. Our outcomes support the idea that a combination of the previously mentioned knowledge sources and type of knowledge is associated with an overall low level of innovativeness in terms of newness in the market place. Most of the innovations are only new to the firm (81%). Nevertheless, a small segment of more innovative firms could be identified. These firms tend to differ from the less innovative ones in various ways. First, they make an overall stronger use of knowledge but rely less on knowledge exchange with buyers (contractors). Secondly, with regard to global sources, they rely significantly more on exhibitions and government agencies than low innovative firms. Third, they make less use of mainly tacit knowledge in processes underlying innovation. Accordingly, support programs that aim to increase the innovative level of small manufacturers need to give due attention to increasing the ways of knowledge achieving as practiced by more innovative firms, provided that financial barriers can be removed.

However, this does not mean that the majority of the small firms - holding a relatively poor position as subcontractors - can easily escape from such a position (Kaplinski et al., 2003) Power relations in the supply chains may hamper upgrading and limit the flow of key knowledge within chains. At the same time, a basic requirement for upgrading is the strategic intent of the firms involved. If this strategic intent exists, various ways of breaking out the poor position may be envisaged (e.g. Humphrey and Schmitz, 2000, 2002). We mention two of them:

- To use the new knowledge for supplying *other* markets than that of the buyer, in which relationships are more balanced, e.g. to modify a design. Small producers cannot easily compete directly with their powerful buyers; therefore, other markets are essential.
- To move to functions that leading firms in the chain leave open, like in logistics and in development of new processes and adaptations to existing design.

How far upgrading can go depends on the type of buyer and type of uneven relationships with them, as well as on the ability of small producers to make individually or collectively the required investment. Here lies a task for future research, i.e. to evaluate which factors - aside from global knowledge - determine success and failure of upgrading strategies. The supply of micro-credits may be one of these factors. The role of regional/local government agencies in a scenario of upgrading would be to improve conditions that enable global knowledge to flow into the local clusters in such a way that it can be accessed and absorbed by all firms in the cluster.

BIBLIOGRAPHY

Afuah, A., (2003), *Innovation Management: Strategies, Implementation, and Profits,* New York, Oxford University Press.

Barney, J., (1991), Firm Resources and Sustained Competitive Advantage. *Journal of Management*, 17, pp. 99-120.

Claver, E., J. Llopis, D. Garcia and H. Molina, (1998), "Organizational Culture for Innovation and New Technological Behaviour", *The Journal of High Technology Management Research*, 9 (1), pp. 55-68.

Beerepoot, N., (2005), "Collective Learning by Artisanal Subcontractors in a Philippine Furniture Custer" *Journal of Economic and Social Geography*,Vol. 96. No. 5, pp. 573-584.

Cohen, W. M. and D. A. Levinthal, (1990), "Absorptive Capacity: a New Perspective on Learning and Innovation", *Administrative Science Quarterly*, Vol. 35, pp. 128-152.

Darroch, J. and R. McNaughtan, (2002), "Examining the Link Between Knowledge Management Practices and Types of Innovation', *Journal of Intellectual Capital*, Vol. 3, No. 3, pp. 210-222.

Geenhuizen, M. van and P. Nijkamp, (2006), "Learning Regions in an Evolutionary Context: Policymaking for High Technology Firms", *International Journal of Entrepreneurship and Innovation Management,* Volume 6 (3), pp. 265-282.

Geriffy, G., (1999), "International Trade and Industrial Upgrading in the Apparel Commodity Chain", *Journal of International Economics* Vol. 48, pp. 37-70.

Hofstede, G., (1991), *Cultures and Organizations: Software of the Mind,* London: McGraw-Hill.

Humphrey, J. and H. Schmitz, (2000), *Governance and Upgrading: Linking Industrial Cluster and Global Value Chain Research*, IDS Working Paper 120, Brighton: Institute of Development Studies.

Humphrey, J. and Schmitz, H., (2002), “How Does Insertion in Global Value Chains Affect Upgrading in Industrial Clusters?”, *Regional Studies* 36 (9), pp. 1017-1027.

Johannessen, J.-A., B. Olsen and G. T. Lumpkin, (2002), "Innovation as Newness: What is New, How New, and New to Whom?”, *European Journal of Innovation Management,* 4(1), pp. 20-31.

Kaplinski, R., O. Memedovic, M. Morris, and J. Readman, (2003), *The Global Wood Furniture Value Chain: What Prospects for Upgrading by Developing Countries, The Case of South- Africa*, Vienna, United Nations Industrial Development Organization.

Kristiansen, S., J. Kimeme, A. Mbwambo and F. Wahid, (2005), “Knowledge Flows and Adaptation in Tanzanian Cottage Industries”, *Entrepreneurship and Regional Development* 17, pp. 365-368.

Loebis, L. and H. Schmitz, (2005), “Java Furniture Makers: Globalization Winners or Losers”, *Development in Practice*, 15 (3/4), pp. 514-521.

Neely, A., R. Filippini, C. Forza, A. Vinelli and J. Hii, (2001), “A Framework for Analysing Business Performance, Firm Innovation and Related Contextual Factors: Perceptions of Managers and Policy Makers in Two European Regions”, *Integrated Manufacturing Systems*, 12 (2), pp. 114-124.

Nonaka, I. and H. Takeuchi, (1995), *The Knowledge-Creating Company: How Japanese Companies Create The Dynamics*, Oxford University Press, Oxford.

Ogawa, S., (1998), “Does Sticky Knowledge Affect the Locus of Innovation? Evidence from the Japanese Convenience-Store Industry”, *Research Policy*, Vol. 26, pp. 777-790.

Porter, M. E., (1995), *Competitive Advantage*, Free Press, New York.

Prajogo, D. I., D. J. Power and A. S. Sohal, (2004), “The Role of Trading Partner Relationship in Determining Innovation Performance: an Empirical Examination”, *European Journal of Innovation Management*, 7(3), 178-186.

Reich, R. B., (1991), *The Work of Nations*, Alfred A. Knopf, New York.

Rogers, E. M., (1995), *Diffusion of Innovations,* The Free Press, New York.

Salleh, Y. and W. K. Goh, (2002), “Managing Human Resources Toward Achieving Knowledge Management”, *Journal of Knowledge Management*, Vol. 6, No. 5, pp. 457-468.

Wissema, J. G., (2007), *Managing the Innovation Pentagon* (forthcoming).

Woodman, R. W., J. E. Sawyer and R. W. Griffin, (1993), “Toward a Theory of Organizational Creativity”, *Academy of Management Review*, 18(2), pp. 293-321.

12

Building the Information Society in Portugal: Lessons from the Digital Cities Program 1998-2000

Jose Luiz Moutinho

INTRODUCTION

Society and technologies co-evolve by both shaping and being shaped by the places and spaces where we live. The variety and intricacy of these recombination processes contribute profoundly to the current diversity of spatial structures and meanings of both cities and country sides. Information and communication technologies (ICT) can offer new scenarios for territorial development and provide a new kind of networked infrastructure that will have a profound and lasting effect on the patterns and shapes of urban areas. One of these scenarios is the concept of "digital cities".

There are five basic design patterns for digital cities: (a) highly wired territory; (b) community network; (c) 3D or 2D representation of physical cities; (d) local government portal; and (e) commercial city guides (Schuler, 2002). While the first pattern concentrates on digital infrastructure, the other four are manifestations of different aspects of urban everyday life on the Internet. Recombinations as well as whole new patterns that emulate the vast diversity of real cities are also emerging.

Building on case studies of selected Portuguese cities and regions which have been engaged in "digital cities and regions" projects, this paper will go beyond the patterns mentioned above and will discuss the main challenges and opportunities for mobilizing the information society in Europe, with emphasis for the conditions affecting less favored regions. Moreover, it will also explore the convergence of telecommunications and computer networks within the urban context as reflecting a new city infrastructure supporting everyday life and mobility.

Existing literature on the subject of digital cities often focuses on technological innovations as the main, if not the only, driver of social transformation. As the new digital technologies become embedded in daily life as an "invisible" infrastructure, mobilization and the capacity of absorption and diffusion of technologies seems to be critical to the development of the information society. Within this context one may ask: *What critical factors can enable a digital city to mobilize individuals, communities and organizations for the construction of an information society in Europe*? *What sort of public policies must be taken into consideration to promote these factors*? These broad questions build on a comprehensive set of data on digital cities presented in a previous work (Moutinho and Heitor, 2004; 2005) and on the need to continuously adapt European and national policies aiming to foster innovation and competitiveness in the information industries. It will also be argued that current broadly target public policies to bridge the digital divide may not necessarily stimulate the construction of the information society. Focusing on communities of practice, interest or proximity might give better results over both the medium and long term.

First, the concept of a digital city and its relationship with physical cities will be discussed. Then, the Portuguese Digital Cities Program will be contextualized within the Portuguese and European strategies and action plans for the development of the Information Society. This will be followed by a brief presentation of the Portuguese Digital Cities Program, focusing on the Aveiro Digital City project. The last section will discuss the existing evidence, present some conclusions, and recommend courses of action for future public policies.

WHAT IS A DIGITAL CITY?

The first known "digital city" was based on Amsterdam's well studied community network experiment in 1994 (DDS – De Digitale Stad), based on the FreeNets and Community Networks in the USA and Canada (Besselaar and Beckers, 1998; Besselaar, 2001). The goal was to provide an electronic space for political discussion and participation in the ten weeks that preceded local elections. But, as one could learn from the examples mentioned in the first section, there is no current unified agreement and understanding concerning the concept of digital cities.

The first Portuguese digital cities' projects were also much diversified. Perhaps their diverse interpretations reflect the relative infancy of this concept. Innovations are often fuzzy at the beginning, and it takes a while before they establish their own evolutionary paths based on pre-existing ones, or simply fade out. Thus maybe these varied readings about digital cities simply mirror the different views about real cities, and will coexist for a long time.

Architects and urban planners usually concentrate on form, patterns, structure and relationships among the various components, both living and non-living, of a city (Kostof, 1991). Engineers mostly give attention to the infrastructures, utilities and artifacts that make cities work. Urban economists prefer to analyze the spatial aspects of decision making of firms and households (O'Sullivan, 2003). Sociologists and geographers are inclined towards studying interactions among people and communities within the urban environment and their effect on the common production of the urban spaces and places (Castells, 1989; Valentine, 2001).

Most of these perspectives tend to enhance one or another aspect of urban life in order to explain or predict, at least partially, the outcome of planning, implementation or living strategies. At one point however, most of the perspectives converge: cities are currently the foci of the creativity and innovation that continuously co-evolve within the economic, social, cultural and political society, institutions and technology. As pointed out by Zook (2001), "despite the space transcending ability of Internet technology, [...] the commercial Internet clustered in a few regions within the United States during the period from 1994 to 2000. The existence of these agglomerations runs counter to expectations that the Internet would bring about the "end of geography".

Mobilization is simultaneously a driver and outcome when building the information society. Cities, and in our case digital cities, are places of dense and energized crowding (using Kostof's (1991) image about the urban processes) that can provide an adequate environment for creativity and innovation and knowledge spillovers through intensive virtual or face-to-face contacts and word-of-mouth (O'Sullivan, 2003).

For the purposes of this paper, the concept employed will be that of the digital city as a new layer of urban networked infrastructure, based on telecommunications and computers, supporting and augmenting everyday life.

EUROPEAN AND PORTUGUESE CONTEXTS

The circumstances that gave birth to the Portuguese Digital Cities Program in 1998 are linked to several agendas and action plans for the development of the Information Society in Europe, starting with the "White Paper on Growth, Competitiveness, and Employment: The challenges and ways forward into the 21st century" (European Commission, 1993).

The development of the Information Society in Europe – or, to use the term currently preferred in the European Union, the 'Knowledge-based Society' – must be understood, as pointed out by Barry (2001), in the context of an "era obsessed by a series of interconnected technological problems: with the maintenance of technological competitiveness and the improvement of research productivity; with the need to patent and protect intellectual property; with the dangers posed by the unintended consequences of technological development; with the public understanding of science; with the prospects of e-commerce and electronic democracy; and with the need for lifelong learning in the face of rapid technical change".

The European Union follows a dual strategy. While member states are stimulated to address national challenges and opportunities with local and/or regional initiatives in a bottom-up approach, the European Commission plays an important role defining top-down strategies, policies and standards for the European Union as a whole (Dearnley and Feather, 2001). Therefore, in order to fully realize the varied aspects of the information society in Europe, it is important to follow both national and European paths, namely by revising the strategic documents that establish visions, goals, demonstration projects and respective evaluation frameworks.

What was particularly significant for the digital cities' projects in Portugal were the areas of application set forth in the "Bangemann Report", prepared in 1994 for the European Council by Members of the High-Level Group on the Information Society as providing the building blocks of the information society, as follows: teleworking (more jobs and new jobs for a mobile society); distance learning (life long learning for a changing society); networking of Universities and research centres (networking Europe's brain power); telematic services for SMEs (re-launching a main engine for growth and employment in Europe); road traffic management (electronic roads for a better quality of life); air traffic control (an electronic airway for Europe); health care networks (less costly and more effective healthcare systems for Europe's citizens); electronic bid tendering (more effective public administration at a lower cost); public administration networks (better government, cheaper government) and city information highways (bringing the information society into the home).

In April 1997, the "Portuguese Information Society Green Book", published by the "Information Society Mission" taskforce (1997) which had been created a year earlier, established a set of key policies for the development of the Information Society in Portugal. It addressed a broad range of issues related with electronic democracy, the digital divide, electronic government, knowledge networks (R&D, Universities and schools), digital business environments, employment, the new economy (software, digital contents, entertainment, telecommunications), institutional and legislative frameworks, security and privacy, as well as others.

The following year, 1998, the Portuguese Digital Cities Program was launched (Veiga, 2000). It proposed four action lines: (a) improve the quality of life in cities, mainly through the provision of electronic local administration services and telemedicine; (b) contribute to the development of peripheral areas; (c) improve the local economy and enhance employment by providing access to new markets, teleworking and e-commerce; and (d) contribute to fight info-exclusion and help citizens with special needs. Several cities or regions were invited to participate in the program using a diverse mix of organizations, mainly led by local governments, universities, or regional agencies. Nonetheless, the basic content of each project was somehow predetermined by previous beliefs and only occasionally would local needs and expectations be addressed.

Figure 1
ICT Intensity and Growth (1992-97)

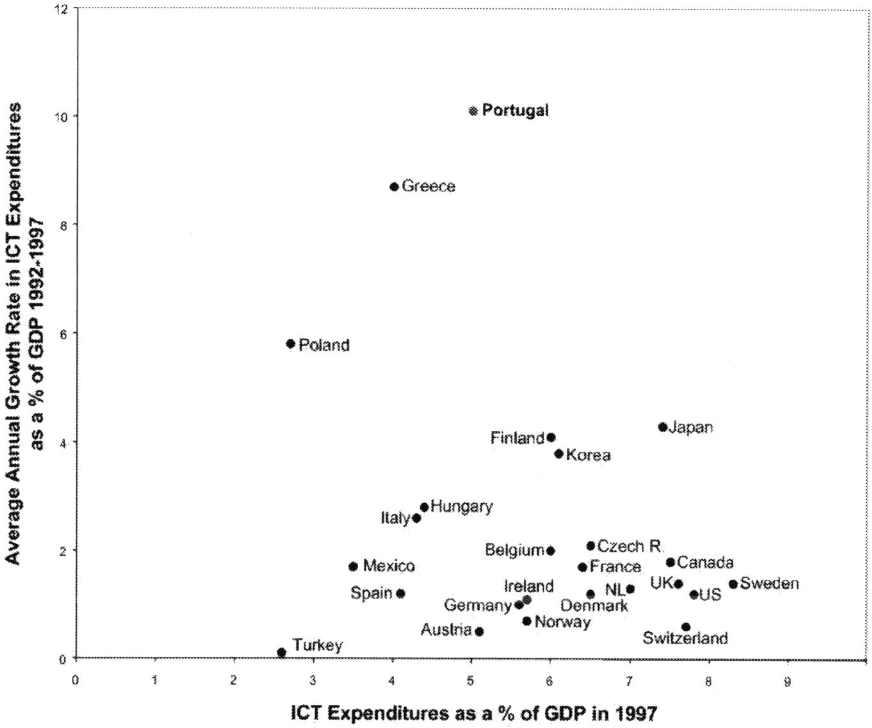

Source: OECD, 2000

In order to better understand the national context it would help to pay attention to the local diffusion of ICTs in comparison with other European countries. Fig 1. presents the intensity of ICT expenditures in 1997 against the growth rate of this intensity from 1992 to 1997. Following a recent analysis for knowledge-based industries (Conceição and Heitor, 2003), the results show that Portugal was the leading OECD country in the rate of growth for ICT expenditures from 1992 to 1997, with a growth rate of more than 10%, mainly accounted for by increases in expenditures for telecommunications (about 9%). Expenditures in IT services and software are particularly low, below 1%, and only Turkey, Greece and Poland have lower shares of expenditure on IT software and services. The growth in this category has been equally dismal, below 2% a year.

Figure 1 shows large variations associated with countries characterized by small absolute values, exhibiting the typical patterns industrialization such as

Portugal. In addition, the results may present indications of a process whereby latecomer countries become engaged in the new techno-economic paradigm (Freeman and Louçã, 2002). Most countries are clustered in the bottom of the figure, with growth rates below 4%. The levels, as indicated by the horizontal distribution of countries, confirm the perception that the US is a leading country. The expenditures on ICT as a percentage of GDP in the US are about 2% above the European average.

Table 1
Expected ICT Expenditure per capita for selected European Regions, 2000-06

Border Midland and Western Region	357.8 €
La Rioja	357.8 €
South Aegean	269.4 €
Ionian Islands	241.4 €
Baleares Islands	238.2 €
Western Greece	151.1 €
Açores	**117.9 €**
Highlands & Islands	98.4 €
Epirus	83.4 €
Alentejo	**44.5 €**
Peloponese	43.1 €
Continental Greece	42.8 €
Algarve	**42.5 €**
Centro	**29.9 €**
Norte	**13.3 €**
Southern Scotland	9.2 €
Lisbon and Tagus Valley	**6.8 €**
Liguria	2.2 €

Source: Tsipouris, 2002

Evidence of the still low absolute value of investments on ICT in Portugal is clearly illustrated in Table 1, which shows values per capita for sample European regions in the census whose programming documents indicated information society actions and that provided the necessary financial information (Tsipouris, 2002). It is clear that the Table refers, above all, to those regions that have attracted European structural funds. As a result it is important to take into consideration the wide diversity of situations and framework conditions for attracting these funds, which clearly influence any analysis undertaken based on their distribution. For the purposes of this paper it is interesting to attempt to define the extent to which the performance of digital networks and cities would depend exclusively on funding limitations, as well as on the capacity to attract them.

Besides its high growth rate in ICT investments, the extent to which the Portuguese society is engaged in the knowledge economy compared to other nations can be analyzed making use of the recently established systematic assessment by the World Economic Forum in collaboration with INSEAD and the World Bank's *info*Dev programme using "networked readiness", as represented in Figure 2 for 2002 (Dutta and Jain, 2003).

Figure 2
Network Readiness Index versus GDP (PPP) per capita, for 2002, with partial Log regression

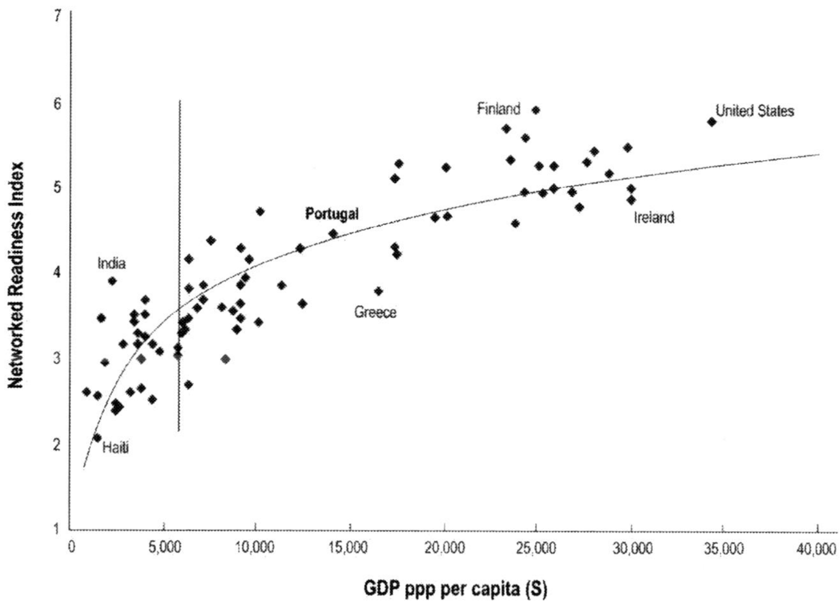

Source: Dutta and Jain, 2003

This indicator offers an aggregated idea of "the degree of preparation of a nation to participate in and benefit from ICT developments" and illustrates the still weak position of Portugal in the European context, above Greece only. The main point to note is that the results for Portugal and for most of the OECD countries appear to be dependent more on each other than the country's overall wealth (as measured in terms of GDP per capita). Considering the partial log regression plot included in the figure, Portugal is in fact entering the cluster of countries where the effect of increasing GDP on network readiness is less pronounced and other factors, namely at the institutional and contextual levels, have been shown to particularly influence the country's competitiveness (Conceição and Heitor, 2003).

The earlier-mentioned challenges associated with latecomer industrialization, and the relative positioning of Portugal in the international scenario at the outset of the 21st century demonstrate that Portugal consistently ranks as one of the poorest performing countries in Europe.

THE PORTUGUESE DIGITAL CITIES PROGRAM

The Portuguese Digital Cities Program was set up by the Portuguese Government through the Ministry of Science and Technology in 1998 and jointly funded by the Portuguese Science and Technology Foundation (25%) and the European Union (75%). There were several simultaneous initiatives related to the diffusion of ICT at a national level, e.g. Internet access in schools (RCTS network), community building portal (Mosaico / Terra-à-vista), and a successful electronic government one stop shop (Infocid). Nonetheless, the digital cities projects were the only integrated initiatives related to the territory and specifically designed for the diffusion of ICT at the local level, covering multi-aspects of society; collective and individual, public and private.

The Program involved 2 rural regions (Trás-os-montes and Alentejo) and 5 small and mid-sized cities (Bragança, Guarda, Castelo Branco, Marinha Grande and Aveiro), as identified in Figure 3. The main objectives of the program were to (a) improve the quality of life in cities; (b) contribute to the development of peripheral areas; (c) improve local economy and employment; and (d) fight info-exclusion and help citizens with special needs (MCT, 1997).

In terms of regional penetration, these projects covered about 11.3% of the total Portuguese population (10.44% of the population under 15 years of age) and about 42% of the total surface of Portugal. All projects involved a broad range of relevant actors and change agents within each of the territories, but nonetheless were always led by local municipalities. Local higher education institutions were particularly involved only in a limited number of projects (Aveiro, Bragança, Trás-os-Montes).

Alentejo, literally "beyond the Tagus River", is an agricultural region famous for its cork plantations. The population density is very low, less than 1/5 of the national average. Its relatively large territory, constituting roughly to one third of the total national area, lies in the south of Portugal. It is limited to the north by the Tagus River, to the South by the Monchique mountain ridge and the Algarve and to the east by the Spanish regions of Extremadura and Andalusia. To the west lies the Atlantic Ocean. The economy relies on forestry, livestock, mining, and agriculture, notably olive oil and wine.

Figure 3
The main projects for the specific development of digital cities and regions, established over the period 1998-2000

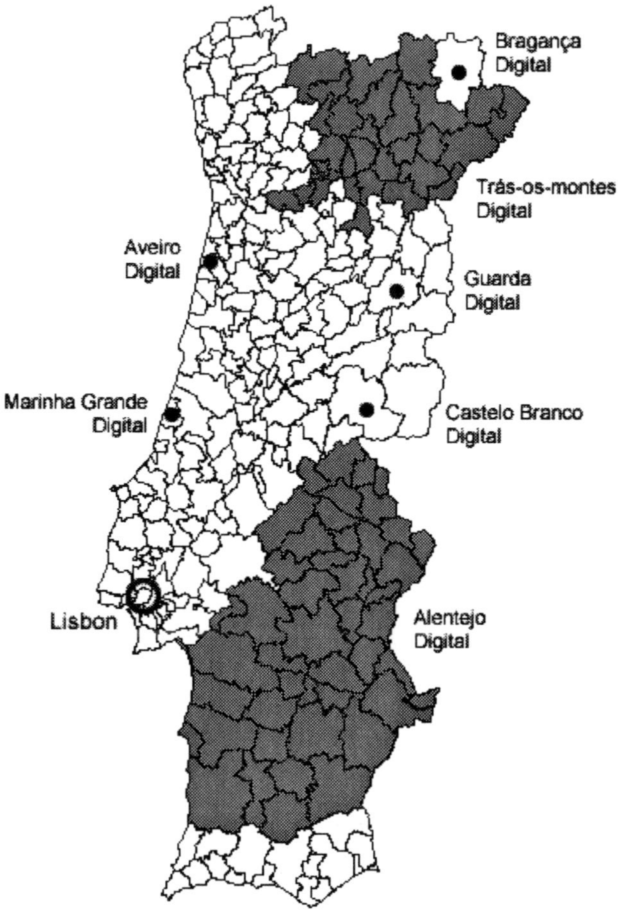

Alentejo Digital brought together 47 municipalities and 3 regional agencies in order to create a regional information network to provide services and territory-related content to citizens and local firms through regional content based web sites. The main objective was to enable local government teams to learn, use and promote new technologies, namely computer network management and digital content production and publishing. An Intranet was set up linking all municipalities and regional agencies, aimed at creating the necessary collaborative work environment. About 50 people were recruited, mostly from local unemployment lists, to work on the project, which lasted until July 2001 (http://www.alentejodigital.pt/).

Table 2
Characterization of digital city projects, in terms of population and area considered in each city/region

Geography	Population	(%) total population	Pop. < 15 aged	(%) total pop. < 15 aged	Area (Km^2)	(%) total area
Aveiro	69.560	0,67%	12.160	0,73%	208	0,23%
Bragança	32.440	0,31%	4.760	0,29%	1.138	1,23%
Castelo Branco	54.260	0,52%	7.440	0,45%	1.440	1,56%
Guarda	38.560	0,37%	6.230	0,38%	709	0,77%
Marinha Grande	33.370	0,32%	5.050	0,30%	186	0,20%
Alentejo	510.690	4,93%	71.930	4,34%	27.227	29,55%
Trás-os-Montes	431.540	4,17%	65.450	3,95%	11.122	12,07%
Total	1.170.420	11,30%	173.020	10,44%	42.030	45,61%

Source: INE, 2001

Trás-os-montes, literally “behind the mountains”, is a remote region in the northeastern corner of Portugal, one of the poorest in Europe, riddled with desertification, an aging population, and socio-economic exclusion. Its population is largely rural and sparsely disseminated. Both agriculture and industry are still underdeveloped in the area. Accessibility is difficult and public transportation is infrequent. There is a noticeable lack of institutions involved in R&D, technical training, education or technological support, as well as regional business networks or associations. Therefore, there are very few opportunities for employment in the technology or information industries (Marques, 2004).

Trás-os-Montes Digital included regionally-based web contents (i.e., www.espigueiro.pt), which aggregated the content and services of 31 municipalities. The portal is managed by the local university and includes business and employment opportunities, geo-referenced information, healthcare facilities and technologies aimed at coordinating medical services in rural areas (Morgado, 2003a, 2003b).

Bragança belongs geographically to the Trás-os-montes region and experiences the same regional constraints. Bragança Digital, led by the local municipality and the local Polytechnical Institute, focused on creating basic ICT infrastructures and a wireless networking environment for local government buildings, health institutions, educational institutions, and the local employment agency, in order to provide information and services to local citizens. Other initiatives included the provision of local products (www.rural.net), and health, educational and e-business activities (Association for the Development of Bragança Digital City, 2001).

Guarda and Castelo Branco are peripheral cities that nonetheless play an important role as regional centralities. The subordinated network of small urban areas, organized along the main accesses to Spain and Europe, are characterized by low-density settlements that rely mainly on agriculture and livestock and some related light industries. Both cities are located in the neighborhood of the Estrela Moutains, the highest peak in continental Portugal.

Guarda Digital was promoted by an organization formed by the municipality, local educational institutes, associations, and the incumbent telecommunication operator. It included pilot projects in healthcare e-business, tele-working and educational initiatives (http://www.ipg.pt/adsi/).

Castelo Branco Digital aimed to connect all public institutions (municipality, social security and health institutions) and local associations (sports, culture and business) to provide an integrated information network to citizens and tourists. For example, it included the provision in rich media of old Portuguese theatre classics (http://www.cm-castelobranco.pt/cb_digital/).

Marinha Grande is a dynamic city, particularly engaged in traditional, labour-intensive industries (e.g. glass and moulding industries) and has been heavily investing in local competitiveness and competence building. Marinha Grande Digital, as managed by the local municipality and the Technological Centre associated with the moulds and plastic injection industries, focused on creating an Extranet to provide business-related (mould, plastics and glass) contents and services and on facilitating communication among companies and clients. Other initiatives included a centre of advanced telecommunications aimed at promoting the use of the Internet (http://www.marinhagrandedigital.com/).

Table 3
Public Funds' Expenditure Per Capita in the Digital Cities Program, 1998-2000

Project	Pop. (a)	Total investment	Investment p.c.
Aveiro	69.560	5,590,000 € (b)	80,39 €
Marinha Grande	33.370	1,200,000 € (c)	35,96 €
Bragança	32.440	1,044,000 € (d)	32,18 €
Castelo Branco	54.260	1,082,000 € (e)	19,94 €
Guarda	38.560	350,000 € (f)	9,08 €
Trás-os-Montes	431.540	1,735,000 € (g)	4,02 €
Alentejo	510.690	1,500,000 € (h)	2,94 €

Sources: (a) INE, 2001; (b) PACD, Final Evaluation Report, 2001; (c); (d) Associação para o Desenvolvimento de Bragança, Final Evaluation Report, February 2001; (e) personal communication http://www.dpp.pt/pdf/info52.pdf; (f) http://www.freipedro.pt/tb/110698/guarda3.htm; (g) personal communication; (h) http://home.telepac.pt/telepac/net/13/regionalismo_2.html.

AVEIRO DIGITAL

Aveiro is a seaport, located at the Vouga estuary, with a population of 69,560. The city's innovative and active character, although of a recent nature, draws from the singular institutional framework established in close collaboration between the local university and the local business environment, mainly driven by the national telecommunication operator.

Following the launch of the first Digital Cities public funding program in Portugal (1998-2000), the municipality, the university and the incumbent operator set up a public-private partnership to develop the idea of Aveiro Digital City focusing on (a) quality of life in the city; (b) democratic participation; (c) extensive access to public and private digital information and services; (d) modernization of local public administration ; (e) inclusive development and sustainable growth; and (f) job creation and lifelong learning (Aveiro Digital City Coordination Team, 2001a,b; Municipality of Aveiro, 1998). The complete funding life cycle was expected to be 8 years, with the first phase of the project starting in February 1998 and lasting until December 2000, totaling an investment of 5,590,000 Euros. The second phase, originally planned to start in January 2001, has only begun on June 2003 and is planned to last until December 2006.

After a troubling start – negotiations over budget allocations and bureaucracy in both the formal approval procedures and the technical implementation schedule caused a lengthy delay of over a year – the first phase included 38 projects covering several different aspects of the use of information and communication technologies. Emphasis was given to infrastructures and digital contents, including local e-government, e-health, and e-business and entertainment, as amplified below (Table 4).

E-business and education related activities accounted for 35.1% of the total number of approved projects and 40.7% of the budget allocated. E-government used up to 20.4% of the available funds. University-based and e-health projects included only two projects and utilized less than 9% of the total budget. On the other hand, projects involving entertainment, culture and the arts accounted for about 30% of the total number of approved projects, although they only received about 8% of the total budget available.

In general, ICT infrastructure – computers, applications, Internet access and basic ICT training – was the most important component of all projects, while investments in activities oriented towards mobilization of the population for the information society were practically nonexistent. Consequently, the evaluation of many activities demonstrated lower levels of public participation than planned, with some of the initiatives falling short of their original objectives (Aveiro Digital City Coordination Team, 2001a,b). E-government and other projects involving basic and secondary schools had more permanent effects, while e-commerce and e-health performed poorly. Budget cuts and uneven financing flows during the implementation phase posed extra difficulties and unnecessarily increased risk.

Table 4
Main digital contents included in the first phase of the Aveiro Digital City Project

Intervention Areas	Activity
Digital Community	Facilitate internet access and active learning support in basic ICT to all citizens: 14 centers in local parishes, and 4 in disadvantaged neighborhoods.
Local Government and Public Services	Aveiro Municipality geographical and administrative information systems
	Water, sewage municipal services: one-stop shop
	Water quality sensing and monitoring system
	Internet and Intranet services for the Aveiro Judicial Court;
Schools and the School Community	Construction of communities involved in education, through training and facilitating communication between families, teachers and students.
	Pedagogical Games and Interactive learning applications for children to play, learn and communicate
University and academic community	Online biology competence centre in the areas of vegetal and animal diversity, molecular biology, local biodiversity and the Aveiro estuary ecosystem
Health Services	Intranet and Internet services for the D. Pedro V District Hospital and the 20 health centers functionally dependent of this Hospital, complemented with ICT training of health professionals
Social Solidarity	Computers, Internet access, ICT training, focused online information and services and design and validation of telework models for people with special needs
Economic Sector	Computers, Internet access and ICT training for 20 SMEs, including basic digital services provisioning
	Sea harbor telecommunication infrastructure, comer-cialization and management information systems. Integrated online services serving various communities
	eCommerce service centre (Internet and public access Kiosks) for local shops
	On-line shopping mall
	Cattle farming sector internet and intranet services for online management, involving 36 farms, with a national expansion
	E-Learning and interactive professional training
Culture and leisure	Interactive Pay-Per-View Pilot Project
	Online edition and publication of news and Internet radio broadcast
	Equipment, training for the production of digital art
	Interactive and pedagogical learning of classical music for all ages
	7 Public access information kiosks (City guides)
	History, culture, art and nature from Aveiro

Source: http://www.aveiro-digital.pt

During 1999-2000, Aveiro Digital City made available 446 personal computers to diverse public and private organizations, published 8 CD ROMs and 32 websites, supplied 73 interactive services, and trained 529 people. The number of Intranet and Extranet users exceeded 3,000 people in different public and private organizations and the Aveiro Digital City Website (http://www.aveiro-digital.pt) registered a monthly average of 4,700 visitors.

The major question raised by the local persons involved in the project has been consistently associated with the structure of public financing and the conditions for long term sustainability, mainly due to the fact that when the limited public funds dried up some of the projects came to a close, while others kept their presence in the Internet but were rarely updated. Moreover, the funding concentrated mostly on the inputs of a long range process, namely infrastructures, information systems and ephemeral content, giving little consideration to mobilizing the population at large. Yet, knowledge networks, especially related to basic education, continued to operate and provided the necessary technology infrastructure for social capital building and information exchange.

Aveiro Digital makes a very interesting case for the need to streamline the processes for the submission, approval and funding of projects. First, the excessive bureaucracy postponed the start of the projects more than a year in most cases. Then the total public funding agreed upon with the Portuguese Government was stretched too thin to fund all the projects approved by the coordination team. On average, the approved incentive was only 50% of the amount requested. The reimbursements were also often delayed. These three factors contributed to cash flow problems in most of the projects, and together with poor management in the early months they created a number of technical and operational difficulties.

After more than a year negotiating the continuation of the project with the Portuguese Government, Aveiro Digital City just entered its second phase. It now includes 11 municipalities. The number of projects has grown to 72, totaling 22,000,000 Euros in investment in community building, local e-government, education, e-health, social inclusion, e-business and digital entertainment (Aveiro Digital City Coordination Team, 2005).

CONCLUSIONS

In previous papers Moutinho and Heitor (2004, 2005) argued that knowledge networks as particular forms of communities of practice, interest or proximity are important drivers of larger communities of users and make both public administration and markets more effective, but that they require infrastructures, incentives and adequate institutional frameworks over time and across space. The analysis was built on the co-evolution of regional development and the endogenous process of technical change, namely in terms of the diffusion of ICT, where the territory is a basic infrastructure that justifies and invites the construction of several layers of information, above all the creation of communication infrastructures and well arranged digital contents within local contexts.

In Portugal, the development of digital cities and regions has become one of the centerpieces of national policies for the development of the information society, playing a critical role for the appropriation of information and communication technologies at the regional and local levels. Therefore, this program must be understood within the context of the increasing focus on national technological competitiveness, global productivity growth, and modernization of public administration. Such a context set the stage for a top-down, technologically deterministic approach that has been constantly challenged by the increasing territorial complexity and heterogeneity of uses, attitudes and skills at the local level.

Taking into account the evidence provided by some of the projects discussed above, namely those at Aveiro, the role of higher education institutions appears to be particularly important in fostering network activities, namely in the form of knowledge-based communities. Following the analysis of Castells and Hall (1994), "it takes a very special kind of university, and a very specific set of linkages to industrial and commercial development, for a university to be able to play the role it often claims to play in the information-based economy". Definitely those technical universities that are pure teaching factories, or work under a bureaucratic structure, are unlikely to act as the generators of an advanced technological milieu. Again, this brings our attention to the role of institutions in planning digital cities and promoting their impact.

Also, against the background of the conditions described above it is clear that market mechanisms do not necessarily work at the level of the issues associated with digital cities, namely the promotion of less favorable zones. At least in the early stages of their development, digital cities require an effective mix of public and private support mechanisms in a long-term perspective, taking into consideration specific regional and thematic aspects. Moreover, digital cities may not be sustainable if promoted independently of a regional innovation system.

As a result, Conceição at al. (2003) have called our attention to the fact that private investment levels (which result from the incentive structure provided by the market to economic agents) in activities with learning or spillover potential tend to be lower than the social optimum, and may even generate what is known in the literature as "low-level equilibrium traps". In principle, these shortcomings in the market mechanism call for some sort of government intervention. Governments are concerned with making sure that societal costs and benefits are endogenized in the decisions of private firms. In a learning environment this may mean subsidizing specific activities, investing in education, or protecting infant industries (Chandler and Hikino, 1996; Lundvall and Johnson, 1994). But government intervention must balance the potential distortions on competition that may result from intervention with the needs to "correct market failures": artificial restraints on competition can also divert profits to activities other than building technological capabilities.

However, incentives and infrastructures do not operate in a vacuum, being shaped by and in turn shaping the particular context in which they operate. In the scope of this analysis, the city or region must have an embedded set of social capabilities that define the context under which digital cities evolve. Consideration of contextual issues in building up network societies have not always been considered in the many different situations which prevail throughout the world, as acknowledged by Castells (2001), among others, and the analysis of the projects earlier considered shows that specific measures to promote adequate contexts have also been scarce.

The evidence demonstrates the need to extend the design of digital cities from a technocratic paradigm of technical change and to look at broader system designs fostering societal developments. In particular, the experience of projects such as Aveiro Digital City shows the importance of mutual relationships that specific project-based communities have on the facilitation of network societies, as well as the fact that the implementation of digital cities may significantly improve the efficiency of those communities.

The complex technology-enabled infrastructures typical of digital cities call for the integration of social and cultural factors in the early design phases in order to mitigate uncertainties, such as sustainability, flexibility and scalability. Moreover, digital cities may develop other unexpected properties, or emergent properties "developed by users of a system" and "often unbeknown to the system designer" (Moses, 2004). This being so, the stakeholders involved in the co-evolution of urban areas and ICT would be better off if, as proposed by Cooley (2000), "the current mechanistic paradigm of technological and societal development [would be substituted by] human-centered systems [that would] provide a powerful alternative philosophy for system design and a broader educational and societal development". He adds that "[this philosophy] regards the social and cultural shaping of technology as central to the design and development of future technological systems and society as a whole", in terms of "knowledge-based adaptive human-centered environments" (Cooley, 2000).

In expanding this conceptual framework to the entire city or even whole regions in order to consider the way millions of people interact with information and communication technologies in their daily life, it is clear that the initial approach to designing digital cities described above needs to be reconsidered. Table 5 illustrates the main implications and requirements of emerging trends, so that the vast majority of potential late adopters are inclusively considered in future digital city projects. As Dertouzos (2001) argues, we can avoid "drowning in information overload and computer complexity only by throwing out last century's model for computing and adopting – indeed, demanding – a new computing philosophy, a new master plan, that lets people interact naturally, easily, and purposefully with each other and the surrounding physical world". And he adds that "to put it in action requires three big steps: changing the mind-set of users and designers; ensuring that our machines are easier to use and make us more productive; and insisting that new technology reach many more people".

Table 5
Emerging trends for a new generation of "Digital Cities"

Layer of Analysis	From	To	Implications and requirements
Institutions and Incentives	Centrally designed programs	Open public / private partnerships	Embedding ICT infrastructures in urban daily life, fostering human-centered systems
	Fixed access	Roaming	Competitive mobile services and improved regulatory framework for increased individual participation
Infrastructure and technologies	Conspicuous objects	Invisible infrastructure	Embedding ICT infrastructures in urban daily life, fostering human-centered systems
	Fixed access	Roaming	Competitive mobile services and improved regulatory framework for increased individual participation
Content and Tools	One-way distribution of information	On-line collaboration and participation	Specific knowledge of institutional and local contexts in order to help develop interactive contents
	Web functionalities	Networked Activities	New competences in content and services development, enhancing user activities and networks
Human and Social Context	Technology supply	Mobilization of users	Mobilizing "change agents" to foster communities of practice, CoP's, and user involvement
	Standards	Interoperability	Building individual and social competences through knowledge-based adaptive human centered environments

This analysis led to the suggestion that while the role of public policy needs to be re-examined (see the recommendations in Table 6), the cultural and social shaping of information technologies requires the specific development of human-centered systems designed to support community building activities ("edge to core" strategies) for the next generation of digital cities.

The reflections listed above were based on the need to take into account the uncertainty in the mobilization of ICT, which requires individuals, firms and organizations to operate in dynamic environments, where markets and technology change rapidly and in unpredictable ways. This calls for the need to combine flexible infrastructures and adequate incentives with institutions which will foster the necessary context for digital cities to succeed. The new paradigm of semantic grids can help ICT complexity to be alleviated and become an invisible infrastructure embedded in urban daily life.

Table 6
The 10 main recommendations for the development of a new generation of digital cities

1.	New financial mechanisms and simpler administrative procedures must be created at the central level to decrease transaction costs and cope with accelerated technological change and the emergent properties of complex systems;
2.	Digital city programmes must be decentralized and focused on fostering public-private partnerships at local or regional level;
3.	The provision of competitive telecommunication services and an improved regulatory framework is a critical success factor for the deployment of public networks;
4.	ICT infrastructures must be embedded in urban daily life, fostering human-centered systems that respond to specific needs and desires;
5.	Project managers must develop an ongoing dialogue with local civic society and accommodate theirs needs and desires;
6.	Projects managers should also evaluate and eventually accommodate or mitigate emergent properties of digital cities;
7.	Inclusion of specific knowledge of local contexts is required in order to facilitate the development of relevant and useful contents and the strengthening of social networks;
8.	Identification and mobilization of local change agents is necessary to foster communities of practice, interest or proximity;
9.	The evolution from one-way distribution of information or two-way transaction (e-government) to real time collaboration and participation in local governance (e-democracy);
10.	Promotion of diversity and inclusiveness.

BIBLIOGRAPHY

Association for the Development of Bragança Digital City, (2001), "Relatório Final Bragança - Cidade Digital". Available from http://www.braganca-digital.pt/

Aveiro Digital City Coordination Team, (2001a), "Relatório Global da 1ª Fase do Programa Aveiro Cidade Digital", http://www.aveiro-digital.pt/arquivo/docs/Rel_GlobalACD1998_2000.pdf

Aveiro Digital City Coordination Team, (2001b), "Relatório de Avaliação do PACD 1998-2000", internal report.

Aveiro Digital City Coordination Team, (2005), "Relatório de Actividades 2003-2006", http://Aveiro digital.pt/ arquivo/docs/RelActPAD2004.pdf

Barry, A., (2001), *Political Machines: Governing a Technological Society,* Atholone Press, London

Besselaar, P. and D. Beckers, (1998), "Demographics and Sociographics of the Digital City", [in:] T. Ishida, (ed.), *Community Computing and Support Systems: Social Interaction in Networked Communities*, Berlin.

Besselaar, P., (2001), "E-community versus E-commerce: The Rise and Decline of the Amsterdam Digital City", *AI & Society*, No. 15 (5), pp. 280-288.

Castells, M., (1989), *The Informational City: Information Technology, Economic Restructuring and the Urban-Regional Process*, Blackwell, Oxford.

Castells, M. and P. Hall, (1994), *Technopoles of the World: the Making of the Twenty-first Century Industrial Complexes*, London, Routledge.

Castells, M., (2001), *The Internet Galaxy: Reflections on the Internet, Business and Society*, Oxford University Press, Oxford.

Chandler, A. and T. Hikino, (1996), "The Large Industrial Enterprise and the Dynamics of Modern Economic Growth", [in:] A. Chandler, F. Amatori, and T. Hikino, (eds.), *Big Business and the Wealth of the Nations*, New York: Cambridge University Press.

Conceição, P., M. Heitor, and F. Veloso, (2003), "Infrastructures, Incentives and Institutions: Fostering Distributed Knowledge Bases for the Learning Society", *Technological Forecasting and Social Change*, 70 (7), pp.583-613.

Cooley, M., (2000), "Human-centered Design", [in:] R. Jacobson, (ed.), *Information Design*, Cambridge, MA: MIT Press, pp. 59-81.

Dearnley, J. and J. Feather, J., (2001), *The Wired World: an Introduction to the Theory and Practice of the Information Society*, Library Association Publishing, London.

Dertouzos, M., (2001), *The Unfinished Revolution: Human-Centered Computers and What They Can Do for Us*, Nova York, Harper Collins Publishers.

Dutta, S. and A. Jain, (2003), *The Networked Readiness of Nations, The Global Information Technology Report 2002-2003*, World Bank's infoDev Programme.

European Commission, (1993), *White Paper on Growth, Competitiveness, and Employment: The challenges and ways forward into the 21st century*, COM(93) 700 final, http://europa.eu.int/en/record/white/c93700/contents.html

European Commission, (1994), "Recommendations to the European Council: Europe and the global information society (Bangemann Report)", Members of the High-Level Group on the Information Society, Brussels, http://europa.eu.int/ISPO/infosoc/backg/bangeman.html

Finnemann, N., (1999), "Thought, Sign and Machine: the Idea of the Computer Reconsidered", Transl. by Puckering, L., http://www.hum.au.dk/ckulturf/pages/publications/nof/tsm/tsm.pdf

Heitor, M. and Moutinho, J. (2004), "Digital Cities and the network society: Towards a Knowledge-Based View of the Territory?", [in:] D. Gibson, M. Heitor, and A. Ibarra, (eds.), *Knowledge and Learning in the Network Society*, New York, Purdue University Press.

INE (Instituto National de Estatistica), (2001), *Census Results,* Lisbon.

Information Society Mission taskforce, (1997), "Livro Verde para a Sociedade da Informação em Portugal", http://www.acesso.umic.pcm.gov.pt/docs/lverde.htm (in portuguese)

Kostof, S., (1991), *The City Shaped: Urban Patterns and Meanings Through History*", Thames and Hudson, London.

Lundvall, B., and B. Johnson, (1994), "The Learning Economy", *Journal of Industry Studies*, 1/2, pp. 23-42.

Marques, T., (2004), "Portugal na Transição do Século: retratos e dinâmicas territoriais", Afrontamento Editions, Oporto.

MCT, (1997), "Iniciativa Internet", Portuguese Ministry of Science and Technology (in portuguese)

Morgado, L., et al., (2003a), "Trás-os-montes Digital and CES: a Case of Social Benchmarking", UTAD, http://home.utad.pt/~leonelm/papers/iadisesociety2003/e-society2003.html

Morgado, L., et al., (2003b), "Rede de Exploração do Projecto Trás-os-montes Digital", UTAD, http://home.utad.pt/~leonelm/papers/ModeloTMD.PDF

Moses, J., (2004), "Foundational Issues in Engineering Systems: a Framing Paper", MIT ESD, Engineering Systems Monograph, http://esd.mit.edu/symposium/pdfs/monograph/framing.pdf

Moutinho, J., and M. Heitor, (2004), "Digital Cities and the Opportunities for Mobilizing the Information Society: Case Studies from Portugal", [in:] P. Besselaar, P. and S. Koizumi, (eds.), *Digital Cities III: Information Technologies for Social Capital: Cross-cultural Perspectives*, Berlin, Springer Verlag.

Moutinho, J. and M. Heitor, (2005), "Building Human Centered Systems Based on Communication Infrastructures: Evidence from Portugal", [in:] S. Marshall, W. Taylor, and X. Yu, (eds.), *The Encyclopedia of Developing Regional Communities with ICT*, Hershey, Idea Group.

Municipality of Aveiro, University of Aveiro and Portugal Telecom, S.A., (1998), "Programa Aveiro — Cidade Digital", Available from http://www.aveiro-digital.pt/

Norman, D., (1998), *The Invisible Computer: Why Good Products Can Fail, the Personal Computer Is So Complex, and Information Appliances Are the Solution*, Cambridge, MIT Press.

OECD, (2000), *OECD Information Technology Outlook*, OECD, Paris.

O'Sullivan, A., (2003), *Urban Economics*, McGraw-Hill, New York.

Schuler, D., (2002), "Digital Cities and Digital Citizens", [in:] M. Tanabe, P. Besselaar and T. Ishida, *Digital Cities: Computational and Sociological Approaches*, Springer-Verlag, Berlin.

Tsipouris, L., (2002), "Final Report for the Thematic Evaluation of the Information Society", retrieved on January 20, 2004, from http://europa.eu.int/comm/regional_policy/sources/docgener/evaluation/doc/information_society.pdf

Valentine, G., (2001), *Social Geographies: space and society*, Prentice Hall, London.

Veiga, P., (2000), "Digital Cities: The Portuguese Experience", PowerPoint presentation, Conference on The Information Society and Economic, Social and Territorial Cohesion 2000-2006: a new opportunity, meeting organized by the European Commission, Lyon.

Zook, M., (2001), The Geography of Internet Industry: Venture Capital, Internet Start-ups, and Regional Development, PhD paper, University of California, Berkeley.

13

Reflections on ICT in World City Development and Development of Remote Regions

Marina van Geenhuizen

ICT AND CHANGING ACTIVITY AND LOCATION PATTERNS

The use of information and communication technology (ICT) has attracted attention from many researchers and policymakers in urban development and urban planning. ICT may influence the performance and growth of the city in many ways. First, by connecting the city with important external networks and by employing adequate internal networks (rings) the potentials of cities may significantly increase for large data transport and use by high-technology companies and research institutes. Secondly, ICT and related technologies in informatics, multimedia etc. constitute the basis for new firm development and enables the city economy to renew itself. Thirdly, the performing of particular tasks by city departments may increase in quality and urban regeneration may be supported using ICT (e.g., Kotval, 1999), like in improving traffic flow and prevention of congestion, watching services to prevent crime and search for criminals, particular tasks in healthcare, supply of education for particular groups, and support in decision-making in spatial planning through GIS-based tools. We may also mention the supply of e-governance services to the public, and in particular areas, support in certain modes of participatory urban planning (Cohen et al., 2005; Conroy and Evans-Cowley, 2006).

If we focus on companies, ICT enables companies to add electronic space to physical space, and to develop the best mix of physical and electronic activity, the latter including the use of the Internet (for example, Li et al., 2001). Through their "distance shrinking" nature, ICT enables interaction over much larger distances than before allowing companies to become more footloose and locate more freely. In addition, based on the latest optical network technology, opportunities are created to facilitate the flow of huge amounts of data. In connecting computer technology, modern communication and science, e-science is created, allowing research institutes and companies to participate in basic and applied research in an interactive manner. Impacts of ICT on production, including research and development, are based on various unique and far-reaching characteristics of the Internet (Kenney and Curry, 2001). An important feature is interactivity, such as in information (feedback) loops in supply chains, leading to an increased efficiency but also to re-configuration of physical segments and the potential removal of some of them (for example, wholesale and retail), also named desintermediation.

Another and probably most powerful feature of the Internet is information intelligence, i.e. the ability to collect information across the network, to store and process the information, and utilise and disseminate the results through the network. This also includes the more sophisticated activities of monitoring as a decision support tool and as an advances business service, grid cluster computing to efficiently connect computer power and data inventory capacity across global networks, and data-analysis like data-mining, and interactive experimentation, development and design (van Geenhuizen, 2006). An example of the latter kind of interaction is the participation of thousands of scientists in research at the accelerator facility of CERN (Geneva, in Switzerland) using a highly sophisticated optic network (Gigaport, 2004). The above developments point to a spread of urban activity in much larger areas than the large cities through virtual connectivity, and to opportunities for peripheral regions to "escape" the digital divide.

However, scholars in regional economy and economic geography increasingly recognise that digital infrastructures disproportionately agglomerate in a few world cities and that high-order activities increasingly concentrate there, with low-order activities being exported to more peripheral regions (e.g. Malecki and Gorman, 2001; Richardson and Gillespie, 1996; Pollard and Storper, 1996). According to evolutionary theory, this seems an inevitable development based on selection and self-reinforcing in cumulative processes. In addition, it is said that the use of ICT is still modest due to various technical limitations in practice (e.g. Bolisani and Scarso, 2000) and due to various behavioural and institutional barriers. For example, substitution of physical meetings by telecommunication is only possible if the communication is standardised, and if there is sufficient trust between the parties at hand. In the case of non-routine communication, such as in learning about context, creation of trust and comfort, negotiation including risk-taking, evaluation of results - all containing high levels of tacit knowledge - telecommunication is facing basic shortcomings (e.g. Howells, 2002). From this point of view, high-level economic activity would remain located in the large cities and the latter would remain growing. In particular evolutionary approaches, it is however also recognised that the phenomenon of chance, and policies of 'grasping' opportunities of unique circumstances and of creativity may allow remote regions to catch-up and narrow the divide by attracting high-order economic activities (e.g. Arthur, 1994; van Geenhuizen, 2006; Grabher, 1993). Thus, in most cases, once started development paths are reinforced by path-dependency and lock-in phenomena, but there are ways to escape from such paths.

In the light of the above ideas, the aim of this chapter is to explore the spatial layout of the digital infrastructure in the Netherlands and the location behaviour of young, innovative companies in urban areas. The following questions will be addressed: To what extent is the most advanced digital infrastructure available in metropolitan areas and remote regions? Are companies that traditionally enjoy agglomeration economies becoming more footloose under the influence of ICT use? What are the potential impacts of changing location behaviour for the large cities and what are the new opportunities for remote regions? We will answer these questions using a mix of study of the literature and empirical work.

AN ORGANIZATIONAL AND INSTITUTIONAL VIEW

Distance as an attribute of space has received a renewed attention in the geography of production since the late 1980s, including an increased recognition of the different dimensions of this concept. We may distinguish between geographical (physical) distance, organizational (social) distance and institutional distance. Geographical (physical) distance means separation in space and in time-space; it is also associated with missing physical connections in transport and telecommunication infrastructures. Note that physical distance may be appreciated differently by actors in countries of a different size. Thus, the same physical distance may be seen as a barrier for interaction in small countries but not as a barrier in a much larger country.

Organizational or social distance refers to separation in interaction and co-ordination. It is based on the logic of similarity and means that actors do not belong to the same "space" of relations (firms, networks) and do not share common knowledge, capacities and aims (Torre and Gilly, 2000). In other words, an emphasis is put on accessibility of socio-economic networks and lack of (perceived) opportunities for interaction (Andersson and Karlsson, 2002). Organizational distance may include geographical distance but this is not necessary. The third type of distance distinguished here, institutional distance, is also based on the logic of similarity. There is institutional distance if the actors do not belong to a common "space" of representation, sets of norms and rules, thoughts and action, or if certain laws prevent interaction. The conclusion may be drawn that even if physical distance is overcome by telecommunication, interaction may still be hampered by organizational and institutional distance. The latter types of distance are quite basic and cannot be changed overnight, but it seems that boundary spanning agents (individuals, organizations) are able to act as relational bridges (e.g. Gertler, 2003).

In the course of the 1990s, optimistic views of the distance-shrinking role of ICT in economic development of remote regions became increasingly questioned, based on a growing recognition of the limiting influence of various institutional factors. The most important limiting factor forwarded was the quality of the *seedbed conditions* in the recipient economy and society in remote areas. There was a growing awareness that ICT cannot simply be transplanted into a remote economy with information directly affecting this economy like a bullet hitting the target. There is no simple linear cause-effect model of the introduction of ICT in the regional economy. Rather, the outcome is the result of a complex interplay between the technologies and the socio-economic field of forces in the recipient economy and policy unit (Graham and Marvin, 1997). The impact of ICT on regional economic growth, for example, varies according to the regional production structure, e.g. the size of local companies and their embeddedness in larger supplier/customer networks. Regions with many large companies linked with metropolitan areas and the global economy seem to benefit more from ICT than regions with a dominance of independent, small and medium-sized companies (SMEs) (Capello, 1997; Cornford and Gillespie, 1993). Thus, a certain amount of networking between companies must exist before it can be mediated through ICT. Networks cannot be created by providing an ICT infrastructure if no networks existed before (Melody, 1991). The point made here is that the competitive position of regions cannot be changed merely through the introduction of ICT in a supply side approach. Much depends on the organizational quality and learning attitude in the recipient economy (e.g. Grimes, 2003). These institutional factors are also known as "soft infrastructure" (Estache and Goicoechea, 2005; World Bank 2006), but the latter may also include elements of social and cultural life. Good quality dining, entertainment, recreation and art facilities may attract and help retain high quality knowledge workers. Tolerance and diversity issues have also been identified as relevant (Florida, 2005). Today, soft infrastructure is seen as just as important as hard infrastructure in the economic development of a region or country.

A MANIFOLD UNCERTAINTY IN ICT POLICYMAKING

In many Western European countries in the 1990s, local policymaking concerning ICT-based development was often rather passive or poorly coordinated. This situation can be attributed to various limiting circumstances. A first constraint was the way in which policy initiatives came into being, that is, without being embedded in a solid structure. In the absence of a clear vision, initiatives often emerged as a result of brainstorm sessions without a structure for continuation of action. Follow-up elaboration was then often based upon optional action, with high risks of lack of commitment. In the Northern part of the Netherlands, six of the ten largest municipalities had developed a clear vision or plans on ICT in 2000, but most of the smaller municipalities had no visions (plans) at all. A second reason for lack of success was that ICT initiatives were derived from other existing policy fields, like transport and economic affairs, leading to a fragmented approach and sometimes overlap.

More importantly, urban policymaking for ICT has been and still is subject to manifold uncertainty (Haynes, 2006) causing a passive attitude and poor coordination. This is stronger than in any other infrastructure policymaking. There is a lack of awareness and knowledge in policymaking about the potentials of ICT use, such as the emergence of new corporate strategies (value chains) and the supply of new services by private and public actors, and the concomitant spatial impacts. Also, the technology is a source of uncertainty itself. A wide range of technologies is already in existence or in the process of research and development, but knowledge about which specific techniques and strategies will be the most effective and beneficial for the tasks they are designed to perform, is only increasing by bits and pieces. Of course, extensive testing, demonstrations and evaluations are continuing to determine the best options. This links into another source of uncertainty, namely market uncertainty. For example, it is not known when wireless facilities offering the highest capacity against affordable prices will be available in the market. And when new technologies are available, public acceptance may be difficult to assess, for example end-users' willingness-to-pay and the private sector's or public/private combination's willingness to provide or to invest. The latter point refers to a further source of uncertainty, namely institutional uncertainty. The design and introduction of new institutions that need to go along with application of the new technologies, like protection of privacy, safety of data transport, authorization, etc., but also new ways of financing and operation of the infrastructures and services (public-private partnerships), often lag behind the introduction of the new technologies. There is also uncertainty about how policy incentives may impact on adoption of the new technologies and new institutional arrangements.

What was often missing in policymaking in the previous years can be summarized as follows (Cohen et al., 2002, 2005; van Geenhuizen, 2004; Gibbs and Tanner, 1997; Sleurink et al., 2001): (1) an explicit policy attention for the development of up-to-date ICT networks as a field on its own, instead of a derived issue from other policies, (2) a sufficient horizontal co-ordination between policy areas affected by ICT, like education, transport, and economics, (3) a coherent policy vision that underpins initiatives and guarantees a systematic follow-up, and (4) a learning approach to deal with the manifold uncertainty, like the development of scenario's, future horizon scanning, and monitoring of actual developments.

We may conclude that the above poor policy circumstances may have hampered a structural improvement of use of ICT both of by city as a whole and its individual actors, like companies and households, thereby reducing the development of new economic activity in the city based on ICT. This situation has, however, not precluded that particular private or public initiatives have closed mental gaps or span boundaries, through which conditions improved, in some cases even drastically. A few well-coordinated projects with positive impacts on different sectors may be the result. This can be illustrated with a project in a relatively remote region in the Netherlands (the medium-sized town Emmen) at the end of the 1990s.

The establishment of ICT learning trajectories at a higher educational institute here has caused various positive impacts on the labor market and on the supply of innovative products (business services) in the area (van Geenhuizen, 2004). As indicated in the previous section, an "escape" from an almost predetermined development path is possible under particular circumstances.

QUALITY OF TELECOMMUNICATION INFRASTRUCTURE

In this section, we will discuss various quality aspects of telecommunication infrastructure. These quality aspects include the spatial layout of the networks and their capacity, safety (failure) of connection, compatibility, supply of services, and safety and protection of information (van Geenhuizen, 2006).

A major quality aspect is the spatial lay-out of the networks and their capacity (bandwidth). Only selected networks can accommodate high-speed Internet that matters today in information transfer and knowledge interaction. Advanced broadband connections and high-speed Internet (a minimum transmission speed of 10Mbit/s) allow for highly complex activities like data-mining and multimedia applications (CBS, 2003). A new development is to facilitate extremely large data flows between fixed points without bothering the Internet traffic by using a hybrid network as a combination of a state-of-the-art Internet network and revolutionary optical technology using light paths. Such a hybrid network in the Netherlands is the currently introduced SURFnet6, including NetherLight as the optical Internet exchange in Amsterdam. This network allows the supply of specific services and applications like in astronomy and high-energy physics research and provides capacity between 1 and 10 Gbit/s (Gigaport, 2005).

Empirical studies in various countries indicate that mostly the largest cities are connected with advanced glass fiber grids allowing for high-speed communication of huge amounts of information (Graham, 1999; Malecki and Gorman, 2001). Extending these networks to remote places in times of liberalized telecommunications markets and a global downturn in the telecommunications sector is problematic. There are signs, however, that a few new "network cities" also benefit from ICT networks (e.g., Malecki, 2002). These may be smaller cities engaged in global knowledge exchange, like at technical universities and R&D laboratories of large companies. Wireless technology affordable by companies could be a solution for non-connected places but seems to stay behind in capacity compared with the gigabits of advanced optical networks. In some cases, however, wireless technology might help to bridge the "last mile" from backbone to customer (or "first mile", seen from the customer).

A shortage of bandwidth causes delay in transmission but also loss of information. In synchronous modes of communication delay is more problematic than in other modes. In video-conferencing and virtual interactive design it is particular annoying, whereas e-mail through its asynchronous nature allows for a time-lag between receipt and response (Mitchell, 1999). Delay is related with our third quality issue, i.e.

connection and its failure (disruption). Failure of connection refers to frequency and duration of disruption, and waiting time to restore connections. Failure is not only a technical matter (a highly intensive use of the network in particular areas) it is also caused on purpose, such as by inserting viruses that spoil software, connections and databases.

Compatibility is the fourth quality aspect to be discussed here. Compatibility refers to hardware and software, and to entire information systems, and it is still problematic to date in advanced applications. Compatibility problems prevent interaction and cause information loss. It is not only a technical issue, but also a matter of willingness of the actors involved. Sometimes, non-compatibility is used on purpose to prevent disclosure of sensitive information to e.g. competitors. Our fifth quality aspect, services, seems to be the most comprehensive aspect facing many different parties and user functions. It embraces hardware and software suppliers, and all kind of service providers, and refers to user interfaces and supporting services (portals, searching machines), storage facilities of messages and data, connections with additional links, and maintenance and upgrading to newest versions such as of software. The last quality aspect to be discussed here, safety and protection, is considered as a separate class of services. Safety and protection are important in business relations for authorisation of electronic transactions – to be sure that these are acknowledged and safe – and in situations of sensitive information, for example, concerning privacy and bank accounts, and news concerning latest innovations.

A CASE STUDY OF THE NETHERLANDS: SURFNET

In this section, we will analyse the spatial pattern of the network for research and education purposes, SURFnet, as it has enrolled in the past few years as a spin-off of a public-private initiative named Gigaport (Gigaport, various years). We will focus on SURFnet4 and SURFnet5. As previously indicated, SURFnet6 is different from the earlier versions because of its hybrid character (additional light paths). The capacity that can be provided ranges from 1 Gbit/s to 10 Gbit/s (SURFnet, 2006). It is generally believed that SURFnet6 is a strong asset in realizing the knowledge economy, not only by serving universities, research laboratories, large libraries, etc. in the Netherlands with high speed Internet, but also by connecting these institutes with similar networks abroad. The current international connectivity using light path exchange (NetherLight) includes 2x 10 Gbit/s wavelength to the Starlight facility in Chicago (USA), the European nuclear research facility in CERN in Geneva (Switzerland), and the MAN LAN facility in New York (USA), aside from other connections at lower capacity wavelength (NetherLight, 2006). In addition, the services developed by Gigaport and related actors in using the potentials of high-speed Internet, are highly innovative and call for a broader use. We mention the following examples: facilities for large-scale analyses and simulations (bioinformatics); toolkits and infrastructure facilitating grid computing covering global computing resources and data storage facilities; and, new services for audio visual infrastructure for metropolitan networks. Note that in terms of performance, light path connections reach higher quality levels compared with traditional Internet (NetherLight, 2006). There is not only higher capacity available (10 Gbit/s), there is also no disturbance of transmission. Besides, protection against intrusion is better due to the point-to-point structure and independence of the traditional Internet.

In our analysis of network connectivity of cities with SURFnet, we use Cluster Leaders and Points of Presence (PoPs) as units of analysis (Table 1). Cluster Leaders are major nodes in the backbone. There are four of them, i.e. SARA in Amsterdam and the three technical universities (Delft, Eindhoven and Enschede/Twente), and each is connected with a number of Points of Presence. Points of Presents denote nodes (mostly on a lower level) where network operators can hand information onto the network of another operator (Grubesic and O'Kelly, 2002). To give an idea of the established capacity of SURFnet4 at the end of 2003: two-thirds of the connected institutes (117 over the country) had a connection with SURFnet of 1 Gbit/s.We may observe the following developments in the metropolitan area in the Western part of the country (Randstad) (Table 1):

- In SURFnet4 and SURFnet5 Amsterdam is the only large city that is Cluster Leader from the start, offering access to the highest speed Internet. The three remaining large cities, The Hague, Rotterdam and Utrecht are Points of Presence from the beginning. Delft as a smaller, academic, city is also Cluster Leader from the start.
- Two Cluster Leaders are outside the metropolitan area in the adjacent areas, i.e. Eindhoven and Enschede.
- In the metropolitan area the number of Points of Presence has gradually increased in medium-sized towns (like in Haarlem and Dordrecht); this is also true for medium-sized towns in the adjacent regions (like in Den Bosch, Arnhem and Zwolle).
- For the Northern provinces we may observe:
- In SURFnet4, there was one Point of Presence, namely in the largest city (Groningen), but no Cluster Leader, meaning that connections at highest speed level were missing.
- Following the shift to SURFnet5, there has been a small improvement in terms of refinement of the grid, with one Remote Point of Presence (Leeuwarden).
- In 2003 the grid has been strongly refined in the province Drenthe by six Remote Points of Presence.

Broadly stated, the construction of the net has followed a city-size principle and the principle of presence of a (technical) university. In the Northern provinces, the picture of SURFnet4 and SURFnet5 initially partly complied with commercial grids, including a basic data transport infrastructure and local access only for the largest town (Groningen) but without access for some smaller medium-sized towns, even those with higher educational institutes and research facilities. Thus, for the remote areas in the North it can be concluded that initially, distance played a substantial role in terms of lack of connections.

However, since 2003 there is an "anomaly" in this pattern due to strong supportive action by the province of Drenthe and the pushing power of an organisation to make the grid better accessible through additional access points, aside from the radio telescope space observatory in Westerbork (ASTRON, 2003). We may notify that an extension of the number of PoP's was a few years later also realized in the province of Zeeland, outside the economic core.

Table 1
SURFnet4 and SURFnet5 networks in the Netherlands

Stage	Cluster Leaders (GigaPoP's)	Points of Presence (PoP's)	Conclusions concerning the Northern provinces
SURFnet4	4 Amsterdam, Delft, Eindhoven, Enschede	12 Groningen, Marknesse, Wageningen, Nijmegen, Maastricht, Tilburg, Breda, Rotterdam, Den Haag, Leiden, Utrecht, Hilversum	One PoP (Groningen) Absence of fastest connections (no Cluster Leader)
Changes in SURFnet5	No change	10 additional PoP's (Remote Points of Presence): Alkmaar, Arnhem, Den Bosch, Deventer, Zwolle, Leeuwarden, Haarlem, Heerlen, Dordrecht, Nieuwegein	One additional Remote PoP (Leeuwarden)
Later extension in SURFnet5 (2003)	No change	6 additional PoPs (Remote Points of Presence): Emmen, Meppel, Assen, Hoogeveen, Dwingeloo and Westerbork	Six additional Remote PoP's in the province of Drenthe
Subsequent extensions (2004/5)	No change	12 additional connections in medium-sized and small towns, mainly in the province of Zeeland	One additional Remote PoP in the province of Groningen

Source: adapted from van Geenhuizen, 2006.

The improved accessibility of SURFnet5 in the province of Drenthe illustrates that regional policy initiatives (subsidies) and individual or collective organisations' power may sometimes change situations in remote areas drastically by using opportunities of facilities that are by nature merely located in empty areas, like radio telescope stations for astronomy research, nuclear research facilities, earth observatories, air-traffic observatories. Accordingly, relatively empty regions may "escape" the digital divide. We may conclude that the previous analysis indicates a general lagging behind for remote regions but also the potential of a sudden jump ahead, causing a differentiated pattern of potentials for integration through high-order economic activities. In addition, the analysis illustrates that any study in the field of telecommunications infrastructure tends to be just a snapshot of one moment in time, thereby facing the danger of overlooking ongoing developments.

As previously indicated, if seedbed conditions are well developed, ICT may attract specific high-technology activity to important nodes because of needs for a large and quick data transport or additional services. However, ICT may also cause other high-technology activity to become more footloose provided that the activity is sufficiently connected to the network. The latter development will be the subject of the next section.

AN INCREASE OF FOOTLOOSENESS?

We now turn to young innovative companies and their location behavior. A concept that is in the heart of the discourse on ICT advantages for companies in remote regions, is footlooseness. Despite a frequent use, it is rather poorly conceptualized so far with regard to companies (van Geenhuizen, 2005). An early conceptualization can be found in the work of Klaassen (1967) as a situation in which the long run profitability of an industry is the same for any location in an economy. Although we agree with the core idea, we prefer to include the notion of different degrees of footlooseness and add other spatial entities than national economies, like larger regions within countries. Note that there is an increasing number of sector-based studies that address the question of footlooseness (or conversely, location-boundness) in terms of changing location patterns (e.g. Gorman, 2002; van Oort and Stam, 2005; Sohn, 2004; Wernerheim and Sharpe, 2003) and these studies either point to an increased agglomeration or modest dispersion; the latter however in an uneven pattern on a local scale, connected with a higher level of IT infrastructure. By contrast, studies on the micro-level of companies and location behavior are scarce. In this section we turn to a study using such a micro-approach, to achieve a greater understanding of location choice.

In an exploratory study in the Netherlands, the valuation of various agglomeration advantages by entrepreneurs of young, innovative companies was used to depict the degree of footlooseness (or location-boundness) (van Geenhuizen, 2005). The following trends can be found in this study:

- The pattern of footlooseness is rather differentiated, even within the economic (sub)sectors studied, that is, biotechnology, ICT producer-services and mechatronics.
- Network characteristics of the company seem the most determining factor of the degree of footlooseness, i.e. related with corporate status (origin), the business model (outsourcing) and maturing of innovative activity. For example, companies that are corporate spin-offs, those that subcontract most activities to other companies (research institutes) and highly innovative companies employing global R&D alliances tend to be somewhat footloose. By contrast,

companies employing a local focus in customer and supplier networks and very young academic spin-offs tend to be location-bound.

- Over time, location-boundness tends to be reinforced, whereas footlooseness tends to be reinforced for some company types but reduced for others.
- Somewhat footloose means that the companies involved prefer to enjoy a limited number of benefits from their location which could be in a larger part of the Netherlands, i.e. a certain (lower) level of agglomeration and centrality, proximity to knowledge and a good knowledge culture, accessibility by car, and easy access to a well-connected airport.

The latter point suggests that the metropolitan area in which young, innovative companies may perform profitably is larger than in previous times. It seems that specific agglomeration advantages that were previously confined to the large city-region have spread over a larger area today, probably including about one third of the country. In terms of size, we guess a shift from the large city region of approximately 20 x 20 km to an area of approximately 100 x 100 km at maximum. We may also notify that particular agglomeration advantages can only be enjoyed in the largest city (Amsterdam) and that these remain exclusive for this city, i.e. a pool of internationally oriented (multilingual) knowledge workers; direct access to the highest capacity telecommunication node and grids and connected services; and proximity to particular customers and suppliers to participate in joint innovation projects based upon learning on a daily basis.

"Somewhat footloosc" in the Netherlands tends to refer to three spatial areas, that is, the Southeast of the country, the Southeast and the highly urbanized West, and a large centrally located region. However, it does *not* refer to the entire country and not to relatively remote regions like the North, except for the largest town here (Groningen). Results on footlooseness so far do not provide support for the idea of increased opportunities of remote regions to attract high-order economic activity from elsewhere, except for some specific company types and maybe for the largest town(s) here. This would mean an important role for endogenous initiatives.

We now turn briefly to a study of a sector that is often associated with footlooseness and with large potentials for remote regions, that is call centers (van Geenhuizen, 2006). Call centers are organizations that answer and distribute telephone calls, using advanced telecommunications hardware and software and computer assistance. Their main goal is to support the own or customers' organizations in terms of public relations, customer services, account handling, technical support, telemarketing, etc. Not all of these tasks are important for all call centers. Aside from very simple centers, providing basic advice by telephone, there exist more advanced centers that are fully functionally integrated in the business processes and serve learning activity of the company involved, using feed-back from marketing and customer management results. Note that in terms of sophistication, there is an upward move from answering simple questions to a full integration in the business processes.

Call centers are either back-office units within existing organizations or newly established entities working for a variety of related organizations. In both cases it is possible that the call center is a separate physical unit (an office) or largely a virtual organization with most employees working from home. Location factors in Amsterdam – as illustrated by an in-depth case study of advanced call centers - indicate that the most important factors cannot be found in remote regions, like the specific labor supply in the capital city (young, flexible, internationally oriented, multilingual), the presence of a cluster of actual and potential clients, and proximity to an international airport (van Geenhuizen, 2006). Also, aggregate patterns (like in the UK) do not point to remote areas as most appropriate for advanced call center development, unless there is a well-developed financial and business service activity and a relatively large population concentration (Beekman et al., 2002; Bristow et al., 2000). The first criterion cannot be satisfied in the North of the Netherlands and the second can be satisfied to a limited extent given the size of the towns of Groningen and Emmen (respectively about 170.000 and 110.000 inhabitants at the municipal level in 2000). With regard to the labor market, the region provides an additional advantage not mentioned in the UK research, i.e. a large availability of bilingual workers due to close proximity of Germany, which is especially important for internationally operating call centers. Given the previous results, we may assume that high-order call centers are an option for the North of the Netherlands only in a limited number of cases, i.e. activity without strong financial and business services links but with an international orientation. Thus, at this stage of our knowledge, advanced call center activity is not likely to contribute to economic growth in the North of the country in a pervasive way. The physical distance to large concentrations of financial and business service activity and the lack of some labor market characteristics - present in the region of Amsterdam - tend to remain a bottleneck.

CONCLUSION

Despite being connected to the best digital infrastructure in early stages, large cities may loose young innovative companies to adjacent areas. Only the largest city is endowed with unique agglomeration advantages, as appeared in a case study of the Netherlands. The potential spread of young innovative companies from the large cities seems not to reach remote regions, except for the largest town here. Similarly, the findings about location behaviour of advanced call center activity provide no ground for the likeliness of economic growth of remote regions through such activity. However, a great deal of work can be done in these regions through a joint learning of local public and private actors and increase of their organising capacity, which might enable sudden jumps in development. Our analysis has illustrated that particular remote regions are able to "escape" from lagging-behind by utilizing their remoteness or emptiness as a positive asset in developing information-intensive research facilities for which emptiness is prerequisite. Such research facilities may justify a connection with the highest band-width in the national digital infrastructure and can help to develop local research companies linked with them on a similarity or a supplying basis following an endogenous growth path. An alternative strategy to improve connections with the highest level digital infrastructure may be effective in regions located close to national political borders. If these regions are included in current EU programs and national policies of cross-border integration, it may be useful to develop and experiment with broadband network projects in existing cross-border programs. However, availability of access to high-speed Internet is a key condition for new economic activity but it is not sufficient. The development of a soft infrastructure, e.g. institution building, good governance, and skills development, seems equally important.

ICT policies and their impacts are subject to uncertainty for many reasons. In this context, the initiative to establish a comprehensive monitoring system of ICT-related policy impacts in the Northern provinces of the Netherlands (Sleurink et al., 2001) is an interesting example of a major step forward, particularly if results from monitoring and analysis serve as an input to a regular re-evaluation of policies and learning. Here lies a field of important research, namely an internationally comparative analysis of how remote regions connect with the highest-level digital infrastructure and how new economic activity is linked with that, particularly which policy strategies are most successful given various specific regional conditions. A lot of research is currently being done, but conditions and actions that allow for a major "jump" in development remain largely unknown.

Note 1

There are no comprehensive standard statistics covering backbones and local connections of digital networks. Data are often only available from private network owners, service providers, private research institutions, etc.

BIBLIOGRAPHY

Andersson, M. and Ch. Karlsson, (2002), *The Role of Accessibility for Regional Innovation Systems*, Jönköping International Business School, Jönköping.

Arthur, W. B., (1994), *Increasing Returns and Path Dependency in the Economy*, The University of Michigan Press, An Arbor.

ASTRON, (2003), www.astron.nl (www.lofar.nl).

Audretch, D., (1998), "Agglomeration and the Location of Innovative Activity", *Oxford Review of Economic Policy*, Vol 14 (2), pp. 18-29.

Beekman, M., F. Bruinsma and P. Rietveld, (2002), "Spatial Diffusion Patterns of Call-centres in the Netherlands", Paper for the 42nd ERSA Congress, Dortmund, August 27th-31st 2002.

Bolisani, E. and E. Scarso, (2000), "Electronic Communication and Knowledge Transfer", *Int. Journal of Technology Management*, 20 (1/2), pp. 116-133.

Bristow, G., M. Munday and P. Gripaios, (2000), "Call Centre Growth and Location: Corporate Strategy and the Spatial Division of Labour", *Environment & Planning* A, 32, pp. 519-538.

Capello, R., (1997), "Telecommunication Network Externalities and Regional Development Policy Implications", [in:] C. Capineri and P. Rietveld (eds), *Network Interconnectivity*, Aldershot, Ashgate, pp. 13-36.

CBS (Netherlands Bureau of Statistics), (2003), *The Digital Economy*. Voorburg/Heerlen, CBS.

Cohen, G., M. van Geenhuizen and P. Nijkamp, (2002), "Urban Planning and Information and Communication Technology: Perceptions, Ideas, and Facts", [in:] Y. Higano, P. Nijkamp, J. Poot and K. van Wijk (eds), *The Region in the New Economy,* Avebury, Ashgate, pp. 47-67.

Cohen, G., M. van Geenhuizen and P. Nijkamp, (2005), "Bytes of Urban Planning: A Dutch Perspective", [in:] M. van Geenhuizen, D. V. Gibson and M. V. Heitor (eds), *Regional Development and Conditions for Innovation in the Network Society*, Purdue University Press, West Lafayette (Indiana), pp. 99-117.

Conroy, M. M. and J. Evans-Cowley, (2006), "E-participation in Planning: an Analysis of Cities Adopting on-line Citizens Participation Tools", *Environment and Planning* C, 24 (3), pp. 371-384.

Cornford, J. and A. Gillespie, (1993), "Cable Systems, Telephony and Local Economic Development in the UK, *Telecomm. Policy*, 17, pp. 589-603.

DDV, (1999), *Spatial Disparities in Telecommunication Infrastructure* (in Dutch), De Meern: DDV.

Estache, A. and A. Goicoechea, (2005), *A Research Database on Infrastructure Economic Performance,* World Bank Policy Research Working Paper 3643.

Florida, R. (2005), *The Flight of the Creative Class: the New Global Competition for Talent*, Harper Collins, New York.

Geenhuizen, M. van (2006), "ICTs and High-order Integration of Remote Regions", [in:] Geenhuizen, M. van, Reggiani, A., and P. Rietveld (eds), *Policy Analysis of Transport Networks*, Aldershot, Ashgate, (in press), pp. 257-273.

Geenhuizen, M. van, (2005), "Footloose or not? A rough set approach to relevance of proximity in the urban e-economy", *Udevalla Symposium Anthology Series 2004*, University of Trolhattan/Udevalla, Udevalla, pp. 389-402.

Geenhuizen, M. van, (2004), "ICT and Regional Policy: Experiences in the Netherlands" [in:] Gibson, D., Heitor, M. and A. Ibarra (eds), *Learning and knowledge for the network society,* Purdue University Press, West Lafayette (Indiana), pp. 227-224.

Geenhuizen, M. van, (2000), "Interconnectivity of Transport Networks: A Conceptual and Empirical Exploration", *Journal of Transportation Planning and Technology* 23, pp. 199-213.

Geenhuizen, M. van, and Nijkamp, P., (1999), "The learning capability of regions: conceptual patterns and policies", [in:] Rutten, R., Bakkers, S., Morgan, K., and Boekema, F. (eds), *Knowledge, Innovation and Economic Growth: Theory and Practice of the Learning Region,* Edward Elgar, London, pp. 38-56.

Geenhuizen, M. van, and Ratti, R. (eds), (2001), *Gaining Advantage From Open Borders. An active space approach to regional development*, Aldershot, Avebury.

Gertler, M. (2003) "Tacit knowledge and the economic geography of context, or The undefinable tacitness of being (there)", *Journal of Economic Geography,* 3 (1), pp. 75-99.

Gibbs, D. and Tanner, K., (1997), "Information and Communication Technologies and Local Economic Development Policies: The British Case", *Regional Studies* , 31 (8), pp. 765-774.

Gigaport, (2000-2006), www.gigaport.nl

Gillespie, A. and Williams, H., (1988), "Telecommunications and the reconstruction of regional comparative advantage", *Environment and Planning A,* 20, pp. 1311-1321.

Gillespie, A., Coombes, M. & Reybould, S. (1994) "Contribution of telecommunications to rural economic development: variations on a theme?", *Entrepreneurship and Regional Development,* 6, pp. 201-217.

Gorman, S. P. (2002) "Where are the web factories: the urban bias of e-business location", *Journal for Economic Geography,* 93, pp 522-536.

Gorman, S. P. and McIntee, A. (2003), "Tethered connectivity? The spatial distribution of wireless infrastructure. *Environment and Planning A",* 35, pp. 1157-1171.

Grabher, G. (1993) *The Embedded Firm. On the socioeconomics of industrial networks*, Routledge, London & New York.

Graham, S and Marvin, S. (1997) *Telecommunications and the City. Electronic Spaces, Urban Places*, Routledge, London.

Graham, S. (1999), "Global Grids of Glass: On Global Cities, Telecommunications and Planetary Urban Networks", *Urban Studies*, 35 (5/6), pp. 929-949.

Grimes, S. (2003), "The digital economy challenge facing peripheral rural areas", *Progress in Human Geography*, 27 (2), pp. 174-193.

Grubesic, T.H. and O'Kelly, M.E. (2002), "Using points of presence to measure accessibility to the commercial internet", *Professional Geographer*, 54 (2), pp. 259-278.

Haynes, K. E. (2006), *Infrastructure: the glue of megacities*. Megacities Lecture 9, Megacities Foundation/NICIS, Amsterdam/The Hague.

Howells, J.R.L. (2002), "Tacit knowledge, innovation and economic geography", *Urban Studies,* 39 (5), pp. 871-884.

Kenney, M. and Curry, J., (2001), "Beyond Transaction Costs: E-Commerce and the Power of the Internet Dataspace", [in:] Leinbach, T.R. and Brunn, S.D. (eds), *Worlds of E-Commerce: Economic, Geographic and Social Dimensions,* Wiley, Chichester, pp. 45-65.

Kinder, T. (2001), "The use of call centres by local public administrations", *Futures,* 33, pp. 837-860.

Klaassen, L., (1967), *Methods of selecting Industries forDepressed Areas: An Introduction to Feasibility Studies,* OECD, Paris.

Kotval, Z. (1999) "A realistic strategy for the revitalization of American cities" *Telecommunications: Cities,* 16 (1), pp. 33-41.

Li, F., Whalley, J. and Williams, H. (2001) "Between physical and electronic spaces; the implications for organisations in the networked economy" *Environment and Planning A*, 33 (4), pp. 699-716.

Malecki, E. (2002) "The Economic Geography of the Internet's Infrastructure" *Economic Geography*, Vol. 78 (4), pp. 399-424.

Malecki, E. and Gorman, S.P (2001) "Maybe the death of distance, but not the end of geography; the Internet as a network", [in:] Brunn, S.D. and Leinbach, T.R. (eds) *The Worlds of Electronic Commerce*, John Wiley, New York, pp. 87-105.

Melody, W. (1991) "The information society: the transnational economic context and its implications", [in:] Sussman, G. and Lent, J. (eds) *Transnational Communications: Wiring the Third World,* Sage, London, pp. 27-41.

Mitchell, W.J. (1999) *E-topia, Urban life, Jim – but not as we know it,* The MIT Press, Cambridge.

Morgan, K. (1997) "The Learning Region: Institutions, Innovation and Regional Renewal", *Regional Studies,* 31(5), pp. 491-503.

NetherLight (2006) www.netherlight.nl

NOM (Northern Development Agency), (2001), The ICT sector in the North of the Netherlands (in Dutch).

OESO (2001) *Cities and regions in the New Learning Economy,* OESO, Paris.

Oort, F. G. van, and E. Stam (2005) Entrepreneurship and Agglomeration Economies in the Dutch ICT Industry. *Udevalla Symposium Anthology Series 2004*, University of Trolhattan/Udevalla, Udevalla, pp. 689-717.

Pollard, J. and Storper, M. (1996) "A tale of twelve cities: metropolitan employment change in dynamic industries in the 1980s", *Economic Geography,* 72, pp. 1-22.

Porter, M.E., (1998) "Clusters and the New Economics of Competition", *Harvard Business Review,* 76 (6), pp. 77-90.

Richardson, R. and Gillespie, A., (1996), "Advanced communications and employment creation in rural and peripheral regions: a case study of the Highlands and Islands of Scotland", *The Annals of Regional Science*, 30, pp. 91-110.

Sleurink, H., Berg, A. Van den, and Nieuwenhuis K., (2001), *The North of the Netherlands in the Information Society* (in Dutch), Koninklijke Van Gorcum, Assen.

Sohn, J. (2004) "Information technology in the 1990s: More footloose or more location bound?", *Papers in Regional Science,* 83 (2), pp. 467-485.

SURFnet (2006) www.surfnet.nl

Torre, A., and Gilly J-P., (2000), "On the Analytical Dimension of Proximity Dynamics", *Regional Studies*, 34 (2), pp. 169-180.

Wernerheim, C.M. and Sharp C.A., (2003), "High-order Producer Services in Metropolitan Canada: How Footloose Are They?", *Regional Studies*, Vol. 37 (5), pp. 469-490.

Worldbank, (2006), Local economic development website (www.worldbank.org).

PART III

RESEARCH & DEVELOPMENT, PARTNERING, KNOWLEDGE MANAGEMENT AND FIRMS' STRATEGIES

14

Surveying the Status of Organizational Culture, Improvement, and Innovation in the Process Assessment and Knowledge Management Communities

Han van Loon

BACKGROUND

ISO/IEC 15504 is a process assessment standard used to evaluate the working processes within an organisation (Van Loon, 2004a, 2004b). A lot of work has gone into refining the process-focussed aspects in the standard, mainly to clarify the process dimension and capability dimension. However, not much work has been done in the ISO/IEC 15504 standard's community about how people aspects impact on the use of process focussed approaches.

During the SPICE 2004 panel discussion, the author raised the issue that people aspects are inadequately considered in the process assessment standards, in part citing the SPICE Trial (Jung et al., 2003) results that show people oriented factors (commitment, politics, focus) have the greatest impact upon whether assessment results are used.

As more organisations move towards an innovation paradigm from a continuous improvement paradigm, the application and relevance of process-focussed approaches have come under increasing scrutiny and debate. At one end of the debate, some pundits believe that good processes can produce consistently high quality results without relying extensively upon individual skills of the person(s) involved, at the other end of the debate some pundits believe that individual skills and experience are paramount and standard processes are disadvantageous or even hinder positive innovation outcomes. Agile programming methodologies such as Extreme Programming emphasize these people aspects while minimising process aspects (Beck, 1999).

The author of this chapter has written in his process assessment books about the need to consider people, processes and products holistically. In addition the Practical Guide book provides some modelling of organisational culture as an introduction to how some people aspects impact on process-focussed approaches in general and process assessment and improvement in particular (see, van Loon 2004a and 2004b). The cultural modelling is based in part upon the work of Trompenaar and Hampden Turner (1998) that use seven cultural dimensions.

Therefore, given the debate concerning process focus versus people focus, a seemingly apparent shift towards innovation focus from continual improvement focus within organisations over the past five years, the author planned to survey the SPICE community, as well as the knowledge management community in an attempt to collect information about the current organisational orientation.

THE SURVEY

The survey was run over three weeks during September and October 2004 at KnowledgeBoard[1], a respected and independent knowledge community web based community. KnowledgeBoard was partially sponsored by the European Commission to set up a knowledge society community portal (www.knowledgeboard.com)

The author devised the survey and received feedback to refine several of the questions asked. One aim of the study is to determine whether preparation for process assessments can be improved by including organisational cultural and people issues in Pre Assessment Questionnaires.

The survey was advertised on the KnowledgeBoard web site and also via email to the SPICE 2004 attendees, and other known process assessment community participants. The survey gathered 124 responses. Next, the response data was cleaned, this consisted of:

- Ensuring that only attributable responses were used (respondents who voluntarily provided email addresses). Each person was emailed to check for a valid email address and to agree to his or her input.
- All responses without email addresses or invalid email addresses were eliminated.
- 4 email responses provided 'out of office' replies. These were subsequently checked separately to the other responses and found to be relatively consistent, so were included in the overall analysis.
- Checking and eliminating double responses (2 people were asked to select their preferred response between similar but not the same responses).

[1] Thanks to Ed Mitchell of KnowledgeBoard for setting up the survey web pages.

- Obviously false data patterns (e.g. all 1 response answer or an escalating pattern response) were eliminated.

After data cleaning, there were 105 valid responses. These responses represent over 100 different organisations (there were a couple of respondents from the same organisation). A few questions produced non-response, so the number of responses varies between 103 and 105.

The questions mostly used a five-point ordinal scale, except for two questions that used a six-point scale.

While the data used for analysis represents attributable data from the respondents, the usual caveats for this type of survey apply. Firstly, the data sample size, while significant, is still a small data set that may be entirely representative of the communities covered. Secondly, as response was voluntary, the data may be biased towards organisations and people who may show more interest in the surveyed areas than normal. This may skew the results (more of less favourable), although data analysis has not indicated any particular bias. Thirdly, many of the questions elicit people's opinions rather than independently verifiable facts. This is typical of cultural issue surveys, but needs to be understood when considering the results.

Finally the author may have unintentionally introduced some bias into the questions asked and the response categories. Data analysis does not indicate any such bias but this does not mean it can be eliminated as a consideration.

THE QUESTIONS ADDRESSED

The survey questions were concerned with the importance of knowledge workers and process specialists in a successful performance, the role of the vision and mission and accessibility of the information concerned, etc. There were also some explicit cultural dimension issues examined (e.g., organisational hierarchy, decision making, gaining management support, rewarding people).

- How important are knowledge workers to the success of the organisation?
- How important are process specialists to the success of the organisation?
- Is the organisation's vision and mission lived by people or more remote?
- To what extent are employees informed about overall business objectives and processes, running projects and their results?
- How easily accessible is this information?
- To what extent are employees informed about innovation and improvement?
- How easily accessible is this information?
- Are there any forms of informal communication (peer to peer communication)?
- Is improvement-focused conduct promoted/supported/rewarded?
- How would you rate the commitment and active participation of personnel doing the work (for example, software practitioners, service staff) in defining and improving practices?
- How would you rate the need for process specialists in helping to define and improve practices?
- How important is process-related training of the employees in your company?

- How would you see the approach to process improvement in your company: is it focused on individuals or teams to proceed?
- Is innovative behaviour promoted/supported/rewarded?
- Is risk of failure recognised and tolerated as part of innovative behaviour
- Are innovations directed from management or can they be proposed and developed by people at all levels of their organisation?
- Does the organisation look outside for new knowledge, improvements or innovations to learn and use?
- Do processes within your organisation work for or against innovation?
- How easy is it to gain resources, support and privileges from management for improvement or innovative activities?
- Are there activities in order to improve the capabilities of employees, for example through training, job rotation (personal development)?
- How would you rate hierarchy in the organisation?
- How would you rate the decision-making approach in your organisation?
- Why are people rewarded?
- How are people rewarded?
- How would you describe your organisation?

THE ANALYSIS

The analysis presented here does not provide the complete analysis of the data, but highlights some general results and also some comparative results (between questions). Appropriate confidence tests were applied to the analysed data to support the conclusions, but this is not presented in this chapter. This chapter focuses on the analysis of process and improvement related questions due to restrictions in acceptable size of the chapter. Future publications will focus on innovation and cultural aspects.

The organisational business covered a wide area of endeavour (Figure 1).

Figure 1
Main business of the organisation

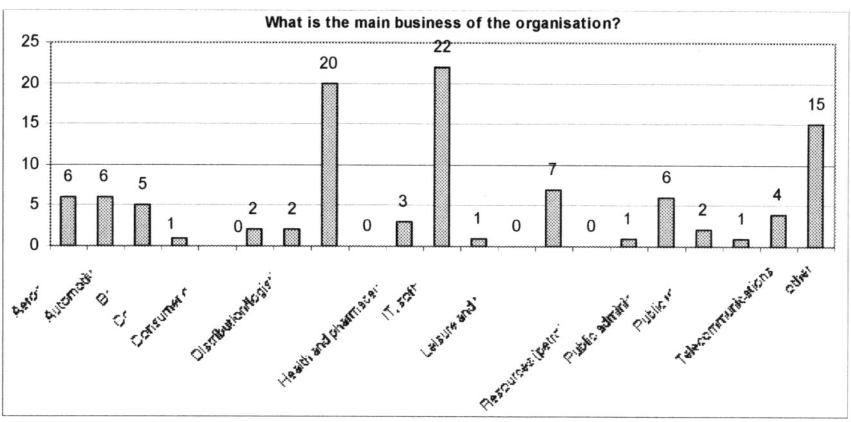

The main business areas were education, and information technology and software development. Both areas are knowledge intensive, so they tend to reflect in part both the KnowledgeBoard community and the SPICE community. Other significant business sectors were aerospace, automotive, banking, manufacturing and public administration.

Figure 2
Type of organisation

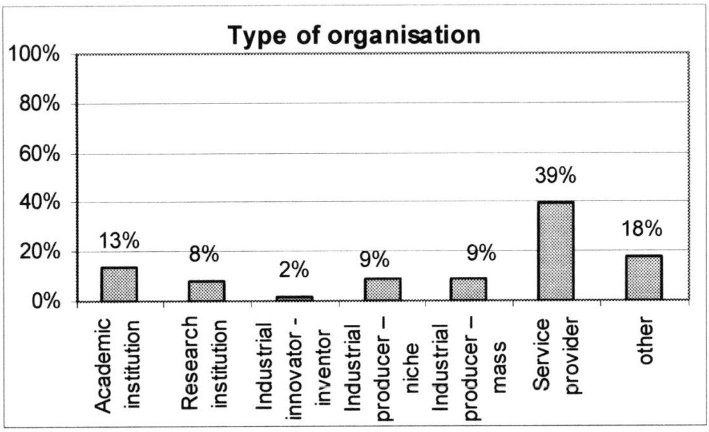

Respondents were also asked to categorise their organisation. The largest group were service providers (Figure 2). The respondents also covered a wide variety of roles and positions, with a small majority of middle-managers (19%) (Figure 3).

Figure 3
Role/position of the respondent

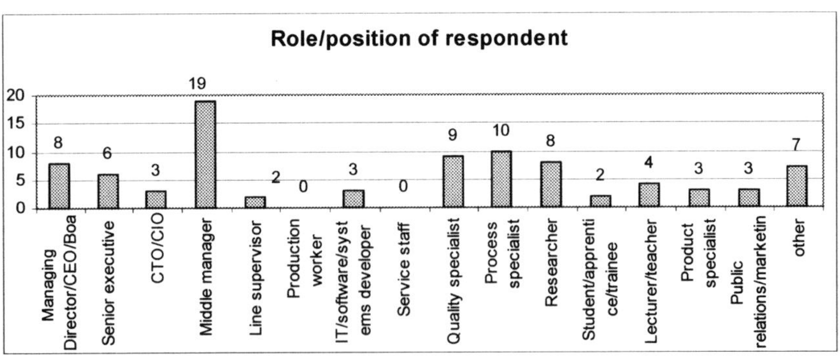

RESULTS AND SOME COMMENTS

The first set of questions deals with the importance of people to the organisation. With regard to importance of knowledge workers and of process specialists, the following could be observed (Figure 4):

- No respondents believed that knowledge workers were not needed or were of low importance to organisational success. 95% of respondents state that people able to apply knowledge for improvement or innovation are important or extremely important to success. 62% stated that knowledge workers are extremely important to create new ideas, improvement and innovation as a basis for success.
- This confirms that the survey target audience is the one expected. It also (unsurprisingly) reflects the KnowledgeBoard community.
- 9% of responses indicated that process specialist were of low or very low importance. 77% of responses indicated that process specialists were important, to both create and tailor processes, as well as to improve and optimise the organisation. 37% of responses saw process specialists as extremely important to processes that aid innovation, improvement and optimisation.

Figure 4
Importance of specific workers

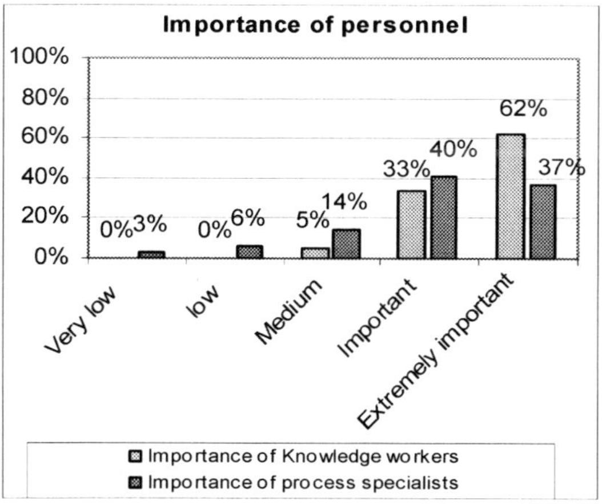

Using individual respondent's data to compare the relative importance of knowledge workers and process specialists, it was found that on average knowledge workers scored 0.56 higher than process specialists. A 99% confidence gives a variance of 0.24 (i.e., one can conclude that overall knowledge workers are considered more important to the organisation by a small but significant amount).

To explore the role of the organisation's vision and mission in the work of the people, the following question was posed: Is the organisation's vision and mission lived by people or more remote? The results are as follows:

- 24% of responses indicated that the organisation's vision and mission did not affect them. Within this group, over 60% (15% of the total responses), the vision and mission were not even recognised as real - meaning that they either did not exist or were just 'public relations' exercises, and management did not follow them. 54% of total of responses saw organisational alignment between the department/team and the overall organisational vision and mission. A smaller percentage of respondents (22% of total responses) indicated that they felt individually responsible for achieving the vision and mission of the organisation.
- A majority (78%) of respondents indicate that the organisation's vision and mission are still seen as non-personal.

In the next paragraphs, the interpretation of results on the following questions is taken together as the intent of the study was to see if corporate information is treated differently to improvement and innovation information: (1) To what extent are employees informed about overall business objectives and processes, running projects and their results? How easily accessible is this information? (2) To what extent are employees informed about innovation and improvement? How easily accessible is this information?

For overall business information the results can be summarized as follows (Figure 5):

- A small share of responses (8%) indicated little or no general information received about the overall organisation, only information within own area/project (this may suggest a relative isolation in organisation information terms).
- 62% of responses indicated that information about overall business objectives, processes and projects is disseminated, but little or no feedback is sought (one-way information dissemination). Within organisations that practice this one-way dissemination, approximately half provide both overall organisation information, as well as own/related projects, the other half only provide very limited information outside their own area/project.
- 31% of responses indicated an active information dissemination and collection approach covering overall business objectives, processes and projects.
- 8% of responses indicated an active organisation-wide information dissemination and collection approach in order to use the information for business purposes (including process related purposes).
- 21% of responses indicated a need to actively search for information, as little was made available.
- 79% of responses found overall organisation information to be available without requiring special activity for the receiver. About 2/3rds of these respondents (50% total) forwarded that most information was available and regularly updated.
- However, only 6% had both push (dissemination) and pull (Knowledge Management systems) access. It appears that such systems are still not widely popular and/or available (although the survey did not investigate the reasons why).

For improvement and innovation information the results suggest the following situation (Figure 5):

- 85% of the respondents indicated that they were either moderately informed (32%), relatively well informed (36%) to very well informed (17%) about innovation and improvement.
- 27% of the respondents needed to actively pursue innovation and improvement information.
- 70% of respondents suggested that innovation and improvement information was regularly provided or available.
- A small percentage of the respondents (5%) indicated that active push and pull (Knowledge Management systems) mechanisms rarely exist.

Figure 5
Information accessibility

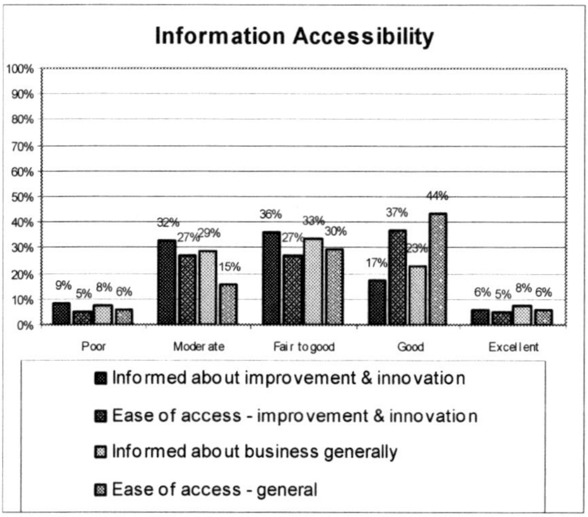

The low percentage of both push and pull information is related to the use of Knowledge Management systems. The low result for both push and pull information mechanisms is somewhat surprising, given the KnowledgeBoard community focus; the author conjectures that this indicates the entire domain is still at a low level of maturity in understanding what are the real requirements for pull-based Knowledge Management in organisations.

In addition, a comparison of general business information versus improvement and innovation information reveals that respondents are better informed about general business information and ease of access is better. Some further analysis is planned about how information access and the way it is provided correlates with support for innovation and improvement.

Next, the study examined whether there are any forms of informal communication (peer to peer communication). The following results were found:

- 83% of the respondents indicated that informal peer-to-peer communication was considered important and/or encouraged by management. A significant percentage

(47% of total) of responses indicated that management formally recognised and supported informal communications.

The last percentage is similar to that concerning how well employees are informed about innovation and improvements. It can therefore be hypothesized that informal peer-to-peer communications are still a major information resource for innovation and improvement.

In the next paragraph, the results concerning improvement related issues are presented. These may be summarized as follows (Figure 6):

- A significant percentage of the respondents (16%) stated that improvement focused conduct was not promoted, supported or rewarded.
- 32% of the respondents indicated that improvement activity is promoted (i.e. it is expected), but not specifically supported or rewarded.
- 29% of the respondents indicated that improvement activities are supported. 11% of respondents indicated that improvement activities are rewarded, while 13% of respondents indicated that active promotion, support and reward for improvement occur.

Concerning the level of commitment and participation in improvement activities, the following results were achieved:

- The majority of respondents (56%) indicated that commitment to defining and improving practices is based on personal interest (or lack of it). This matches quite well with the result that 53% of improvement activities are supported and/or rewarded. It is therefore possible to suggest that personal commitment is related to the degree of support/reward from management. This seems to be connected to ways in which people are rewarded.
- A significant percentage of respondents (30%) indicated that senior (21%) or all personnel (9%) actively define and improve practices. On the one hand, this means that a significant percentage of respondents indicates an active improvement culture, on the other hand, this can also be interpreted to mean that less than 1/3rd of respondents have an organisation-wide commitment to continual improvement activities, and the majority operate on a more ad-hoc basis.

With regard to the use of process specialists and use of process training, the results were as follows:

- 39% of the respondents indicated that they make use of process specialists as needed. 50% of the respondents indicated that process specialists are frequently or always used.
- A small percentage (7%) of the respondents indicated that they receive no process training. Half the respondents (49%) indicated that only some people are given process training and that is only for specific processes related to their roles.
- A little over a quarter (27%) of the respondents indicated that everybody is given process training related to their roles, while 14% of the respondents indicated that some people are trained on every process. Only 4% indicated that everyone receives extensive process training on all processes.

Figure 6
Improvement support, commitment and resources

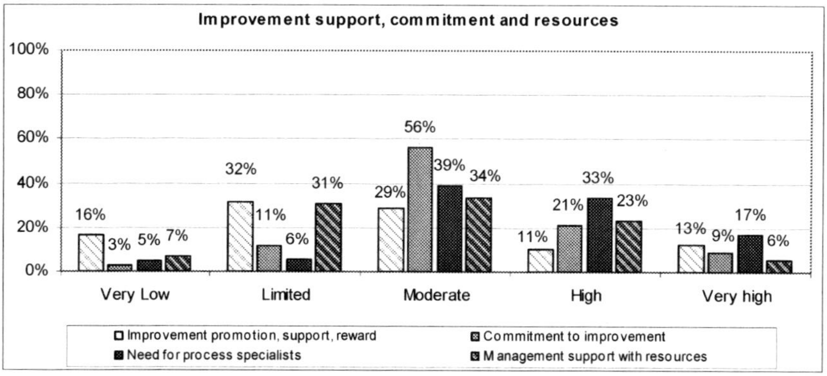

As a next step, the results concerning individual or team based improvement will be discussed:

- 43% of the respondents indicated that individuals mainly drive process improvement.
- A significant percentage of the respondents (34%) indicated that improvement is driven equally by a team or an individual, while 20% of respondents indicated that teams mainly drive improvements, but that some individual effort is still involved.

The combined results suggest that improvement is still driven by individuals, even inside teams. Hence the concept of an improvement champion is still very important to drive improvement in an organisation. Overall, the respondents indicated that process related training is still a significant activity in organisations, witness the very small percentage of people who receive no training, although in half of the cases (49%), it is limited to their role.

One of the correlated questions deals with how people gain resources from the management. This obviously impacts not only upon the ability to improve, but also upon the level of commitment to improve.

- 31% of the respondents indicated that it was difficult to get support for improvement or innovation unless it aligned with their manager's desires.

The author believes the above response confirms the result that 'improvement is only promoted'. Where improvement is only promoted and not always supported and rewarded, there exists a commitment gap (shown as the 'limited' set of results in the above figure). It also seems to connect with the lack of process specialist use.

- 34% of the respondents indicated that obtaining support was based upon clear guidelines but still subject to manager approval.
- 29% of the respondents indicated that support is generally based upon the merits of the activity, 1/4 of which was specifically given a high level of positive management support as opposed to the remainder that were approved without resistance. This difference between the two sub-categories is more to do with the level of active management support based upon merit.

How the respondents evaluated the need for and importance of process specialists, together with the need for process training is shown in Figure 7. Respondents saw the importance of process specialist as generally important to very important, but not quite the same level of need. Training people to become process specialists (most or all people to be trained in all processes) was considered less important, witness almost an 80% score for limited and moderate importance (in fact, 14% are trained on every process, while only 4% indicated that everyone receives extensive process training on all processes).

Figure 7
Process specialists and training

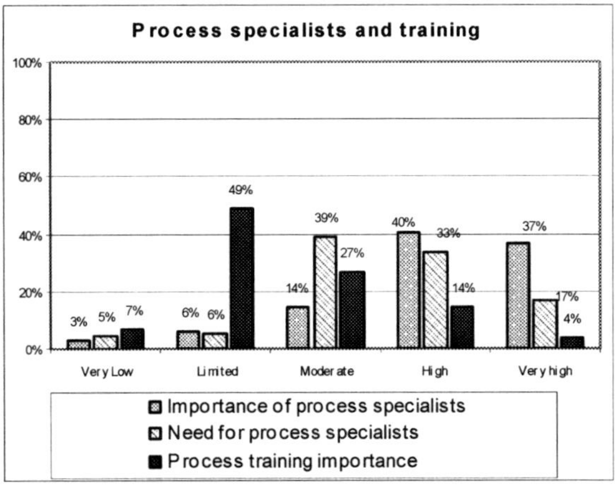

There may also be a relation with the way in which people are rewarded, as shown in Figure 8:

- The majority of respondents (52%) indicated that rewards are based purely on performance. Of these, more than half of respondents (28% of total) indicated this was on an individual success basis, the rest upon a team success basis.
- A significant percentage of respondents (33%) indicated that behaviour as well as performance is considered, while a minority but still significant number of respondents (15%) indicated that reward is based upon behaviour.
- More respondents (38%) indicate that reward based upon individual criteria takes precedence over reward based upon team criteria (29%).

Figure 8
Reasons for rewarding people

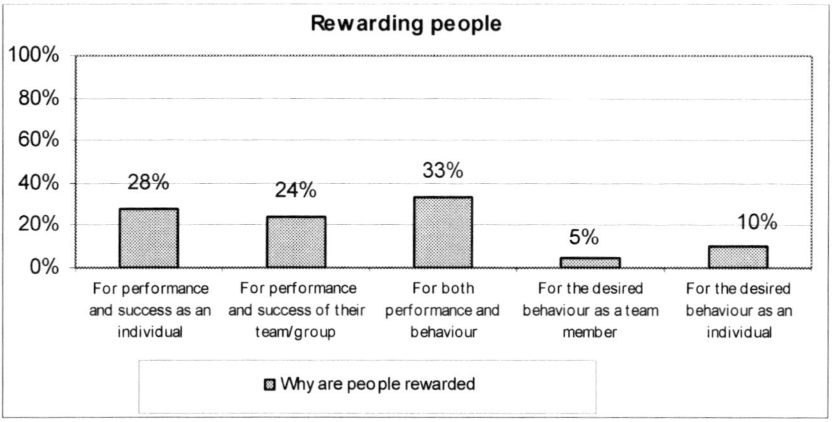

SOME SHORT COMPARISONS ON INNOVATION

The survey included several questions about innovation and organisation cultural dimensions. It is not possible to provide all the details here due to restrictions in size of the chapter, but a couple of the more interesting results are briefly described here. There were complementary questions about the level of promotion, support and reward for improvement and for innovation activities.

Figure 9
Improvement and innovation support

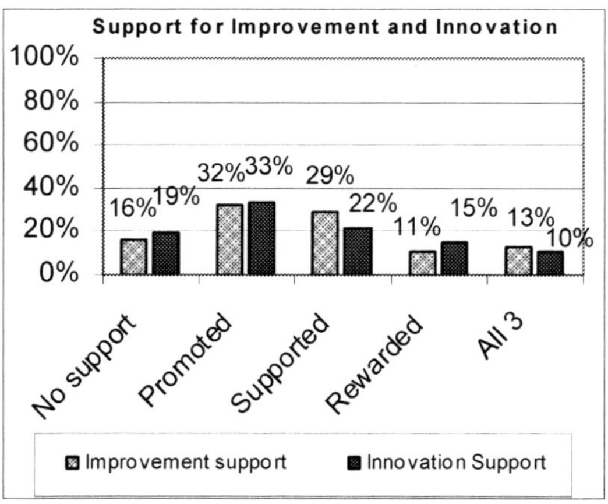

Figure 9 illustrates that overall there was a similar level of support for innovation and improvement. A one-to-one correlation is not yet made to see if the same organisation responded to both improvement and innovation in a similar way, or whether there was difference (one innovation focussed, the other improvement focussed).

Another issue that was addressed was the impact of standard processes upon innovation (Figure 10).

- 39% of the respondents indicated that standard processes have a negative effect upon innovation, mostly slowing down rather than preventing innovation. On the other hand, 27% indicated that their organisations have processes to support innovation.
- 24% of the respondents stated that standard processes have neither a negative or positive effect on innovation, while a small percentage stated that they had no standard processes and therefore had a high level of freedom to innovate.

Figure 10
Standard processes and innovation

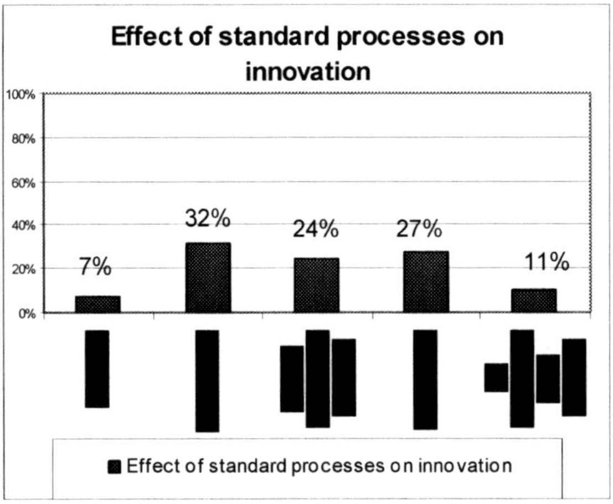

The above outcomes suggest that a greater share of the respondents believe standard processes adversely affect innovation than support it (32% versus 27%). Note that the amount that processes slow down innovation (32% group) was not surveyed in further detail.

However, if one concludes that greater innovation is a trend, then care must be taken that process-focussed organisations do not unnecessarily adversely affect innovation. On the other hand, the process support to innovation shows that this can be adequately resolved if care is taken to handle innovation issues correctly.

CONCLUDING REMARKS

The survey results show the importance of people to the success of the organisation, both generally as knowledge workers and as process specialists. This confirms the need to consider people in process-focussed approaches, and is further supported by a majority of the respondents indicating that commitment to defining and improving practices is based upon personal interest (and that is in part driven by how people are rewarded).

Furthermore, if management only promotes improvement and does not support and reward it, the results support the hypothesis that this situation impacts negatively on the level of improvement performed. This is further supported by the results on the ease of gaining management support and resources.

One of the arguments for process assessment has been that determining the status quo in terms of process capability and improvement opportunities provides a good basis to promote improvement. The author concludes that promotion in itself does not lead to improvement; it is support and reward that leads to improvement.

The general conclusion is that process assessment itself and the associated standards need to be expanded to take people and culture into account. For further information, visit the author's web site: www.starswebworx.ch/starsweb.

BIBLIOGRAPHY

Beck, K., (1999), *Extreme Programming Explained: Embrace Change*, Addison Wesley, Reading, Massachusetts.

Jung, H. W., R. Hunter, D. Goldenson, and K. El-Emam, (2003), SPICE Project. Phase 2 Trials Final Report. Version 1.01. January 2003. Vol 1. Executive Summary and detailed reports. Vol 2. Findings from Phase 2 of the SPICE Trials. www.csi.strath.ac.uk/research/spice/

Loon, H. van, (2004a), *Process Assessment and ISO/IEC 15504*. A Reference Book. Springer.

Loon, H. van, (2004b), *Process Assessment and Improvement. A Practical Guide For Managers, Quality Professionals And Assessors.* Springer.

Trompenaar, F. and Ch. Hampden-Turner, (1998), *Riding the Waves of Culture. Understanding Diversity in Global Business* - 2nd edition, McGraw-Hill.

15

The Role of Knowledge Management in Improvement of Management Systems

Maciej Urbaniak

INTRODUCTION

The essential ingredient for benefiting from knowledge management is the establishment of a learning culture in an organization. If a company believes that it already knows the best way or that the 'best practice' is not appropriate to their circumstances, then sustainable improvement will just not happen. The development of a learning culture of an organization does not, of course, happen overnight; it takes time, and requires that the appropriate infrastructure to be in place.

The term "organizational learning" has been applied to so many different aspects of corporate management, from human resources management to technology management strategies, that it has become a particularly vague concept (Trott, 2005).

An organization's success depends increasingly on the diverse backgrounds, knowledge, skills, creativity, and motivation of all its employees and partners. This is the reason why many organizations realize that their employees' success also depends increasingly on having opportunities for personal learning and on practicing new skills. Organizations invest in employees' personal learning through education, training, and other opportunities for continuing growth and development.

Managing information and knowledge can require a significant commitment of resources inasmuch as the sources of information grow dramatically each year. Information from internal operations, from the Internet, and from business-to-business (B2B) and business-to-consumer (B2C) communications challenges the organizational capabilities to provide the information that people need to do their work, keep current, and improve their management system (Evans, Lindsay, 2002).

This chapter is an attempt to pinpoint the role of knowledge management in the improvement of a management system. The author discusses the very concept of knowledge management, as well as the notions of knowledge conversion process and knowledge assets. This is followed by the discussion of the tools of learning and a comparison of the continuous improvement of management with knowledge management. The author concludes that an integrated approach to management is required in order to underpin high levels of innovativeness and the success of companies.

THE CONCEPT OF KNOWLEDGE MANAGEMENT

According to Webster's Dictionary, knowledge is the fact or condition of knowing something with familiarity gained through experience or association (Basu, Wright, 2003). The concept may also be described as a set of models that describe various properties and behaviors within a domain and that may be recorded in an individual brain or stored in organizational processes, products, facilities, systems and documents.

The notion of knowledge management has been defined in different ways and from different perspectives (Cobb, 2003). Inasmuch as it is an evolving technology area, it is very difficult to find a standard definition of what it is. For example, the concept is described as follows:

"Knowledge management is the strategy and processes to enable the creation and flow of relevant knowledge throughout the business to create organizational, customer, and consumer value".

"Knowledge management is the broad process of locating, organizing, transferring, and using the information and expertise within an organization. The overall knowledge management process is supported by four key enablers: leadership, culture, technology, and measurement" (American Productivity and Quality Center).

"Knowledge management is the process of continually managing knowledge of all kinds to meet existing and emerging needs, to identify and exploit existing and acquired knowledge assets and to develop new opportunities".

Thus knowledge is distinguished from information and data in that an interpretation is applied to it. This also means that knowledge is something that is developed by individuals, and that its quality depends on this individual's insights and expertise (Von Stamm, 2003).

Knowledge management is concerned with the exploitation and development of the knowledge assets of an organization with a view to furthering the organization's objectives. The label 'knowledge assets' refers to the accumulated intellectual resources that an organization possesses, including information, ideas, learning, understanding, memory, insights, cognitive and technical skills, and capabilities. They are held not only by an organization but also reside within its customers, suppliers, and partners as well, and may be described as the "know-how" that an organization has available to use, to invest, and to develop. Employees, software, patents, databases, documents, guides, technical drawings, policies and procedures are repositories of the knowledge assets of an organization. Building and managing them are key components in order for an organization to create value for its stakeholders and to help sustain competitive advantage. Customers, suppliers, and partners may also hold key knowledge assets. All in all, in many organizations knowledge assets have become more important than financial and physical assets.

Knowledge management also invests in the most valuable resource of competitive advantage; that is, people. The support structure for such a knowledge-sharing process should include:

- a regular best practice forum to learn from each other and to network in order to achieve continuous improvement of an organization,
- internal and external benchmarking to assess targets and gaps,
- continuous communication with the stakeholders of an organization.

Knowledge management helps an organization to gain insight and understanding from its own experience. Specific knowledge management activities help an organization focus on acquiring, storing and utilizing knowledge for such things as problem solving, dynamic learning, strategic planning and decision making. It involves the process of identifying, capturing, organizing, and using knowledge assets to create and sustain competitive advantage and it differs from information management in that the latter is focused on data whereas knowledge management is focused on information. A knowledge management system allows intangible information to be managed as an organizational asset in a manner similar to tangible assets.

Knowledge management and learning systems have a practical application when an organization wants innovation from everyone and everywhere, especially where "everywhere" is spread over five continents and tens of thousands of people. Knowledge management refers to how a firm creates, collects, stores, and shares knowledge. An effective knowledge management system can provide most people in an organization with access to knowledge and information about innovation, as well as related topics (Johnson, 2005; Snyder, Duarte, 2003).

The knowledge to be managed includes both explicit, documented knowledge, and tacit, subjective knowledge. Explicit knowledge includes information stored in documents or other forms of media. Tacit knowledge is information that is formed around intangible factors resulting from an individual's experience, and is personal and content-specific. These two aspects represent the above mentioned "know-how". Management entails all of the processes associated with the identification, sharing and creation of knowledge. This requires systems for the creation and maintenance of knowledge repositories as well as the cultivation and facilitation of knowledge sharing and organizational learning (Okes, Westcott, 2001).

An organization's knowledge management system should provide a mechanism for sharing the knowledge of employees and the organization to ensure that high-performance work is maintained through transitions. A culture and value system that support the collaborative sharing of knowledge across functions encourages the full participation of all employees in the process. Therefore, each organization should determine what knowledge is critical for its operations, and then implement systematic processes for sharing this information.

Armistead explores knowledge management from the perspective of operational processes, that is, the basic input-output transformation process (Fig. 1). At the input end, we have a combination of knowledge of a customer's needs and expectations, knowledge of raw materials and resources to be used, knowledge of products and services to be delivered, as well as data information or knowledge (Armistead, 1999). The knowledge conversion process is actually a changing and/or improving process. It consists of preserving, embedding and enhancing knowledge of a process, products and services. The knowledge conversion process can also be seen as one of knowledge creation, transferring and sharing, and a process of knowledge access improvement as well. Fostering a knowledge environment that is conducive to knowledge development, use, and transfer is vital in the knowledge conversion process.

Figurel
Input-output transformation knowledge process

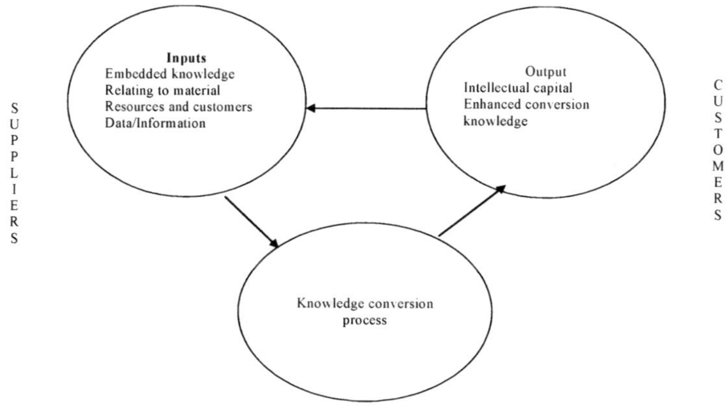

Source: Adapted from Armistead, 1999, pp.143-154.

In the Information and Technological Age which we have no entered, knowledge embedded in products and services, intellectual capital, as well as improved knowledge and understanding of customer needs are among the most important outputs of the knowledge conversion process. The process clearly indicates that knowledge management takes information, knowledge and people as its basic inputs, and applied knowledge and intellectual capital as its desired outputs. Knowledge management emphasizes knowledge creation, transfer, and embedding to serve different organizational purposes. These may include the enrichment of knowledge of customers, the building of knowledge capital, or developing enhanced access to knowledge.

Process improvement requires that new knowledge results in better processes and procedures. Increasing the knowledge of the organization, both in an individual sense as well as for the organization as whole, is the essence of learning. Knowledge can easily be lost if information is not documented, or when individuals are promoted or leave the organization.

The organization's knowledge base is made up of several dimensions (Trott, 2005). The following five dimensions can be considered:

- *Individual assets* - the skills and knowledge of the individuals that form an organization, the application of which significantly influences corporate success.
- *Technological assets* - the most immediately visible elements of the technological base, the set of reproducible capabilities in *product, process* and *support areas.* Technological assets can be more or less reliably reproduced; the other elements are, by contrast, fundamentally relational, which makes them much more difficult to replicate.
- *Administration assets* - the resources that enable the business to develop and deploy individual and technological assets. These are specifically the skill profile of employees and managers, the *routines, procedures and systems* for getting things done, the organizational structure, the strategies that guide action, and the culture that shapes shared assumptions and values.
- *External assets* - the relations that the firm establishes with current and potential allies, rivals, suppliers, customers, political actors and local communities, e.g. joint ventures, distribution channels, etc.
- *Projects* - the means by which technological, organizational and external assets are both deployed and transformed. Projects should be considered as part of the knowledge base insofar as the organization's *modus operandi* is a learned behavioral pattern that can contribute to or detract from technological and business performance.
- *Lagging measures* (outcomes) refer to what has happened; *leading measures* (performance drivers) predict what will happen. For example, customer survey results about recent transactions might be a leading indicator for customer retention (a lagging indicator); employee satisfaction might be a leading indicator for turnover, and so on. These measures and indicators should also establish cause-and-effect relationships across perspectives (Rampersad, 2003).

TOOLS OF LEARNING

There are clearly other factors that will also dramatically impact performance:

- Competence management encompasses the process of the continuous development of human potential within an organization. The goal of competence management is the continuous deliverance of top performance with a motivated and developed community. It focuses on the maximum development of employees and makes optimal use of their potential in order to achieve the goals of the organization. Competence management involves the development of job-related competences; the collection of information, capabilities, experience, skills, attitudes, standards, values, views, and principles (knowledge) that is focused on the expert fulfillment of a job. The development cycle is a critical focus. It consists of the following phases: result planning, coaching, appraisal, and job-oriented competence development.
- Kolb's learning cycle - those processes of instinctive learning (learning by experience) - is seen as being at the core of the management concepts mentioned. Together with the process of conscious learning (learning by education) these processes result in individual and collective behavioral changes. These two learning forms, as well as individual learning, PDCA-learning, and collective learning are all important. They are used to create conditions for effective organizational change. Kolb's learning cycle contains the following four phases: gaining hands-on experience; observing this

experience; drawing conclusions from this experience; and testing these ideas in experiments; which in turn will result in new behaviors and experiences. This philosophy encompasses a composition of closely-related management concepts, which together form a harmonious whole (Rampersad, 2005).

- Kaplan and Norton's Balanced Scorecard is only one version of performance measurement systems that have emerged as companies have come to recognize the need for a broad set of performance measures that provide a comprehensive view of business performance. These performance indicators span the entire business operation, from suppliers to customers, and from frontline workers to the top levels of management (Kaplan, Norton, 2001). The Balanced Scorecard provides a framework for designing a set of measures for activities which are deemed to be the key drivers of business. By having four distinct perspectives (financial, customer, internal process and innovation, and learning) it promotes a more holistic balanced view of an organization. By creating one's own measures under each of these headings, no important area would be overlooked (Gupta, 2004).

Implementing a balanced performance measurement system can have far-reaching effects on an organization. It can be used for:

- deciding what the key drivers of performance are;
- refocusing and stimulating activity on these key business drivers;
- drawing attention to goals and targets;
- creating a culture of achievement;
- recognizing in advance the trends affecting a business.

Figure 2
Relations between Balance Score Card, PDCA/PDAC cycle and Kolb's learning cycle

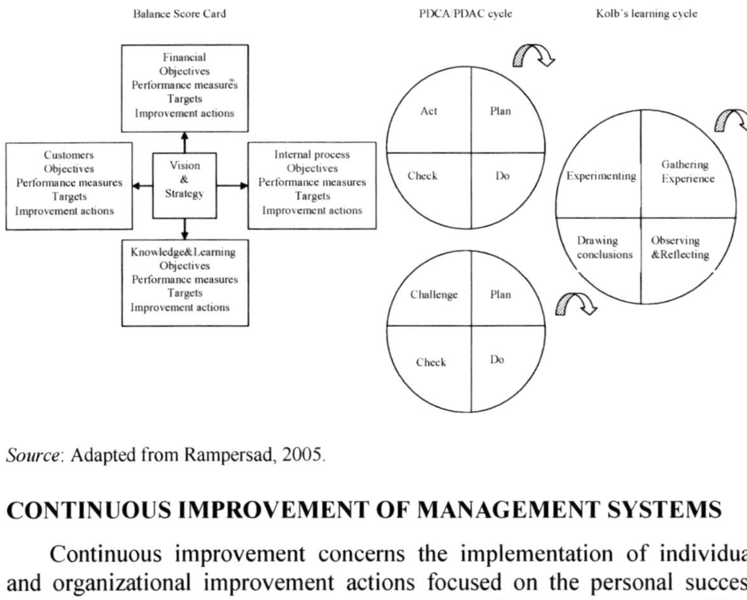

Source: Adapted from Rampersad, 2005.

CONTINUOUS IMPROVEMENT OF MANAGEMENT SYSTEMS

Continuous improvement concerns the implementation of individual and organizational improvement actions focused on the personal success and competitive advantage to the organization, respectively. When implementing the improvement process, the following five phases are identified:

- Process selection and definition: this involves selecting and defining a critical process related to the improvement actions eligible for continuous improvement.
- Process evaluation and standardization: this entails the description, evaluation, and standardization of the selected process.
- Process improvement: this is based on the continuous improvement of management of the evaluated processes according to the PDCA cycle.
- Personal improvement: spurring continuous individual improvement according to the Plan-Do-Act-Challenge cycle (PDAC-learning).
- Development: this entails continuous development of the job-related competences of individual employees based on the development cycle.

The emphasis lies on the job-related development and growth of individual employees through the absorption of knowledge, as well as the optimal use of their abilities, which enables them to fulfill their jobs more effectively. The development cycle is continuously followed through and consists of the following phases: result planning (drafting the competence profile and making related appointments regarding the results to be obtained); coaching (helping to gain the agreed upon results and giving feedback); appraisal (judging the functioning, checking if and how all the appointments are met and the results are obtained, giving feedback, and reviewing the results); and job-related competence development (development of employee competence through courses, on-the-job training, personal guidance, coaching, etc.).

Continuous improvement and learning are integral parts of management system. This requires analysis and problem solving everywhere within the company. Emphasis on continuous improvement encourages change, innovation, and creative thinking as regards how work is organized and conducted. Customer-driven quality emphasizes the "positive side of quality", that is enhancement, new services, and customer relationship management. Success with the "positive side of quality" depends heavily on creativity, more so than on steps to reduce errors and defects, which rely on well-defined techniques. Human resource focus stresses employee involvement, development, and recognition, and encourages creative approaches to improving employee effectiveness, empowerment, and contribution (Feigenbaum, Feigenbaum, 2003).

Definitions and descriptions of the concept of continuous improvement system/Business Excellence are often vague (Evans, Lindsay, 2002). It is therefore useful to provide a brief profile of continuous improvement of management system concepts by reviewing the vital principles:

- Customers include internal and external customers.
- Meeting and exceeding customer needs is a clearly stated aim.
- Leadership of a continuous improvement system/Business Excellence concept stems from top management and requires both individual and team commitment.

- The highest levels of integrity, honesty, trust and openness are essential ingredients of the continuous improvement system/Business Excellence concept.
- Mutual respect, mutual trust and mutual benefit for all stakeholders are important factors.
- The continuous improvement system/Business Excellence concept offers each individual the opportunity to participate, contribute and develop a sense of ownership.
- The continuous improvement system/Business Excellence concept involves continuous and measurable improvement at all levels of an organization.
- The continuous improvement system/Business Excellence concept requires consistent and precise performance to reach high standards in all areas of an organization.
- The aim of the continuous improvement system/Business Excellence concept is to use better resources in order to achieve effectiveness and efficiency.

Table 1
Similarities and differences between knowledge management and the continuous improvement system/Business Excellence concept

Similarities	
Knowledge management	**Continuous improvement system/Business Excellence concept**
– Continuous improvement and learning from the others	– Continuous improvement and measurement to achieve satisfaction of customers and other stakeholders (employees, owners, suppliers, society)
– Valuing employees/intellectual capital	– Valuing employees/intellectual capital
– People's competence development	– Employee training/education-competence development
– Empowerment involvement	– Empowerment involvement
– Teambuilding/collaboration	– Teambuilding/collaboration
– Acquiring knowledge of competitors, customers, suppliers and partners	– Acquiring customers and market knowledge
– Facilitating/improving access to knowledge	– Selection and use of data, information and opening communication channels
– Improving the quality and efficiency of decision-making	– Improving the quality and efficiency of decision-making
Differences	
Knowledge management	**Continuous improvement system/Business Excellence concept**
– Embedding knowledge of staff, customer, product and process	– Better use of resources to achieve effectiveness and efficiency
– Regarding knowledge as a source of competitive advantage	– Striving for excellence through benchmarking
– Achieving greater productivity through the use of knowledge	– Consistent and precise performance to reach high standards in all areas of organization
– Creating / disseminating new knowledge and embedding it in new technologies and products	– Effective leadership and team commitment throughout the focus on results
– Searching for new sources of information	– Decision-making process based on data facts
– Adapting knowledge to market needs	– Mutual respect, mutual trust and mutual benefits for all stakeholders

In terms of the input-output process, the continuous improvement system/Business Excellence concept is, similar to knowledge management, a process of transformation of a set of inputs, including plant equipment and raw materials, procedures and methods, information and knowledge, as well as people and their skills. The outputs of the transformation are products, services, information/paperwork and any results that meet customer needs and expectations.

The concept of *knowledge management* has been getting a lot of attention in many areas of industry, and this raises questions about how it affects existing efforts to define and implement management systems (based on ISO 9001, a business excellence concept).

A large number of companies introducing the business excellence concept through adopting ISO 9000:2000 series standards, or working out their own solutions in this field, try to assess the degree of their implementation in order to receive quality awards such as the Malcolm Baldridge Award, the European Quality Award, the Deming Prize, or national quality awards.

In the past 10-15 years Six Sigma has evolved as a methodology for implementing business process improvement initiatives (radical improvements as well as more incremental approaches). However Six Sigma has been implemented in many cases as a "program" and is not frequently integrated with other approaches to quality management, such as ISO 9001 and quality awards models (Pyzdek, 2003).

Figure 3
An integrated approach to management

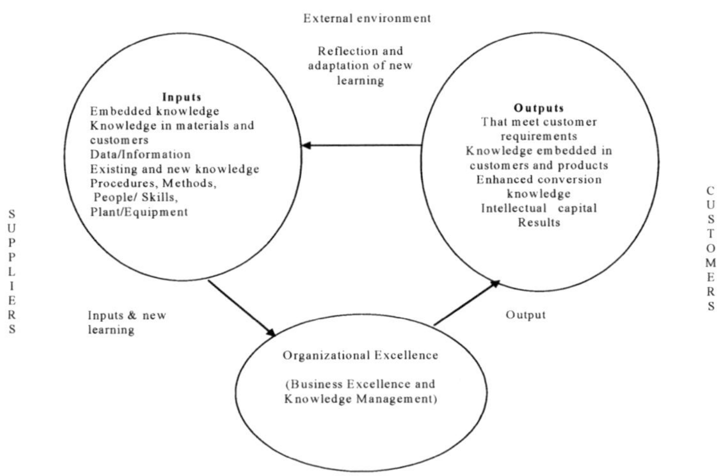

Source: Adapted from Oakland, 2004.

The comparison (Table 1) of the continuous improvement of management systems and knowledge management processes indicates their strength and complementariness. The framework (Fig. 3) shows that organizational excellence can be achieved through incorporating knowledge management concepts into the continuous improvement of management system processes whilst interacting with changes in the business environment.

CONCLUSIONS

In today's ambiguous and uncertain environment, organizations face critical issues of adaptation, survival and competence. By creating, acquiring, embedding and using knowledge those organizations can address these critical issues, as well as obtain competitive advantage. Searching for and acquiring new sources of information and new technologies helps organizations to stand out in the struggle for market share, in terms of both their products and services. Knowledge management and continuous improvement of management systems are complementary. A synergistic combination of knowledge management and continuous improvement of management systems forms a cycle of improvement and development, leading to organizational excellence.

Knowledge of, and understanding of customer needs and requirements are the prerequisite for customer satisfaction. The knowledge embedded in quality products and services is vital to the achievement of customer satisfaction. Moreover management by facts, a core value of the Malcolm Baldridge National Quality Award, is based on an organization's capability for obtaining, processing, disseminating and using data and information. While continuous improvement of a management system is focused on a result, which emphasizes the optimization of resources, and on greater productivity, a better use of intellectual capital and knowledge assets hold the key to achieving the desired results. As elaborated above, in order to be successful it is necessary to take an integrated approach to management. In other words, the continuous improvement of a management system should address environmental changes and deal with them through improving knowledge management capacities and skills. Although the proposed model can be described as a series of steps, it is intended to involve a series of spiraling , self-reflective cycles commencing with the knowledge and other inputs known at the time being fed into a holistic, total management process that produces the required output and stimulates individual and collective reflection, which in turn leads to new learning and adaptation of such into the inputs and the management process.

BIBLIOGRAPHY

Armistead, C., (1999), "Knowledge Management and Process Performance", *Journal of Knowledge Management*, Vol. 3, No. 2, pp.143-154.

Basu, R., J. N. Wright, (2003), *Quality Beyond Six Sigma*, Elsevier Butterworth Heinemann, Oxford.

Cobb, Ch. G., (2003), *From Quality to Business Excellence- A System Approach to Management*, ASQ Quality Press Milwaukee, Wisconsin.

Evans J. R., W. M Lindsay, (2002), *The Management and Control of Quality*, South-Western Thomson Learning, Cincinnati, Ohio.

Feigenbaum, A. V., D. S. Feigenbaum, (2003), *The Power of Management Capital –Utilizing the New Drivers of Innovation, Profitability, and Growth in a Demanding Global Economy*, McGraw-Hill, New York

Gupta, P., (2004), *Six Sigma Business Scorecard- Ensuring Performance for Profit*, McGraw-Hill, New York.

Johnson, J. D., (2005), *Innovation and Knowledge Management*, Edward Elgar Publishing, Northampton, Massachusetts.

Kaplan, R. S, D. P. Norton, (2001), *The Strategy-Focused Organization. How Balanced Scorecard Companies Thrive in the New Business Environment*, Harvard Business School Press, Boston, Massachusetts.

Oakland, J. S., (2004), Oakland on Quality Management, Elsevier Butterworth-Heinemann, Burlington MA.

Okes, D., R. T. Westcott, (2001), *Certified Quality Manager Handbook*, ASQ Quality Press Milwaukee, Wisconsin.

Pyzdek, Th., (2003), *Six Sigma Handbook*, McGraw Hill, New York.

Rampersad, H. K., (2003), *Total Performance Scorecard; Redefining Management to Achieve Performance with Integrity*, Elsevier Science, Burlington, MA.

Rampersad, H. K., (2005), *Personal Balanced Scorecard: The Way to Individual Happiness, Personal Integrity and Organizational Effectiveness*, McGraw-Hill, New Delhi.

Snyder, N. T., D. L. Duarte, (2003), *Strategic Innovation – Embedding Innovation as a Core Competency in Your Organization*, Jossey-Bass A Wiley Imprint, San Francisco.

Trott, P., (2005), *Innovation Management and New Product Development*, Financial Times Prentice Hall, London

Von Stamm, B., (2003), *Managing Innovation, Design Creativity*, John Willey and Sons, London.

16

The Effect of R&D Capital on Firm Productivity - Research Policy Implications

Ebbe Krogh Graversen
Michael Mark

INTRODUCTION

The potential for economic growth is often explained by the ability to be innovative and dynamic, and the present chapter analyses whether R&D, together with knowledge and the ability to use knowledge, i.e. absorption capacity, improves firm productivity. Firms investing in different types of innovation are assumed to generate more economic growth than comparable firms not investing in innovation. At the same time, the community return from investments in R&D is theoretically larger than the firm specific return, since the generated knowledge can be partly shared and used by other firms. The community return from investments in R&D is also generally found to be much larger than the firm specific return; see for example David et al. (2000) or Griffith et al. (2001) for a discussion of this point. In the literature, returns above 20 percent at the firm level and 50 percent at the community level are mentioned. Hence, the community has a large interest in supporting the framework and conditions for private sector R&D; cf. Sørensen et al. (2003). Guellec and van Pottelsberghe de la Potterie (2003) find an optimal subsidy of close to 15 percent.

The return on R&D capital and the R&D related factors that influence growth in firms is found to be positive. The firms examined in this analysis were collected from the official Danish R&D Statistic 2001 from The Danish Centre for Studies in Research and Research Policy, c.f. AFSK (2003). The Danish Centre for Studies in Research and Research Policy collects and delivers data on R&D activities in Danish firms to the OECD MSTI statistics. This data set is the most comprehensive and representative sample of the R&D activities of Danish firms so far used in this type of analysis. The data material for 2001 includes more than 2200 firms and broadly represents all private sector firms with more than 5-10 employees.

THEORETICAL AND EMPIRICAL MODEL

The analysis presented herein is based on a modified Cobb-Douglas production function as found in Solow (1956) and modified as in Hall and Mairesse (1995), among others. Present values of firms' accumulated R&D investments, together with interaction effects, are included directly in the production function, such that the influence from R&D on production is directly verifiable in the empirical model[1]. Similarly, a number of firm specific R&D related and general characteristics is used to identify observable differences in return and the level of firms' productivity. The empirical model is estimated sequentially to test whether the parameter estimate of R&D capital is robust.

In a logarithmic version of the Cobb-Douglas production function, where production is estimated as a function of R&D capital, other assets and labour, the estimation coefficient to R&D capital equals the output elasticity of R&D capital. Production is the firm specific value added. A split of the assets and labour into a part that is caused by R&D and a part that is not modifies the production function. Furthermore, level differences and interaction variables are included in the model in order to improve its explanatory power and to test for these differences[2]. Hence, the empirical modelling of the interaction effects of R&D capital is in line with a number of other analyses[3].

---

[1] Firms' accumulated R&D capital and R&D personnel are included in their assets and number of employees. To prevent double counting, the R&D capital is subtracted from the assets and the R&D personnel are subtracted from the number of employees in the production function. Hall and Mairesse (1995), Wakelin (2001) and Dilling-Hansen et al. (1999) point out that the problem of double counting in the innovation indicators used in the model is an important problem that changes the results significantly. See also Schankerman (1981) and Cuneo and Mairesse (1984) for early explorations of this specific bias problem in the empirical approach.

[2] The interaction variables are, for example, business sector indicators, size indicators etc. Especially business sector indicators are of interest in this study, since it is the first study where firms from all private business sectors are included. Usually only firms from the industry sectors or some part of it have been used.

[3] The modeling of growth rates are usually set up in an extended standard production function written as $Y=AK^{\alpha}N^{\beta}C^{\gamma}e^{\varepsilon}$ and in logarithmic form $y=a+\alpha k+\beta k+\gamma n+\varepsilon$, where y measures output, n measures labour input, k measures assets, and c measures the R&D

Firm data

The primary data set is the survey data collected for the Danish R&D statistics 2001. Additional information comes from the corresponding material from 1991, 1993, 1997, 1998 and 1999; cf. descriptions in AFSK (2003). The data material includes survey data on the R&D of private sector firms and other innovation activities in the years examined. The initial data set consists of the firms that are included in the 2001 data collection and it is weighted to represent the firm population of 18381 firms in 2001. The data broadly represents all private sector firms with more than 5 employees[4]. The R&D data has been merged with economic data and information on employees in the firms, using firm specific data from Statistics Denmark[5]. The firm units in the register and the number in the survey did not match exactly since, by law, they have different informational purposes. Furthermore, 50 firms had either none or a negative value added, so the data set usable for the analyses was reduced to 2228 firms representing 2929 firms with a positive R&D capital, and 14025 firms without.

A few variables in the survey data have a small fraction of empty or missing values. Some of them are substituted by logical values, such as 0 as number of R&D employees if the firm is R&D inactive. Other empty values are caused by missing or inconsistent answers in the survey. The missing values have been estimated using other forms of information in the statistics or other sources. For example, a firm is deemed to have a positive R&D capital if it earlier had R&D expenses, but not any longer.

The firm specific R&D capital is constructed in the same way as most studies in the area, following the pioneering work by Griliches (1979). The present value of a firm's R&D capital is found as described in the appendix, based on the data material for the previous ten years. However, in order to test the correlation of the results with the definition used and the construction of the R&D capital, six other R&D capital measures have been constructed and used in the estimates; see the appendix description. Bernstein and Mamuneas (2006) find R&D depreciation rates in R&D intensive industries at 18-29 percent. As was done in Hall and Oriani (2004), we also use depreciation rates of 15 percent, accumulating over 5 and 10 years respectively. We also use a 20 percent depreciation rate as was done in Smith et. al. (2000) and Dilling-Hansen (1999). Furthermore, we use different depreciation rates for different R&D cost types;

---

capital. This modelling is the basis for a number of empirical studies of the relation between innovation and productivity. Studies use the production model, with various adjustments in the definition of output and specifications of how innovation influences production; cf. Hall and Mairesse (1995), Singh and Trieu (1996), Dilling-Hansen et al. (1999), Smith et al. (2000), Wakelin (2001), Criscuolo and Haskel (2002), Rouvinen (2002), Medda et al. (2003).

[4] More specificly, it is representative of all private sector firms in Denmark with 2 or more employees in the IT, R&D and other knowledge service business sectors, 50 or more employees in the primary industry, retail trade and lawyers' business sectors, and 6 or more employees in the remaining business sectors, c.f. AFSK (2003).

[5] It was not possible to create a panel data set due to limited financial resources.

see Table A.1. Finally, we use the results from Bayer (2003), with a depreciation rate over three years and an immediate depreciation of 40 percent in period 0.

The difference in average R&D capital is small. Naturally the average R&D capital is larger when the accumulation period is 10 years compared to 5 years, but only around 5-10 percent. Similarly, the average is larger when the depreciation rate is 15 percent instead of 20 percent, but again only 10-15 percent. Hence, the average R&D capital fluctuates less than 10 percent, except for the capital measure as constructed by (Bayer, 2003), where it is one-third the size of the others.

FIRM CHARACTERISTICS

One out of six firms was R&D active in 2001; cf. Graversen and Mark (2005); one out of four was innovative, and one-third of the R&D active firms had an R&D department. Naturally, the shares vary in differing business sectors and different firm size groups[6]. For example, more than half of the firms in the medical industry and more than half of large firms are R&D active. These firms also more often have R&D departments and research-educated employees[7].

There is a high consistency between the R&D active firms in 2001 and the firms with a positive R&D capital in 2001. The latter group is a bit larger, since a few R&D inactive firms in 2001 had been R&D active in at least one of the ten previous years. Hence, there is almost no difference in firm characteristics in the two firm groups and there will be no distinction between them in the following sections. The term 'R&D active firms' will cover both groups in the remaining sections. This means that general differences that relate to business sector and firm size can be found in Table 1 as well as in Table 2. Table 1 shows that there is a significant difference in the levels between the sectors and firm groups, which indicates a need for these variables to be included as explanatory variables in the estimation model.

The concept of value added is used to measure production in the present analyses. The results in Table 2 show that R&D active firms have a 40 percent larger value added per employee than R&D inactive firms. R&D active firms have close to 50 percent of the total value added in the private sector, although they only constitute 17 percent of all firms included in the data set.

---

[6] See the definition on business sectors in appendix Table A.2. The sectors are defined according to the Eurostat definition of technology level sectors. The medical and ICT sectors are separated out due to their high R&D intensity and policy interest. As a consequence, the remaining firms in the high and medium-high technology industry sector were so few that they are merged in the sector "Other high and medium-high technology industry".

[7] Research-educated employees are defined as employees with a degree/job at the Ph.D. level.

Table 1
Characteristics among firms with R&D capital

Business sector	R&D personnel	Share of R&D personnel	R&D costs/ turnover	R&D costs/ R&D personnel	R&D Capital/ R&D personnel	Share of firms	Share of employees
------- Average conditional on positive R&D capital ------							
Industry	Number	%	%	1000 DKK	1000 DKK	%	%
Low tech	5	3	1	269	1895	13	22
Medium-low tech	5	6	2	281	1889	11	11
Medical	71	32	15	514	2465	3	7
ICT industry	24	23	7	412	2109	7	6
Other high and medium-high technology industry	13	16	6	352	2314	20	24
Service							
ICT Service	13	35	25	564	1916	16	12
Other knowledge service	13	49	22	302	1464	19	9
Other sectors	3	9	2	334	1201	11	9
Firm size							
Under 50 employees	4	29	13	314	1126	65	10
50-99 employees	8	11	6	351	1343	13	8
100-249 employees	16	10	8	554	2100	13	18
Over 249 employees	74	9	4	519	2415	9	65
All firms	12	23	11	364	1996	100	100

Table 2
Average value added per employee in 2001: by business sector and firm size, 1000 DKK and shares in percentage

Business sector	R&D inactive firms		------------ R&D active firms --------------				All firms
	Value added per employee		Value added per employee	Difference 1)	Share 2)	R&D capital per employee	Value added per employee
Industry	1000 DKK		1000 DKK	%	%	1000 DKK	1000 DKK
Low tech	361	<<	477	32	14	38,5	416
Medium-low tech	382	<	412	8	17	50,0	397
Medical	353	<<<	922	161	64	554,5	909
ICT industry	423	<<	546	29	29	341,6	490
Other high and medium-high technology industry	397	<	434	9	39	178,7	427
Service							
ICT Service	600	<	706	18	43	222,3	676
Other knowledge service	325	<<<	486	50	19	292,5	372
Other sectors	366	<	425	16	6	41,7	373
Firm size							
Under 50 employees	424	<	448	6	13	216,0	428
50-99 employees	374	<	437	17	26	112,3	390
100-249 employees	355	<	420	18	47	147,7	385
Over 249 employees	319	<<<	564	77	60	176,8	461
All firms	371	<<	517	39	17	170,6	431

Note: Number of unweighted observations is maximum 2228.[1)] Percent higher average value added per employee between R&D active and inactive firms. [2)] Share of firms that have a positive R&D capital.

The average value added per employee is also given in Table 2. Differences in average values over business sectors and firm size groups show the differences in levels between these groups. Similarly, the differences in value added per employee among R&D active and inactive firms show that the R&D active firms have a relatively higher average value added. The numbers in Table 2 do not show whether firms have a positive return from R&D capital, only the value added per employee. The differences can be caused by other factors, such as differences in capital intensity between business sectors, differences in employee education levels, etc.

As the numbers in Table 2 show, there are considerable differences, indicating that the medical and ICT business sectors and large firms have the highest value added per employee. The R&D active firms in all sectors and groups have a higher value added per employee than among the corresponding R&D inactive firms. Generally this indicates that R&D capital correlates positively with a firm's value added per employee.

The relatively small difference in value added per employee among small firms with and without R&D capital is caused by the relatively many ICT firms with a high value added per employee, which increases the average for R&D inactive firms in this group. Similarly, the fact that firms are small and young when they do not have R&D activities to support the growth seems to increase value added. The correlation between value added per employee and firm size is small and insignificantly negative for R&D inactive firms, but positive although insignificant for R&D active firms. For these firms there is instead a significant positive correlation of 0.4 between value added per employee and R&D capital per employee. Hence, the basic statistics support the results in earlier studies, in for example Lundvall (2002), finding that there are several typologies of firms: small and large firms with low value added per employee; small firms on their way up with a large value added per employee; and large firms with a large R&D activity and a large value added per employee.

Less than 4 percent of the employees in the R&D inactive firms are highly educated, while more than 9 percent in the R&D active firms were highly educated[8]. The overall average share is 6 percent highly educated employees among the overall number of employees in the firms.

PRODUCTIVITY AND R&D CAPITAL

The influence of R&D capital on productivity was empirically tested on the firms included in the analyses. The estimated results, the variation in the results, and their robustness is presented in Tables 3 and 4. The general result is that the return on R&D capital is positive, large and significant. This result confirms a long list of earlier empirical studies finding a positive correlation between R&D investments and productivity growth; cf. for example Griliches (1998), Smith et al. (2000) and Salter and Martin (2001). A new finding here is that the return on R&D capital increases with the share of R&D personnel and research educated employees.

---

[8] Highly educated employees are defined as those with a bachelor degree or higher.

Generally, the parameter estimate on R&D capital is robust with regard to changes in the model specification and in the data sample[9]. However, a difference is that the parameter estimate of R&D capital increases by a factor of five when the model is estimated on the R&D active firms alone; see Table 3. We do not see this as an indication of a model misspecification, i.e. that the model cannot capture both R&D active as well as inactive firms. Instead, we interpret the result as an indication of a high value added among the R&D inactive firms, as indicated in Table 2. There is a high return on R&D among R&D firms, but a large number of the R&D inactive firms do well as knowledge receivers or users, i.e. imitation through absorptive capacity, as was mentioned in Cohen and Levinthal (1989).

Table 3
Estimation of the production function

Model	All firms				R&D active firms			
	1[1)]	2[2)]	Basis[3)]	4[4)]	1[1)]	2[2)]	Basis[3)]	4[4)]
Constant	**9.308**	**9.366**	**9.664**	**9.813**	**8.294**	**8.102**	**8.846**	**10.390**
Labour	**0.626**	**0.625**	**0.644**	**0.642**	**0.504**	**0.517**	**0.573**	**0.585**
R&D labour	-	-	-	**0.226**	-	-	-	**0.209**
Assets	**0.289**	**0.286**	**0.264**	**0.256**	**0.384**	**0.314**	**0.242**	**0.234**
R&D capital	-	**0.016**	**0.018**	**0.019**	-	**0.094**	**0.114**	**0.023**
Indicator for R&D capital > 0	**0.205**	-	-	-	-	-	-	-
Number of firms	2228	2228	2228	2228	662	662	662	662
R^2	0.90	0.90	0.89	0.90	0.91	0.92	0.92	0.92

Note: All variables are measured in logarithms except the 'indicator variable for R&D'. R&D capital is accumulated over ten years and depreciated by 20 percent per year. Parameters in **bold** are significant at a 1% level. [1)] Simple model with an indicator for R&D capital>0. [2)] Incl. R&D capital, but without correction for double counting. [3)] Incl. R&D capital and correction for double counting in assets. This model is the basis model in Section 4. [4)] Incl. R&D capital and R&D personnel and correction for double counting in both assets and labour.

In the calculation of R&D capital the present value of the firms' salaries to R&D personnel is included. For the year 2001 it is also included in the R&D capital, which results in a double counting of R&D personnel through the labour variable and the R&D capital variable; cf. Model 2 in Table 3. Hence the labour

[9] A truncation of the data set that removes the upper and lower five percent of the firms sorted by value added per employee does not change the results in any significant way. Hence there is no indication of extreme values or firms in the data set. Similarly, the model is estimated in a version where the R&D cost in 2001 is subtracted from the R&D capital in order to test the causality between cost and productivity in the same period. This does not influence the results, so it is not pursued further in this paper, although for example Gu and Tang (2004) on Canadian manufacturing data find that it takes 1-3 years for innovations to generate a positive impact on productivity.

variable is fully reduced by the R&D personnel, cf. Model 3 in Table 3. In the analysis where R&D personnel constitutes a separate variable, the salary costs for 2001 are also removed from the R&D capital in order to prevent a similar double counting;, cf. Model 4 in Table 3. Table 3 shows the estimated results from a sequential extension of the standard production function. This was estimated both for all firms and for R&D active firms, respectively. Model 1 is the standard model, with a single dummy variable for whether a firm is R&D active or not. The extension in the remaining Models 2, 3 and 4 was briefly explained above. Model 3 is the model that is used in our further analysis.

In Table 3 it can be seen that the coefficients to labour and other assets are very stable and robust. In column 1 under all firms the coefficient for the dummy variable measures the difference in productivity between R&D active and inactive firms; here the coefficient means that the R&D active firms have a 20 percent higher productivity on average than R&D inactive firms. The results also show that the correction for double counting between Models 2 and 3 only slightly increases the coefficient for the R&D capital. Including labour reveals the high significance of the R&D personnel factor. In the case of all firms, their inclusion does not change the coefficient for the R&D capital, but in the case of R&D active firms alone, the coefficient for the R&D capital is reduced by a factor of five. Hence, when R&D personnel is seen as a short term R&D indicator and the R&D capital measure in Model 4 as a long term R&D indicator, there seems to be a large short term effect of R&D on firm productivity, and a smaller although still highly significant long term effect. In the basis model, Model 3, the R&D capital measures both the short and long term effects.

As described in the appendix, seven different R&D capital measures were constructed, and Graversen and Mark (2005) show that the coefficient for R&D capital is more or less independent of how the R&D capital is constructed, which results correspond to those found in Hall and Mairesse (1995)[10]. Hence, a brief conclusion would be that it does not matter, for the results, whether R&D capital is accumulated over 5 or 10 years, whether a 15 or 20 percent depreciation rate is used, or whether or not R&D cost type specific depreciation rates are used. However, Hall and Mairesse (1995) do conclude that the longer the accumulation period the better is the robustness of the results, while the choice of depreciation rate does not matter.

Interaction and level effects

Table 4 shows the estimated results from the basis model for R&D active firms, extended with level effects and R&D interaction effects. There is insufficient space to comment on all the results, so what follows will be a summary of the main findings and their corresponding implications. In several of the models in Table 4 the interaction coefficient is positive and the

---

[10] The coefficient for R&D capital constructed by Bayer (2003) halves the coefficient for R&D capital.

corresponding level effect is negative. Basically, this just means a higher payoff per R&D capital unit to those firms with these characteristics, and not a lower level for the same firms, since the level refers to the crossing at zero R&D capital, where no R&D active firm is situated.

As Models 1 and 2 (column 1 and 2) in Table 4 show, neither a business sector nor a size effect exists among the R&D active firms. There is a higher effect on the residual group called 'other sectors', but this is caused by a few atypical firms in this small group of R&D active firms. Hence, in contrast to Kafouros (2005) using UK data, we do not find a higher coefficient for high tech firms. The level coefficient for the residual group is also negative and a negative level effect is found among the small firms in column 2 as well, indicating an increasing productivity level with firm size. Hence, contrary to, for example, Tsai and Wang (2005), who find a u-type relation between R&D productivity and size, and Kafouros (2005), who finds a larger coefficient for large firms, there are no interaction differences across firm size in the present study.

There is no additional influence resulting from foreign ownership, R&D departments, or employment of researchers (columns 3 and 4) among the R&D active firms. The finding concerning R&D departments and the employment of researchers seems robust, since it is invariant to inclusion of business sector dummies; cf. Graversen and Mark (2005).

Firms receiving external financing for their R&D activities have a higher coefficient for R&D capital; cf. column 5. This result was also noted in Bönte (2003). There seems to be a positive correlation between external finance and productivity. The type of R&D also seems to influence the coefficient for R&D capital; cf. column 6, where applied R&D increases the R&D capital coefficient. This result is also invariant to inclusion of the business sector and size group dummies.

The most impressive results are found in columns 7 and 8, where the share of R&D personnel and the share of researchers are included as interaction variables[11]. However, inclusion of the share of R&D personnel reduces the common coefficient for R&D capital by 50 percent. This indicates that the R&D personnel variable is a useable substitute for R&D capital. The model estimate in columns 8 shows a similar significant additional effect of employing researchers. Hence, the coefficient for R&D capital increases both with the share of R&D employees and the share of researchers. In the latter case, some firms do not have researchers employed, while all R&D firms by definition have R&D employees.

---

[11] This effect is also invariant to inclusion of business sector and size group dummies (not shown here).

Table 4
The basis model estimated with different interaction coefficients for R&D capital and level; R&D active firms only

		Basis model with different interaction and level indicators							
		1	2	3	4	5	6	7	8
Constant		**8.772**	**10.055**	**8.879**	**9.118**	**9.099**	**9.118**	**9.455**	**9.780**
Labour		**0.585**	**0.497**	**0.566**	**0.577**	**0.570**	**0.563**	**0.675**	**0.576**
Assets		**0.249**	**0.229**	**0.245**	**0.237**	**0.246**	**0.242**	**0.234**	**0.253**
R&D capital		**0.103**	**0.092**	**0.109**	**0.100**	**0.091**	**0.104**	**0.049**	**0.103**
Interactions with R&D capital									
Industry	Low tech	0.017							
Industry	Medium-low tech	-0.038							
Industry	Medical	-0.006							
Industry	ICT industry	-0.019							
Industry	Other high and medium-high technology industry	Ref.							
ICT service		0.032							
Other knowledge service		-0.009							
Other sectors		***0.059***							
< 50 employees			0.041						
50-99 employees			-0.011						
100-249 employees			-0.052						
> 249 employees			Ref.						
Foreign ownership				0.023					
R&D Department					0.005				
Researchers employed					0.033				
Externally financed R&D						**0.106**			
Public financed R&D						-0.024			
Basic R&D							-0.001		
Applied R&D							**0.061**		
Development R&D							0.005		
Share of R&D personnel								**0.056**	

Share of researchers									**0.136**
Level interactions									
Industry	Low tech	- 0.183							
	Medium-low tech	0.601							
	Medical	0.320							
	ICT industry	0.249							
	Other high and medium-high technology industry	Ref.							
ICT service		- 0.335							
Other knowledge service		0.347							
Other sectors		***- 0.820***							
< 50 employees			**-1.088**						
50-99 employees			-0.175						
100-249 employees			0.580						
> 249 employees			Ref.						
Foreign ownership				- 0.214					
R&D Department					- 0.134				
Researchers employed					- 0.335				
Externally financed R&D						***- 1.524***			
Public financed R&D						0.389			
Basic R&D							0.191		
Applied R&D							***- 0.979***		
Development R&D							- 0.072		
Number of firms		662	662	662	662	662	662	662	662
R^2		0.92	0.92	0.92	0.92	0.92	0.92	0.92	0.92

CONCLUSIONS

The present analysis has documented the positive influence of R&D on firm productivity. The immediate conclusion is that R&D active firms have a much higher value added per employee than R&D inactive firms. Similarly, there is a significant positive return on R&D capital in terms of productivity. The following conclusions may be offered:

- R&D active firms have a 40 percent higher value added per employee. The difference is 20 percent when corrected for differences in labour and capital input.
- R&D active firms have 50 percent of the total value added in the private sector, even though they constitute only 17 percent of all private firms.
- The return on R&D capital is 34 percent among R&D active firms. This return is significantly higher than the return on other firm specific capital.
- The higher the share of R&D employees or researchers in the firms, the higher is the elasticity of R&D capital.

The estimated effects are minimum estimates for the social or community effect, since there is a positive multiplication effect of such increases in R&D activities. The conclusions concerning the private sector are also found in other more or less comparable studies. In Smith et al. (2000) it is concluded that the return on R&D capital is positive and large, and in the DISKO report no. 7 from 1999 it is concluded that increased competitiveness causes more innovation, but often decreases the employment growth, and last but not least in the IDA report (IDA, 2004) it is concluded that innovation increases growth and that the employment of highly educated employees increases the probability of innovation in firms.

Research policy implications

The results found in our analysis give some explanations as to how R&D activities influence private sector productivity. The findings show that the private sector has a group of firms that are more competitive and create a higher value added because they are R&D active. The present research policy in Denmark aims to increase private sector competitiveness through an increased knowledge stock imbedded in the production. This seems to be an effective strategy, since there obviously is an economic benefit from R&D activities.

In the light of the conclusions above, the future development of research policy must be to further support the possibilities of sectors to become and stay innovative, or even better R&D active, and to increase the amount of R&D activities overall. Furthermore, the private sector should be supported in their aims to increase the knowledge imbedded in their production processes. The analyses contained in this report show that an effective way to support both intentions is to increase the education level in firms by employing R&D competent employees or by up-qualifying the competences among the existing stock of employees in firms.

The analysis yields results that support the findings in other studies using data from other countries; cf. footnote 3. Hence, the generality of the results indicates that the policy conclusions above are globally valid. In particular the findings regarding the importance of absorptive capacity in the form of R&D personnel can and should influence future international research policy and be of interest for all countries aiming to become more competitive.

However, the general applicability of the results should be confirmed by studies in other countries using the same model and data type, since the generality builds on indices. OECD, among others, is currently working on such a project using the 4th Community Innovation Survey, CIS4, and the results will hopefully confirm the general applicability of the results of this study.

BIBLIOGRAPHY

AFSK, (2003), *R&D in the Private Sector. Statistics for 2001.* (In Danish: Erhvervslivets forskning og udviklingsarbejde. Forskningsstatistik 2001) R&D Statistics Series 2003, The Danish Institute for Research and Research Policy.

Bayer, M., (2003), *The Market Valuation of R&D Intensive Firms and the Capitalization of R&D Expenditures* (In Danish: Aktiemarkedets prisfastsættelse af FoU-intensive virksomheder, herunder aktivering af FoU-udgifterne). AFSK Report 2003/1, The Danish Institute for Studies in Research and Research Policy.

Bernstein, J. and T. Mamuneas, (2006), "R&D Depreciation, Stocks, User Costs and Productivity Growth for US R&D Intensive Industries", *Structural Change and Economic Dynamics*, Vol. 17, pp. 70-98.

Bloch, C., (2003), "The Effect of R&D Expenditures on Stock Market Returns for Danish Firms", *AFSK WP 2003/6*, The Danish Institute for Studies in Research and Research Policy.

Bönte, W., (2003), "R&D and Productivity: Internal vs. External R&D - Evidence form West German Manufacturing Industries", *Economics of Innovation and New Technologies,* Vol. 12, pp. 343-360.

Chan L. K. C., J. Lakonishok and T. Sougiannis, (2001), "The Stock Market Valuation of Research and Development Expenditures", *Journal of Finance*, Vol. 56, pp. 2431-256.

Cohen W. and D. Levinthal, (1989), "Innovation and Learning: The Two Faces of R&D", *Economic Journal,* Vol. 99, pp. 569-596.

Criscuolo C. and J. Haskel, (2002), "Innovations and Productivity Growth in the UK", *CeRiBA Discussion Paper.*

Cuneo P. and J. Mairesse. (1984), "Productivity and R&D at the Firm Level in French Manufacturing", [in:] Z. Griliches (ed.) *R&D, Patents, and Productivity*, University of Chicago Press.

David P., B. Hall and A. Toole, (2000), Is Public R&D a Complement or Substitute for Private R&D? A Review of the Econometric Evidence. *Research Policy*, Vol. 29, pp. 497-529.

Dilling-Hansen M., T. Erikson, E. S. Madsen and V. Smith, (1999), "The Impact of R&D on Productivity: Evidence from Danish Manufacturing Firms", *AFSK WP 1999/1.* The Danish Institute for Studies in Research and Research Policy.

Graversen E. K. and M. Mark, (2005), "The Effect of R&D Capital on Firm Productivity, *CFA Working Paper 2005/3.* The Danish Centre for Studies in Research and Research Policy.

Griffith, R., S. Redding and J. Van Reenen, (2001), "Measuring the Cost Effectiveness of an R&D Tax Credit for the UK", *Fiscal Studies*, Vol. 22, No. 3, pp. 375-399.

Griliches, Z., (1979), "Issues in Assessing the Contribution of Research and Development to Productivity Growth", *Bell Journal of Economics*, Vol. 10, pp. 92-116.

Griliches, Z. and J. Mairesse, (1984), "Productivity and R&D at the Firm Level", [in:] Z. Griliches (ed.), *R&D, Patents and Productivity*, University of Chicago Press, Chicago.

Griliches, Z., (1998), "R&D and Productivity: Economic Results and Measurement Issues", [in:] P. Stoneman (ed.), *Handbook of the Economics of Innovation and Technological Change*, Blackwell, pp. 52-89.

Gu, W. and J. Tang, (2004), "Link Between Innovation and Productivity in Canadian Manufacturing Industries", *Economics of Innovation and New Technology*, Vol. 13, pp. 671-686.

Guellec, D. and B. Van Pottelsberghe de la Potterie, (2003), "The Impact of Public R&D Expenditure on Business R&D", *Economics of Innovation and New Technologies*, Vol. 12, pp. 225-244.

Hall B. (1993). The Stock Market's Valuation of R&D Investment During the 1980's. *American Economic Review*, vol. 83, pp. 259-264.

Hall, B. and J. Mairesse, (1995), "Exploring the Relationship between R&D and Productivity in French Manufacturing Firms", *Journal of Econometrics*, Vol. 65 (1), pp. 263-293.

Hall, B. and R. Oriani, (2004), "Does the Market Value R&D Investment by European Firms? Evidence from a Panel of Manufacturing Firms in France, Germany and Italy", *NBER WP 10408*.

IDA, (Rambøll Management and The Danish Society of Engineers), (2004), *The Value of High Educated Employees in Small and Medium Sized Enterprises*. (In Danish: Højtuddannedes værdi for små og mellemstore virksomheder). IDA Report.

Kafouros, M. I., (2005), "R&D and Productivity Growth: Evidence from the UK", *Economics of Innovation and New Technology*, Vol. 14, pp. 479-497.

Lundvall, B.-Å., (2002), "Growth", *Innovation and Social Cohesion: the Danish Model*, Edward Elgar.

Medda, G., C. A. Piga and D. Siegl, (2003), "On the Relationship Between R&D and Productivity: A Treatment Effect Analysis", *Working Paper 02/2003*, Nottingham University Business School.

Rouvinen, P., (2002), "R&D-Productivity Dynamics: Causality, Lags and «Dry Holes»". *Journal of Applied Economics*, Vol. V, No. 1, pp. 123-156.

Salter, A. and B. Martin, (2001), "The Economic Benefits of Publicly Funded Basic Research: a Critical Review", *Research Policy*, Vol. 30, pp. 509-532.

Schankerman, M., (1981), "The Effect of Double Counting and Expensing on the Measured Returns to R&D", *Review of Economics and Statistics*, Vol. 63 (3), pp. 454-458.

Singh, N. and H. Trieu, (1996), "The Role of R&D in Explaining Total Factor Productivity Growth in Japan, South Korea and Taiwan", *Working Paper No. 361*, Department of Economics, University of California, Santa Cruz.

Smith, V., M. Dilling-Hansen, T. Eriksson og E. Strøjer Madsen, (2000), "R&D and Productivity in Danish Firms: Some Empirical Evidence", *AFSK WP 2000/4*, The Danish Institute for Studies in Research and Research Policy.

Solow, R., (1956), "A Contribution to the Theory of Economic Growth", *Quarterly Journal of Economics*, Vol. 70, pp. 65-94.

Sørensen, A., H. C. Kongsted and M. Marcusson, (2003), "R&D, Public Innovation Policy, and Productivity: The Case of Danish Manufacturing", *Economics of Innovation and New Technology*, Vol. 12, pp. 163-178.

Tsai, K. and J. Wang, (2005), "Does R&D Performance Decline with Firm Size? - A Re-examination in Terms of Elasticity", *Research Policy*, Vol. 34, pp. 966-976.

Wakelin, K., (2001), "Productivity Growth and R&D Expenditure in UK Manufacturing Firms", *Research Policy* 30, pp. 1079-1090.

APPENDIX
Construction of R&D Capital measures

1) **R&D capital** is the present value of all firm specific R&D costs in the past., here calculated as the present value of the R&D costs over the last ten years, following Griliches and Mairesse (1984), Hall (1993) and Hall and Oriani (2004)
   a) Alternatively period t is not included in the sum, i=0, 1, ... is changed to i=1, 2, ... under point 4.a. In this way a clear time dependent causality from R&D costs to productivity is secured.
2) **R&D Costs in the years previous to 2001**. The costs are collected in real values, i.e. 2001-prices. The firms R&D costs are used if reported. Otherwise, R&D costs are estimated from the other firms in the years the R&D cost is not reported. The exact procedure is as follows:
   a) The firm's own reported R&D costs are used if they exist.
      i) Values in the years between two valid observations are linearly approximated
   b) In case of missing information on R&D costs in previous years, it is assumed that:
      i) There is no R&D cost in previous periods if there is no R&D cost in 2001 or in the first valid observation after a missing observation period.
      ii) Sector specific growth rates are calculated in the respective years. The R&D costs are calculated as $R\&D_{t-1} = R\&D_t * (1+r_t)^{-1}$, where r_t is the sector growth rate.
      iii) The estimated R&D costs are corrected for the non-existence of firms in previous years.
3) **R&D costs** are assumed to have a **real depreciation rate**, d_{t-i}, which may depend on the R&D type, c.f. table A.1. Hall and Oriani (2004) use a depreciation rate of 15 percent and accumulates over 5 and 10 years respectively. They do not consider R&D type. Dilling-Hansen et al. (1999) and Smith et al (2000) use 20 percent, while Bloch (2003) and Chan et al (2001) use a linear depreciation of 20 percent per year over four years (20, 40, 60 and 80 percent).

Table A.1.
R&D cost type and depreciation rates used in K5a and K10a

R&D cost type	Depreciation rate, d_t	R&D cost type	Depreciation rate, d_t
Operational costs:	15%	**Investment costs:**	10%
Salary	14%	Buildings	7%
Other operational costs	20%	Apparatus and instruments	12%

4) **R&D capital,** K_t, is calculated as the present value of the R&D costs in the firms[12]:
   a) As $K_t = \Sigma_i \Sigma_u R\&D_{u,t-i} * \Pi_i (1-d_{u,t-i})$, i = 0, 1, ..., J, u = R&D type. In simple terms, this is just the sum of the present value of all the annual R&D costs, adjusted to 2001 prices.
5) **Different measures of R&D capital,** K_t. Seven different R&D capital measures are calculated and used to test the parameter stability in the empirical model. They are calculated over 5 and 10 years respectively, with depreciation rates as shown in Table A.1, being 15 and 20 percent respectively. Finally, the K_Bayer_t, with a depreciation rate over three years and an immediate depreciation of 40 percent in period 0 is used. See also Graversen and Mark (2005).

Table A.2
Business sector definitions

Business sector	NACE code for R&D activity
Other high and medium-high tech industry	24, 29-35.3 (except medico and ICT industry)
Medium-low tech industry	23, 25-28
Low tech industry	15-22, 35.4-37
Knowledge service	64, 72-74, 92.2 (except ICT service)
Other sectors	1-14, 40-41, 45, 51-52, 60-63, 65-67, 75.25, 90 (except ICT trade)
Medical (industry)	24.4, 33.1
ICT industry and trade	30.01-30.02, 31.3, 32.1-32.3, 33.2-33.3, 51.43.2, 51.64.1, 51.84, 51.86
ICT service	64.2, 71.33, 72.1-72.6

---

[12] The depreciation rates could be estimated at the sector level using the method explained in Bayer (2003). He estimates a model on the ordinary operational result explained by material assets, sales costs and R&D costs, together with three lags of R&D costs, and finds decreasing depreciation rates over three years and an immediate depreciation on 40 percent in period 0. Unfortunately, the depreciation rates are relative (sums to 1) and not absolute. Hence, the corresponding coefficient for the R&D capital is not absolute either.

17

Innovative & Research and Development Activities of Firms in Luxemburg

Vincent Dautel

INTRODUCTION

Research and Development (R&D) and innovation are important for economic success. This has led the European Union Member States to decide, at the Barcelona European Council (2002), that public and private R&D inputs have to increase so as to reach 3% of GDP in 2010.

Moreover, Research and Development, and also innovation, are highly important topics on the agenda of Luxembourg public authorities. We can illustrate this by the steep increase in the public support of R&D activities since 2000. Initially, we can note that Government funding has tripled between 2000 and 2005 (Government Budget Appropriations or Outlays on R&D have increased by 193%). Secondly, public research centres have been developed (R&D expenses of these centres have increased by 89% between 2000 and 2004). Thirdly, the University of Luxembourg, whose aim is to be a "Research university", was created in 2003.

Concerning these changes, it seems interesting to study the effect of public and private R&D investments. In a more restrictive view, it also seems opportune to look at the impact of R&D impact on the innovative performance of firms. This second theme is the subject of this analysis.

In order to study this, we will use the micro data from the Community Innovation Survey (CIS3) carried out in Luxembourg.[1] This survey has many advantages for the purpose of our study. Firstly, it provides micro data of R&D input, external knowledge adoption and co-operation agreements, including co-operation with the "science base" (i.e. scientific partners, including public research centres). And secondly, it includes measurements of innovative results (i.e. share of turnover from innovative products).

Based on this data, our analysis will examine the following issues:

- Do firms involved in R&D activities benefit from a greater innovative performance?
- Do firms engaged in R&D activities above a critical level have greater abilities to take advantage of existing knowledge external to the firm? (i.e. evaluation of the absorptive capacity hypothesis outlined below)
- Will co-operation with scientific actors (including Luxemburgish public R&D centres) generate greater innovation outputs?

This chapter is organised as follows. Section 1 describes our theoretical background for studying the influence of Research and Development. Section 2 focuses on our reference model. The data (CIS3) and the variables used are presented in the third section. Finally, the last section is devoted to an empirical analysis.

ASPECTS AND ISSUES OF RESEARCH AND DEVELOPMENT

Firms are launching Research and Development activities so as to create new knowledge (the first aspect of R&D). In this perspective, the new knowledge provided by R&D becomes an essential aspect of a firm's innovation ability. In turn, the results of innovation contribute to the improvement of firm performance.

This has been demonstrated empirically by Griliches (1986). This has also been confirmed recently by empirical studies (see for example Mairesse and Mohnen, 2004 or Lööf and Heshmati, 2006) based on the model carried out by Crépon, Duguet and Mairesse (1998 and 2000), taking into account the different steps of the innovation process in order to deepen the link between research and innovation.

Nevertheless, the last few years have been characterized by an increasing interest in external knowledge for the innovation process (Arora *et al.*, 2001). This is reflected in the assertion that we are moving toward a "knowledge economy". In this respect, Cohen and Levinthal (1989 and 1990) have shown that external knowledge is not easy to internalize and that firms have to invest resources so as to be able to absorb it. Among these resources, they point out the importance of a previous R&D investment.

---

[1] A large amount of complementary work has already been done with previous CIS data collection carried out in Luxembourg (some of them in addition to the European surveys). The main results and analysis of this process are included or quoted in Allegrezza (2001).

As a result, in addition to a direct contribution based on the creation of new knowledge, research can indirectly contribute to firms' innovative performance through the development of absorptive capacity (the second aspect of R&D). These absorptive capacities are defined as "the ability to recognize the value of new information, to assimilate it, and apply it at commercial ends" (Cohen and Levinthal, 1990).

The role of these absorptive capacities has been broadly confirmed by macroeconomics analysis, such as the one prepared by Griffith, Redding *et al.* (2004) on a panel of manufacturing firms from 13 OECD countries. Veugelers (1997) and Becker and Peters (2000) have carried out some of the few empirical analysis of absorptive capacities using micro data. Simultaneously, the conceptual aspects of the absorptive capacity construct have been fine-tuned (see Van Den Bosch, Van Wijk *et al.*, 2003, for an overview).

Moreover, in order to deepen the analysis of the importance of external knowledge, it seems necessary to distinguish among that acquired by co-operation agreements, especially when these collaborations engage scientific partners (the third aspect of R&D corresponds to co-operation with scientific partners). These scientific partners include universities, public research centres and firms specially dedicated to R&D activities.

Indeed, given that the R&D activities of these scientific partners incorporate more basic research, a fruitful collaboration with them seems to require a certain degree of absorptive capacity (Cockburn and Henderson, 1998), or, in other words, "a sufficient expertise within these firms to use the results of externally performed research" (Mowery and Rosenberg, 1989, quoted by Veugelers and Cassiman, 2005). The success of these collaborations seems also to require the "connection" of scientific actors, especially those from universities, with the economy (see Scott *et al.*, 2001, for a literature review of the transmission channels between universities and the economy).

This kind of distinction a priori supports the retention of R&D co-operation, because it selects co-operations involving actors specifically dedicated to R&D. The insertion of those co-operations has a major advantage: it identifies the existence of complementarities between R&D of firms and the R&D of public actors, as those public actors include most of the firms' scientific partners.[2]

ANALYTICAL APPROACH

The goal of this analysis is to identify the impact of different R&D aspects on innovation performance. As indicated earlier, these different R&D aspects are: new knowledge creation, the mediating effect of R&D, and co-operation with scientific partners. It has to be stressed that the scientific partners used in this analysis are aimed at engaging in R&D co-operation.[3]

---

[2] While 34 firms from the sample note co-operation with at least one scientific partner, among them 14 declare having such a co-operation with at least one commercial laboratory or one R&D firm, 27 with at least one university, and 14 with at least one public or non-profit research centre.

[3] Nevertheless, R&D co-operation can also be realized with other partner types such as the suppliers, customers, or firms belonging to the same group.

However, this identification cannot be made without any difficulties. Indeed these different aspects are intrinsically linked: the commitment to research activities ensures new knowledge creation and also the development of absorptive capacity, and this absorptive capacity induces the establishment of co-operation with scientific partners.

This leads us to set up three distinctive models. Each one aims at identifying one of the different R&D aspects and its relation to innovation output. Even if it seems opportune to insert variables in this analysis which are specifically connected with each of the different aspects of R&D, we will take into account an identical selection of control variables in the different models.

Knowledge creation and innovative performance

In order to take into account the impact of new knowledge creation, we will estimate the resources devoted to R&D by its different measures: intensity, volume, and involvement in R&D above a critical level (i.e. an R&D personnel of at least five persons). This leads to the presentation of three variants of the identification model of the new knowledge creation effect. Consequently, we will successively consider the benefits on innovation performance of variant; R&D intensity, R&D volume, and involvement above a critical level in R&D.

The mediating effect of R&D and innovative performance

Following Cohen and Levinthal (1989 and 1990), we assume that firms with absorptive capacities, based on their involvement in R&D above a critical level, benefit more from external knowledge than others. This leads to our assumption that R&D has a mediating effect.

To our knowledge, there is no common practice that takes into account absorptive capacity that shapes the mediating effect of R&D. Becker and Peter (2000) consider, successively, a permanent R&D activity and a R&D department. Veugelers (1997) considers only a R&D laboratory. As a proxy for a R&D department is lacking in our sample, we assume the presence of potential absorptive capacities by the existence of a R&D personnel of at least five persons: in Luxembourg, this encompasses 31% of the innovative firms that declare that they have performed a R&D activity. In addition, we will also assess absorptive capacity through a more sustained engagement in R&D by taking into consideration ten R&D employees.

As the two faces of R&D are intrinsically linked, we will adopt the following strategy in order to empirically test our hypothesis about the mediating effect of R&D. In the first step, we will distinguish between firms that have absorptive capacities and those that do not. In the second step, we will estimate for each of them the effect of acquired external information on their innovation performance. In this second step we also simultaneously take into account the direct effect of R&D, that is the creation of new knowledge (this is measured by the intensity of resources devoted to R&D). The introduction of this effect is intended to circumscribe the mediating effect of R&D.

Consequently, while controlling the direct effect of R&D, the identification of the mediating effect of R&D corresponds to a test of the following hypothesis: all together, is the effect of acquired external knowledge on innovative performance positive, significant, and more important for firms with absorptive capacities?

Co-operation with scientific partners and innovation performance

Co-operations with scientific partners are partially linked with a given level of R&D engagement, and also with a firm's knowledge resources. Consequently, as could be expected, co-operation practices are connected with a firm's absorptive capacities. Indeed, we observe a strong relationship between co-operating with scientific partners and employing at least five R&D employees for manufacturing firms on one hand (phi=0.503; $p<0.01$), and the proportion of higher educated people in the service sector on the other hand (Cramer's V=0.317; $p<0.01$).[4] In addition, these co-operations are concentrated in those manufacturing or services sectors with higher technological or knowledge intensities (Cramer's V=0.376; $p<0.01$).

The identification of the specific effect induced by co-operation with scientific partners on innovation output leads us to consider in a distinctive manner the other effects of R&D (i.e. knowledge creation and absorptive capacities). We have to mention that the sample size does not allow us to draw a distinction between national (mainly actors in national public research) and foreign partners in our analysis.[5] Consequently, it will not be possible to effectively take into account complementarities between national public R&D actors and firms.

Other aspects influencing innovation performance

In each model, we insert the following identical selection of control variables: firm size, firm sector, and firm type (whether or not a subsidiary of a group and the origin of the group), firm creation in the 1998-2000 period, and market focus of the firm (i.e. national or international market). We also include in these models market pull, usage of protection methods for innovation, and also the lack of technological competencies or skilled employees.

We expect a positive and significant effect of market pull, external knowledge importance, and the use of protection methods on innovative performance. On the contrary, a negative and significant effect is expected with regards to the lack of technological competencies or skilled employees.

---

[4] The proportion of higher educated people is distinguished by the following categories : 'less than 1%' 'from 1% to less than 10%', 'from 10% to less than 20%', 'from 20% to less than 50%', 'from 50% to less than 80%', 'at least 80%'.

[5] Only 34 firms from our sample acknowledged co-operation with at least one scientific actor, national, or firm from abroad, and among these firms 22 also take part in a co-operation with at least one national partner.

Figure 1
Synthetic presentation of the overall model

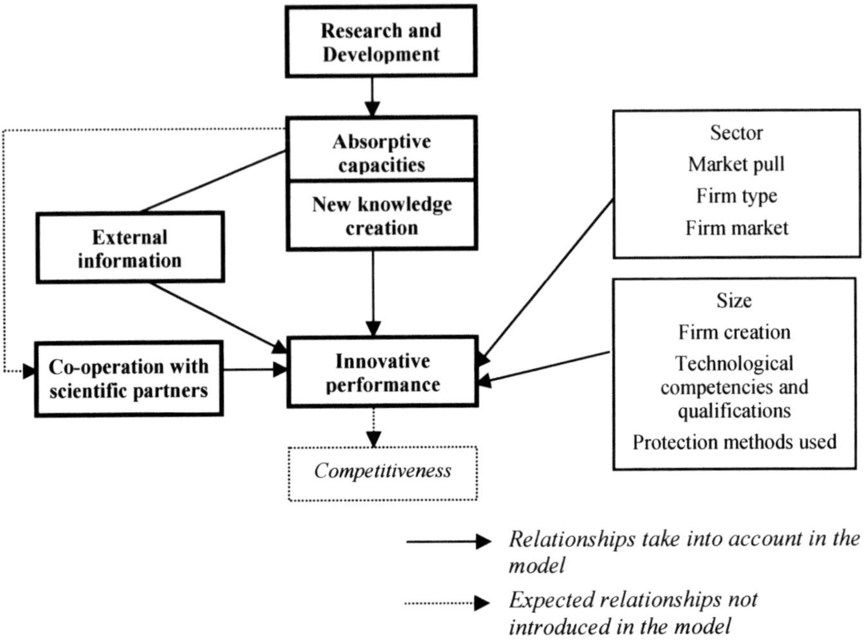

DATA USED AND MEASUREMENT OF VARIABLES

Data used

The micro data used in this chapter comes from the third Community Innovation Survey (CIS3) carried out in Luxembourg. In this survey, 603 firms were polled in order to form a representative sample of the 1393 firms operating, with at least ten employees, in the manufacturing sector or in a selection of the service sectors.[6] Among these 603 firms, 440 are subject to this empirical analysis.[7]

The data collected in this survey includes two types of crucial measures for our analysis. First, they describe resources devoted to R&D (i.e. R&D employees and R&D expenditures) by R&D active firms, and also external knowledge acquired and types of co-operation engaged in. Secondly, they

---

[6] The sectors surveyed are the following: the 'manufacturing sector', 'electricity, gas and water supply', 'wholesale trade', 'transport, storage and communication', 'financial intermediation', 'computer and related activities', 'research and development', 'architectural and engineering activities', 'technical testing and analysis'.

[7] The response rate was higher than 70% and no systematic bias could have been found by size or sector.

include the measurement of the innovative performance of firms that innovate in products (i.e. share of turnover generated by innovative products).

Indeed, we do not have any measurement for the process of innovation output. However, within the manufacturing and the service sector we note a strong link between product and process innovations: 40% of the innovators associate product and process innovation, and only 21% are only innovative in process.[8] Moreover, restricting our analysis to product innovators does not exclude the main R&D performers: firms only process innovators account for less than 1% of total R&D expenditures.

The information collected in this survey comes mainly from innovative firms (45% of the firms). For these firms we describe the effects of innovation, innovation co-operation, protection methods for innovation, and factors hampering of innovation. As a result, the restriction of our measurements to innovative firms could introduce a selection bias in the estimates. In order to correct this potential problem, we have take into consideration the two step Heckman procedure (1979). As no difference was observed in the sign and significance of our estimates, we have not included it in our presentation of our final results.

Measurement of variables

In order to collect data on innovation output, firms were asked to specify, our of their total turnover in 2000, the share of new or improved products introduced during the period 1998-2000. This proportion constitutes our dependent variable. As it is restricted between 0 and 1, we have applied a logit transformation $[\ln(Y_i/(1 - Y_i))]$ so that the estimate will be efficient. Nevertheless, so as not to have extreme values, values below 0.01 and above 0.99 (before the transformation) have been replaced, following Mairesse and Mohnen (2004), by 0.01 and 0.99.

Furthermore, according to the different R&D aspects identified in this chapter, separate variables have been included to take into account the importance of R&D activities.

On one hand, we consider traditional R&D measures such as volume of resources devoted to R&D, represented by the amount of the R&D personnel (logarithm of the number of R&D employees), and the intensity of these resources, quantified by the R&D expenditure divided by turnover. Additionally, we take into account a sustained engagement in R&D as depicted by a R&D personnel of at least five employees (yes=1; no=0).

On the other hand, we insert variables that take place in the R&D aspects defined above. In order to do so we consider firstly the external information that has contributed to innovation projects. These external information sources are divided into the following main types: institutional sources (universities, government, and private non-profit research institutes), market sources (suppliers, clients or customers, competitors or other firms from the same industry) and other sources (professional conferences or meetings, journals, fairs

---

[8] We obtained the following results for the manufacturing sector (Phi=0.451; $p<0.01$) and the service sector (Phi=0.392; $p<0.01$).

or exhibitions). Firms were asked to assess, on a 4-point scale (from 'not used' (0), to 'very important' (3)), the importance of these sources for innovation success. In order to build an index of the importance of theses sources we aggregate, following Cassiman and Veugelers (2002), each item score and we rescale the sum between 0 and 1.

Secondly we take into account cooperation agreements. Innovative firms declare whether or not they have been involved in co-operations in support of their innovation activities. These co-operations (38% of the innovative firms) have taken place with the following partners: suppliers, clients or customers, competitors or other firms from the same industry, consultants, commercial laboratories or R&D firms, universities or higher education institutes, and the government or private non-profit research centres. In order to take into account and measure these co-operations, we first set up a co-operation variable equal to 1 when firms declare that they had at least one co-operation agreement with any type of partner, and otherwise 0. Then, we established the co-operation with a scientific partner variable at 1, 0 otherwise, if the firms have collaborated with the following partners: commercial laboratories or R&D enterprises, universities or higher education institutes, or government or private non-profit research institutes. This distinction is aimed at targeting co-operations with institutes specifically dedicated to R&D. We assume that such a co-operation requires some knowledge resources so as to assimilate their fruits and to apply them to commercial ends. Moreover, this distinction permits us to proxy collaborations between public research and firms. Indeed, as it was pointed out above, firms declared that they co-operate most of the time with universities and government or non-profit research centres.

Prior to conclusion, we introduce other control variables. Firstly we consider the usage of protection methods for innovation. Firms were asked to report the usage of formal methods (patent applications, registration of design patterns, trademarks and copyright) or strategic methods (secrecy, complexity of design, lead-time advantage on competitors) in order to protect their innovation. Unfortunately, they did not have to rate the importance of these methods. As a consequence, we cannot create an index of protection degree, as set up by Cassiman and Veugelers (2002). In other ways as well, the usage of formal methods does not seem to match the practice in the service sector. Therefore, we incorporated into our analysis the two following strategies: secrecy practice (yes=1; no=0) and lead-time advantage on competitors (yes=1; no=0).

The other control variables are related to the environment of the firm or to the firm itself. Referring to the environment of the firm, we consider the firm sector (9 different sectors), whether the firm operates mainly within the international market (yes=1; no=0), membership in a national group (yes=1; no=0), a European group (yes=1; no=0) or a supra-European one (yes=1; no=0).[9] In addition, we take into account the market pull. Firms have to report

[9] These sectors are categorized by the following distinction: high and medium high-tech manufacturing industry, medium low tech manufacturing industry, low tech manufacturing industry, whole sale and commission trade, transport and communication, financial intermediation, computer activities, R&D – engineering activities and consultancy – technical testing and analysis.

the importance of the effects of innovation that they have encountered, using a 4-point scale (from 'not relevant' to 'high'). Among these effects, we selected the following three: increased range of goods or services, increased market or market share, and improved quality in goods or services. We consider that a firm has encountered a market pull if it reports that one these three effects have been very important to the innovations introduced in the 1998-2000 period: 63% of innovative firms were to meet this criterion. With reference to the firm, we note the size of the firm (=logarithm of the number of employees), having started its business in the 1998-2000 period (yes=1; no=0) and the degree of hampering factors to innovation stemming from a lack of technological competencies or lack of skilled personnel (=sum of the score of these two hampering factors measured on a 4-point scale (from not relevant(0) to high(3), rescaled between 0 and 1).[10]

EMPIRICAL RESULTS

In order to consider the effect of R&D activities on innovative performance, we have applied a multivariate analysis. This permitted us to take into account the influence on innovation performance of variables related to firm's characteristics and firm's environment. Moreover, these methods enable us to take the relationship between the different aspects of the analyzed phenomenon into consideration. This constitutes a major advantage for our analysis because, as indicated previously, the different aspects of R&D are intrinsically linked together: engagement in R&D activities ensures both the creation of new knowledge creation and the development of absorptive capacities, and those capacities in turn support the development of co-operation with scientific partners.

In order to take the relationship between the different aspects of the analyzed phenomenon into consideration, we introduce three different regression models based on the share of turnover due to products new for the firm.[11]

New knowledge creation

The first regression (1.1) measures the importance of R&D activities, taking into account the intensity of resources devoted to R&D as quantified by R&D expenditures divided by turnover (Table 1). This regression does not include other R&D measures such as R&D volume included in regression (1.2), nor a sustained engagement in R&D as measured by a R&D personnel of at least five persons.

---

[10] We observe a relationship between the two variables relating to a lack of internal competencies (Spearman rho=0.68; $p<0.01$).

[11] The examination of errors term verified that regression assumptions were met, and that there was no severe multi-collinearity.

Table 1
New knowledge creation and innovative performance

	Regression 1.1			**Regression 1.2**			**Regression 1.3**		
	Est		T-test	Est		T-test	Est		T-test
Intensity of R&D input	**3.30**	***	3.01	-		-	**2.48**	**	2.33
Volume of R&D resources	-		-	**0.28**	*	1.81	-		-
At least five R&D employees	-		-	-		-	**0.81**	**	2.56
External knowledge	**1.69**	***	3.62	1.66	***	3.54	1.70	***	3.68
Firm characteristics									
Firm creation	**1.26**	***	2.78	1.42	***	3.00	1.19	***	2.68
Lack of technological competencies and qualifications	**-1.54**	***	-3.94	-1.60	***	-4.17	-1.61	***	-4.17
No R&D activity	0.04		0.17	0.18		0.56	0.23		0.85
co-operation in innovation activity	0.08		0..37	0.04		0.17	0.00		0.01
Lead-time advantage on competitors	0.10		0.42	0.13		0.57	0.14		0.63
Secrecy practice	-0.17		-0.73	-0.07		-0.28	-0.17		-0.72
Firm size	**-0.40**	***	-5.55	-0.47	***	-5.55	-0.48	***	-5.90
Firm environment									
Market pull	**0.63**	***	2.14	0.64	***	2.74	0.62	***	2.13
Market mainly international	0.11		0.45	0.08		0.33	0.07		0.23
Subsidiaries of Luxemburgish groups	**0.89**	***	3.00	0.85	***	2.76	0.85	***	2.85
Subsidiaries of European groups	0.31		1.08	0.27		0.95	0.34		1.22
Subsidiaries of supra-European groups	**0.98**	***	2.82	0.89	***	2.64	0.90	***	2.69
High and medium high-tech manufacturing industry	-0.27		-0.39	-0.40		-0.60	-0.16		-0.24
Medium low tech manufacturing industry	-0.50		-1.26	-0.62		-1.56	-0.46		-0.46
Low tech manufacturing industry	**-1.23**	***	-3.45	-1.26	***	-3.46	**-1.16**	***	-3.30
Transport and communication	**-0.84**	**	-2.11	-0.85	**	-2.11	**-0.74**	*	-1.87
Financial intermediation	**-0.72**	*	-1.78	**-0.81**	**	-1.98	-0.69		-1.70
Computer activities	-0.33		-0.91	-0.28		-0.72	-0.26		-0.71
R&D – Engineering activities and consultancy – Technical testing and analysis	0.85	*	1.75	0.89	*	1.79	**0.92**	*	1.88
Constant	-0.79		-1.57	-0.60		-1.16	-0.68		-1.37
Number of observations		180			180			180	
Adjusted R^2		0.26			0.27			0.28	

Notes: references = firms not subsidiary of a group; whole sale and commission trade sector.
* significant at the 10% level; ** significant at the 5% level ; *** significant at the 1% level.

Sources: CIS3 survey – CEPS/INSTEAD, STATEC, Ministry of Culture, Higher Education and Research (MCESR); own calculations by author.

As expected, a positive and significant effect of the intensity of resources devoted to R&D (regression 1.1 and 1.3), the volume of R&D resources (regression 1.2), and the existence of sustained R&D activity (regression 1.3) is found. The main positive effect on the proportion of the turnover due to products new for the firm is the size of the firm: small firms seem to be more efficient in innovation than bigger ones. The main negative effect arises from the lack of skilled employees or technological competencies.We also observe in each regression the positive and significant effect of market pull, external knowledge importance, or of the creation of the firm in the 1998-2000 period. We also note that, in comparison with "independent firms", i.e. not belonging to a group, the positive and significant effect of subsidiaries of supra-European groups (72% of these firms belong to a US group), as well as of Luxemburgish groups.

On the contrary, belonging to the 'low tech manufacturing industry' and to a lesser extent to the financial intermediation or the transport and communication sectors have a negative impact on innovation performance. It has to be pointed out that we have previously unexpectedly discovered that the 'low tech manufacturing industry' tends to be more innovative than the 'medium low tech manufacturing industry' (Dautel, 2005b). Yet, if the former seems to be more innovative than the latter, a part of its innovation seems to be of a lower importance, at least for product innovations.

We have also observed that, for the financial sector, accounting for innovation through new or significantly improved products and processes is not appropriate (Dautel, 2005a). In order to improve innovation valuation, it appeared necessary to include other important changes achieved by firms, especially those an organizational nature. The result presented in this analysis, showing a lower output for product innovation in this sector (significant at the 10% level in regression 1.1 and at the 5% level in the regression 1.2), tends to confirm the inadequate nature of "traditional" innovation measurement for the financial sector.

Finally, we have also previously found that the transport and communication sector was weakly innovative: the percentage of innovation firms in this sector (34%) is twice lower than in the computer activities sector (75%). Consequently, in addition to the low rate of innovation in this sector, those accomplished appear to be of a low importance.[12]

The co-operation insertion, without any distinction as to the different partner types, does not have any significant effect on product innovation performance. This seems to be linked to the various nature of co-operation which, following Schmidt (2004), can be presented as: cost-sharing or risk-sharing, knowledge development, and motives related to firm characteristics. The knowledge development that should positively impact innovation performance is only of the many motives for co-operation.

In the same way, we do not observe any significant effect of the following protection methods used: lead time advantage and secrecy practice. Nevertheless, we should not conclude based on this finding that protection innovation modes do not have any effect on innovation output. Indeed, we do not have full information regarding the use of protection methods. No

[12] Given the small number of observations, a grouping of the transport sector (90% of these firms) and the communication sector (10%) has been made.

information is available on their degree of importance for the innovation process.

Mediating effect of R&D

Our main point is to consider empirically if a certain degree of R&D involvement provides absorptive capacities to a firm. Absorptive capacities are defined as the ability to recognize the value of new information, to assimilate it, and apply it to commercial ends. In this way, R&D will have a mediating effect by materializing the absorptive capacities that new knowledge will generate.

In order to approximate a degree of involvement that should confer absorptive capacities, a R&D personnel of at least five persons was taken as a measurement. The testing of our hypothesis leads to the following consideration: taking into account the direct effect of R&D (using R&D intensity), do firms with absorptive capacities better profit from external information sources (i.e. have a larger innovation output) than firms without absorptive capacities.

In a first regression, not including the importance of R&D activities (2.1), we observe a positive and significant effect of external information for firms both with or without absorptive capacities (Table 2). Yet this effect is more important for firms with absorptive capacities than those without. In a second regression (2.2), including R&D activities' intensity, we note that the importance of external information once again has a positive and significant effect on innovation performance. We notice this for firms with or without absorptive capacities, but the external information effect is obviously bigger for firms with absorptive capacities. This seems to corroborate the absorptive capacities' function, which is to ensure a better use of external knowledge in commercial terms.

Furthermore, so as to strengthen the validity of this result, we shape a more sustained absorptive capacity using a R&D personal of at least ten persons. In that configuration (regression 2.3) including the intensity of R&D activities, we notice that the effect of external knowledge has increased for those with this more sustained absorptive capacity.[13]

After studying the situation, a double importance stemming from an involvement in R&D above a certain threshold becomes apparent: it brings new knowledge and absorptive capacities to the firm, each of which positively influences innovative performance.

[13] It must be noted that in that configuration the effect of external knowledge for firms with less than ten R&D employees is not significant due to the interaction with the control variable 'no R&D activity'.

Table 2
Mediating effect of R&D and innovative performance

	Regression 2.1			Regression 2.2			Regression 2.3		
	Est		T-test	Est		T-test	Est		T-test
Intensity of R&D input	-		-	**2.81**	**	2.2	**2.59**	**	2.53
External knowledge with absorptive capacity	**2.86**	***	3.67	**2.29**	***	3	**2.08**	***	4.07
External knowledge without absorptive capacity	**1.23**	**	2.29	**1.32**	**	2.44	0.04		0.07
Firm characteristics									
Firm creation	1.45	***	3.09	1.23	***	2.7	1.19	***	2.81
Lack of technological competencies and qualifications	-1.48	***	-3.83	-1.42	***	-3.63	-1.42	***	-3.63
No R&D activity	0.05		0.18	0.11		0.41	-0.61	*	-1.9
Co-operation in innovation activities	0.01		0.07	0.03		0.14	0.05		0.24
Lead-time advantage on competitors	0.19		0.79	0.13		0.55	0.09		0.42
Secrecy	-0.1		-0.42	-0.17		-0.69	-0.24		-0.99
Firm size	-0.46	***	-5.59	-0.43	***	-5.42	-0.46	***	-5.98
Firm environment									
Market pull	0.57	**	2.46	0.59	***	2.52	0.61	***	2.66
Market mainly international	0.14		0.58	0.13		0.55	0.09		0.38
Subsidiaries of Luxemburgish groups	0.81	***	2.66	0.84	***	2.81	0.89	***	2.93
Subsidiaries of European groups	0.2		0.69	0.25		0.88	0.23		0.83
Subsidiaries of supra-European groups	0.79	**	2.21	0.87	**	2.38	0.82	**	2.51
High and medium high-tech manufacturing industry	-0.44		-0.67	-0.3		-0.45	-0.13		-0.19
Medium low tech manufacturing industry	-0.7	*	-1.76	-0.54		-1.38	-0.41		-0.97
Low tech manufacturing industry	-1.36	***	-3.76	-1.26	***	-3.59	-1.11	***	-3.06
Transport and communication	-0.93	**	-2.36	-0.85	**	-2.19	-0.89	**	-2.26
Financial intermediation	-0.86	**	-2.06	-0.74	*	-1.8	-0.64		-1.55
Computer activities	-0.27		-0.72	-0.29		-0.81	-0.22		-0.59
R&D – Engineering activities and consultancy – Technical testing and analysis	0.91	*	1.9	0.89	*	-1.87	0.91	*	1.75
Constant	-0.37		-0.67	-0.63		-1.18	-0.04		-0.08
Number of observations	180			180			180		
Adjusted R^2	0.25			0.26			0.28		

Notes: references = firms not subsidiary of a group; whole sale and commission trade sector.
* significant at the 10% level; ** significant at the 5% level ; *** significant at the 1% level.

Sources: CIS3 survey – CEPS/INSTEAD, STATEC, MCESR; own calculations by author.

Co-operation with scientific partners

We have found above that co-operation, without any distinction between the different partner types, does not have any significant positive effect on product innovation performance. This can be attributed to the various motives underlying co-operations. Thus it seems appropriate to make a distinction between these co-operations, so as to consider those that take place with scientific partners. It can be expected that the knowledge development motive initiates this kind of co-operation, that specifically includes R&D, and consequently that these co-operations will favour an increase of the innovation output. Nevertheless, it has to be pointed out that the survey does not include any question concerning the reasons for recourse to co-operation.

In order to consider co-operation with scientific partners, we take into account those that take place with universities, commercial laboratories, R&D enterprises, and government or private non-profit research institutes. As seen above, these scientific partners are mainly public R&D actors from Luxembourg or abroad. In addition, a link between co-operation with a national and a foreign partner is observed: innovative firms cooperating with a national/foreign scientific partner are often simultaneously engaged in at least one co-operation with a foreign/national scientific partner.[14]

In order to estimate the influence of co-operation with scientific partners, we consider two variants of the model. These variations differ with the insertion of R&D activities and external knowledge. In a first regression (3.1), we take into account the intensity of R&D resources devoted to R&D and external knowledge importance (Table 3). Nevertheless, so as to take into account the absorptive capacities' effect, a second regression (3.2) includes a distinction between external knowledge with or without absorptive capacities.

In these regressions a positive and significant effect of co-operation with scientific partners on innovation performance is found. In addition, with the insertion of the co-operation, the positive and significant effect of external knowledge remains stable, with or without absorptive capacities. Consequently, if knowledge acquired with co-operation matters, other types of information have also a great interest for innovation performance, with or without R&D engagement above a certain threshold.

Therefore, this result speaks in favour, on the one hand (from a managerial perspective) of firms hand (from a technology policy perspective) of the development of public research actors. Indeed, these are the persons who have the skills to develop the national stock of scientific knowledge and the ability to offer opportunities for further R&D co-operation. However, this conclusion has to be considered cautiously because no distinction is made between either cooperation with national or foreign scientific partners, nor between public and private research.

---

[14] Phi=0.463; $p<0.01$.

Table 3
Co-operation with scientific partners and innovative performance

	Regression 3.1			Regression 3.2		
	Est		T-test	Est		T-test
Intensity of R&D input	**3.33**	***	3.01	**3.01**	**	2.58
External knowledge	**1.41**	***	2.90	-		-
External knowledge with absorptive capacity	-		-	**1.88**	***	2.70
External knowledge without absorptive capacity	-		-	**1.35**	***	2.72
Co-operation with scientific partners	**0.67**	***	2.64	**0.63**	**	2.45
Co-operation with other partners	-0.11		-0.46	-0.13		-0.52
Firm characteristics						
Firm creation	1.25	***	2.74	1.24	***	2.73
Lack of technological competencies and qualifications	-1.50	***	-3.86	-1.50	***	-3.87
No R&D activity	0.04		0.17	0.09		0.34
Lead-time advantage on competitors	0.05		0.22	0.06		0.28
Secrecy	-0.19		-0.81	-0.19		-0.80
Firm size	-0.41	***	-5.71	-0.44	***	-5.57
Firm environment						
Market pull	0.61	***	2.62	0.60	**	2.57
Market mainly international	0.07		0.31	0.07		0.29
Subsidiaries of Luxemburgish groups	0.82	***	2.76	0.80	***	2.67
Subsidiaries of European groups	0.26		0.91	0.25		0.90
Subsidiaries of supra-European groups	0.92	***	2.69	0.88	**	2.51
High and medium high-tech manufacturing industry	-0.30		-0.45	-0.30		-0.29
Medium low tech manufacturing industry	-0.59		-1.48	-0.58		-1.46
Low tech manufacturing industry	-1.33	***	-3.62	-1.32	***	-3.62
Transport and communication	-0.87	**	-2.19	-0.84	**	-2.13
Financial intermediation	-0.69	*	-1.73	-0.68	*	-1.71
Computer activities	-0.47		-1.25	-0.44		-1.17
R&D – Engineering activities and consultancy – Technical testing and analysis	0.56		1.14	0.59		1.20
Constant	-0.49		-0.92	-0.42		-0.77
Number of observations	180			180		
Adjusted R^2	0.28			0.27		

Notes: references = firms not subsidiary of a group; whole sale and commission trade sector.
* significant at the 10% level; ** significant at the 5% level ; *** significant at the 1% level.

Sources: CIS3 survey – CEPS/INSTEAD, STATEC, MCESR; own calculations by author.

CONCLUSIONS

Based on the data from the third Community Innovation Survey carried out in Luxembourg, we have investigated, in this national context, the role played by Research and Development on innovation performance. A positive influence has been observed for the three following aspects of R&D: knowledge creation, the development of absorptive capacities, and R&D co-operation (i.e. collaboration with scientific partners). Therefore, a triple impact of R&D aspects is found.

As expected, a first repercussion of R&D is that new knowledge resulting from R&D increases product innovators' output. As a result, firms actively engaged in R&D tend, in comparison to other innovative firms, to derive a more important proportion of their turnover from their new products.

In other ways as well firms' external knowledge importance for innovation outcomes has been put to the fore. Henceforth, a second repercussion of R&D is that external knowledge appeared all the more fruitful in those firms engaged in R&D above a certain threshold. Consequently, this finding gives empirical support to the absorptive capacities' hypothesis, that postulates that firms with absorptive capacities have, in commercial terms, a more fruitful usage of external knowledge than firms lacking strong absorptive capacities.

Finally, a third repercussion of R&D lies in the importance of the co-operation with scientific partners that appears as R&D co-operation. These co-operations stimulate product innovation output.

In order to take advantage from each of the beneficial R&D effects, a dual importance becomes apparent, from a managerial point of view, of investing in R&D above a certain threshold for firms in the service sector, and triple for those belonging to the manufacturing sector. Indeed, this R&D engagement permits both added knowledge creation (in comparison to firms with a lower investment in R&D) as well the development of absorptive capacities. Moreover, for the manufacturing firms it generates opportunities for co-operation with scientific partners (in the service sector it is rather the share of higher educated employees generates these co-operations). These different advantages have to be in a manager's mind at the launching of new projects seemingly unrelated to ongoing activities.

Therefore, these results highlight the primacy of an R&D investment above a critical level, and argue in favour of stimulating firms' R&D activities. In this respect, public R&D activities have a role to play: the development of public R&D activities should lead firms, especially in the high-tech sectors and in knowledge based services, to increase their R&D investment so as to benefit from new knowledge creation.

In order to derive more precise conclusions, two important issues have to be considered in future studies. First, some new and more appropriate measures of absorptive capacities have to be considered before any strong generalization can be made. Based on this criterion, the replicating of empirical analyses across countries and times would be welcome. Second, some additional distinctions would be needed so as to improve our understanding of the public research contribution. Indeed, owing to proximity and the type of research performed, Luxemburgish and foreign research actors may play different roles.[15] In addition, a contrast may appear between co-operation engaged with universities,

[15] Before the creation of the University of Luxembourg, public research activities were strongly oriented toward technological transfers.

mainly involved in basic research, and those with other public research institutes. Finally, the contextual situation may also have a role to play: whereas Lööf and Broström (2006) have found strong empirical support regarding the positive influence of university collaborations for Swedish large manufacturing firms (at least 100 employees), Medda *et al.* (2006) have not found such evidence for Italian manufacturing firms.

BIBLIOGRAPHY

Allegrezza, S., (2001), "Les déterminants de l'appropriation et des modes d'appropriation de l'innovation technologique : le cas des PMI au G.D. Luxembourg", Thèse doctorale, Université de Liège.

Arora, A., A. Fosfuri and A. Gambardella, (2001), "Markets for Technology and Their Implications for Corporate Strategy", *Industrial & Corporate Change*, Vol. 10, pp. 419-451.

Becker, W. and J. Peters, (2000), "Technological Opportunities, Absorptive Capacities, and Innovation", Discussion Paper Series 195, Universitaet Augsburg, Institute for Economics.

Cassiman, B. and R. Veugelers, (2002), "R&D Cooperation and spillovers: some empirical evidence from Belgium", *American Economic Review,* 92 (4), pp. 1169-1184.

Cockburn, I. and R. Henderson, (1998), "Absorptive Capacity, Coauthoring Behavior, and the Organization of Research in Drug Discovery", *The Journal of Industrial Economics*, Vol. 46, pp. 157-182.

Cohen, W. and D. Levinthal, (1989), "Innovation and Learning: the Two Faces of R&D", *The Economic Journal*, 99, pp. 569-596.

Cohen, W. and D. Levinthal, (1990), "Absorptive Capacity: a new Perspective on Learning and Innovation", *Administrative Science Quarterly* 35, pp. 128-152.

Crépon, B., M. Duguet and J. Mairesse, (1998), "Research and Development, Innovation and Productivity: An Econometric Analysis at the Firm Level", *Economics of Innovation and New Technology*, 7(2), pp. 115-158.

Crépon, B., M. Duguet and J. Mairesse, (2000), "Mesurer le rendement de l'innovation", *Economie et Statistique* 334, pp. 65-78.

Dautel, V. (2005a), "La mise en œuvre d'innovations selon les caractéristiques intrinsèques des PME et PMI luxembourgeoises", Cahier Economique du Statec n°97, pp. 129-150.

Dautel, V., (2005b), "Quelles entreprises ont innové au Grand-Duché de Luxembourg entre 1998 et 2000?", Cahier Economique du Statec n°97, pp. 151-178.

Griffith, R., S. Redding and J. Van Reenen, (2004): "Mapping the Two Faces of R&D: Productivity Growth in a Panel of OECD Industries", *The Review of Economics and Statistics*, MIT Press, Vol. 86(4), pp. 883-895.

Griliches, Z., (1986), "Productivity, R&D, and Basic Research at the Firm Level in the 1970's", *American Economic Review 76* (1), pp. 141-154.

Heckman, J., (1979), "Sample Selection Bias as a Specification Error", *Econometrica*, 49, pp. 153-161.

Lööf, H. and A. Heshmati, (2006), "On the Relationship Between Innovation and Performance: A Sensitivity Analysis", *Economics of Innovation and New Technology*, Vol. 15, Issues 4/5, pp. 317-344.

Lööf, H. and A. Broström, (2006), "Does Knowledge Diffusion Between University and Industry Increase Innovativeness", Working Paper Series in Economics and Institutions of Innovation 21, Royal Institute of Technology, CESIS - Centre of Excellence for Science and Innovation Studies.

Mairesse, J. and P. Mohnen, (2004), "The Importance of R&D for Innovation: a Reassessment Using French Survey Data", working paper 10897, National Bureau of Economic Research.

Medda, G., C. Piga and D. Siegel, (2006), "Assessing the Returns to Collaborative Research: Firm-level Evidence from Italy", *Economics of Innovation and New Technology,* 15(1), pp. 37-50.

Mowery, C. and N. Rosenberg, (1989), *Technology and the Pursuit of Economic Growth,* Cambridge University Press.

Scott, A., G. Steyn, A. Geuna, S. Brusoni and E. Steinmuller, (2001), "The Economic Returns to Basic Research and the Benefits of University – Industry Relationship: a Literature Review and Update Findings", Report for the Office of Science and Technology, SPRU, University of Sussex, Brighton.

Schmidt, T., (2004), "Knowledge Flows and R&D Cooperation: Firm-level Evidence from Germany", DRUID Summer Conference 2004 on Industrial Dynamics, Innovation and Development.

Van Den Bosch, F., R. Van Wijk and H. Volberda, (2003): "Absorptive Capacity: Antecedents, Models, And Outcomes", [in:] M. Easterby-Smith and M. Lyles (eds), *The Blackwell Handbook of Organizational Learning and Knowledge Management.*

Veugelers, R., (1997), "Internal R&D Expenditures and External Technology Sourcing", *Research Policy* 26, pp. 303-315.

Veugelers, R. and Cassiman, B., (2005), "R&D Cooperation between Firms and Universities: Some Empirical Evidence from Belgian Manufacturing", *International Journal of Industrial Organization*, Vol. 23, pp. 355-379.

18

Factors Governing a Firm's Choice of R&D-Cooperation Partners

Ebbe Krogh Graversen
Michael Mark

INTRODUCTION

Firms have to continuously develop and increase their knowledge base in order to strengthen their market position through innovation. From the point of view of a firm, it is not always optimal to possess the necessary R&D competences internally. It may be better, less costly, and more efficient to gain access to necessary R&D related knowledge and competences through cooperation with partners carrying out R&D. Furthermore, R&D cooperation and other interactions with R&D partners may facilitate access to new markets through reduction in entry costs, thereby improving the market position of the firm. Cooperation may also lower financial risk in R&D projects by cost splitting, which is a particular obstacle for SMEs (small and medium sized enterprises). As such there are numerous economic incentives to cooperate on R&D.

The positive economic synergies of R&D cooperation, i.e. transfers of knowledge between firms and between firms and public research institutions, are among the main reasons why the EU and OECD, among others, choose to put R&D cooperation at the top of their research policy agenda.

Initially, it is assumed that firms' choice of R&D cooperation partners is motivated by a particular need for R&D in the firms themselves. Therefore, the analysis in the present chapter will focus on the firm specific R&D profile, R&D competences, and absorption capacity as factors that determine a firm's need for and choice of R&D cooperation partners. R&D cooperation in the present analysis is defined to include active contributions to R&D projects together with other firms or R&D institutions. This means that all participating firms gain some kind of benefit, such as increased sales or reduced costs, from the project. Firms that only buy R&D are also included in the analysis, since purchasing R&D is considered to be a complex action that requires absorption capacity and R&D knowledge, i.e. an active input.

THEORETICAL MOTIVATIONS FOR R&D COOPERATION

There are numerous pros and cons for R&D cooperation among firms.[1] From a rational choice point of view one should expect that a firm would cooperate whenever there are economic advantages in either the short or long term. Hagedoorn and Duysters (1997) state, for example, that the accumulation of knowledge through R&D cooperation may create a long-term positive economic return, and Vinding (2002) claims that technology and the use of technology is crucial for the competitiveness and economic returns of firms. Similarly, D'Aspremont and Jacquemin (1988) point out that the return on R&D cooperation is higher than the return of individual firm based R&D. Becker and Dietz (2004) show that joint R&D enhances both innovation input and output. Likewise Cassiman and Veugelers (2002) state that cooperating firms maximize their common return by cooperation on R&D, even in a competitive market for end or near market products.

R&D cooperation may also provide a firm with additional resources to create, build and sustain its competence stock, and it may also help the firm to gain new knowledge and competences, as well as increasing the firm's technology level. The additional or new knowledge gained from R&D cooperation can be stored inside the firm and will improve its ability to implement and develop complex knowledge in future R&D projects. Thus, a considerable incentive for the firms to participate in R&D cooperation is the faster and easier access to new knowledge and technology; cf. Robertson and Gatingnon (1998) and Polt et al. (2001). This faster and easier access to complex knowledge will minimize the development costs of the firm, and thereby give the firm competitive advantages inasmuch as it can more efficiently utilize and develop new knowledge into end products or cheaper production processes; cf. Hagedoorn (1993).

A precondition for firms' participation in R&D cooperation is that they have something to contribute: either R&D inputs or the competences to

---

[1] An extensive overview can be found in Caloghirou et al. (2003) as well as in Schmidt (2005). See also Graversen (2002), Polt et al. (2001), Lundvall (2002), Belderbos et al (2004a, 2004b), Cassiman and Veugelers (2002), Fritsch and Lukas (2001), Laursen and Salter (2004), Genua (1998), Oerlemans and Meeus (2001), Fontana et al (2003), Tether (2002), among many others.

understand and implement the results of the particular R&D project. Free riders will usually only participate once in R&D cooperation, since the abuse of their partners' trust will preclude them from future cooperation; cf. Cassiman and Veugelers (2002). It is therefore of utmost importance that the participating firms have something to offer their cooperation partners. This in turn means that the firm must posses R&D competences to offer as inputs into the common R&D project. A firm that has sufficient competences is also better suited to commercialize the new knowledge and the results of the improved knowledge base; cf. Oerlemans and Meeus (2001), Cohen and Levinthal (1990), Negassi (2004) and Fritsch and Lukas (2001). Hence, a vital requirement for a participating firm must be sufficient absorption capacity; cf. Cohen and Levinthal (1990), who find that the absorption capacity of the firm positively influences its capability to explore the quality and value of new external knowledge, to understand and employ the knowledge, and in the end to commercialize this knowledge onto the market.

Knudsen et al. (2001) finds that the openness of a firm towards R&D cooperation positively correlates with the firm's utilization of the new knowledge obtained. They find that sequential R&D cooperation enables a firm to secure the needed absorption capacity and knowledge that is crucial for the new knowledge to pay off. Thus it may be expected that firms with a large stock of basic competences and a high absorption capacity will participate in R&D cooperation more often than other firms, due to a *ceteris paribus* larger return from such R&D cooperation; see also Laursen and Salter (2004). Absorption capacity also influences the utilization over a longer perspective, and it is therefore important that firms detect and attract the specific partners that are interesting for them, i.e. have a solid and trustful reputation, not only today but also in the future. Also, the R&D personnel in a firm have their own network of past and present R&D cooperation partners, which often are based on previous R&D cooperation; cf. Polt et al. (2001) and Mora-Valentin et al (2004). Therefore, the employees' network is important for the firm, because the existence of research employees or firm competences make the specific firm attractive as a partner for R&D cooperation. The absorption capacity in turn is the foundation upon which a firm accumulates this new knowledge and, in turn, further participates in new R&D cooperation projects based on the accumulated new knowledge; cf. Kastelli et al. (2004), Knudsen et al. (2001) and Vinding (2002).

Another way to increase the attractiveness of a firm is to secure that its organizational foundation and culture are geared toward participation in R&D cooperation; cf. Simon (1999) and Graversen (2002). Although desirable, it is not enough for the firm merely to possess the physical framework if it does not possess the right organization and culture, i.e. one that motivates, allows and stimulates R&D cooperation.

Differences in the choice of private and public R&D partners

The results in Table 1 show that firms more often cooperate on R&D projects with other private firms than with public research institutions. The innovation life cycle reveals a straightforward explanation for this fact, as illustrated in Figure 1.

Often a single firm or a consortium of partners carries out the fundamental research behind an initial innovation, typically building on basic research. There is a large input of expert knowledge and this affects the possibility for cooperation with public research institutions versus private firms. Public research institutions are strong on basic research, while the average private R&D firm is strong on development and implementation. Hence, initial R&D cooperation more often involves public research institutions than private firms. Over time the initial innovation spreads through the networks, including employees who have worked with it and those who see the potential of the innovation in terms of market interests. More and more firms will work with the innovation, trying to make it fit market demands, or alternatively commercialize it, believing that it will create a change in demand, adopting the blue ocean strategy. Many firms are at this point working with the innovation while further development still requires some research dependent input, although not basic research. At this stage the input from public research institutions will typically be from non-profit independent research and technology providers. Simultaneously there is an increasing necessity for complementary knowledge from other private firms, i.e. access to markets, production capacity, etc.

Figure 1
R&D as a source for innovation in the Innovation Life Cycle, ILC

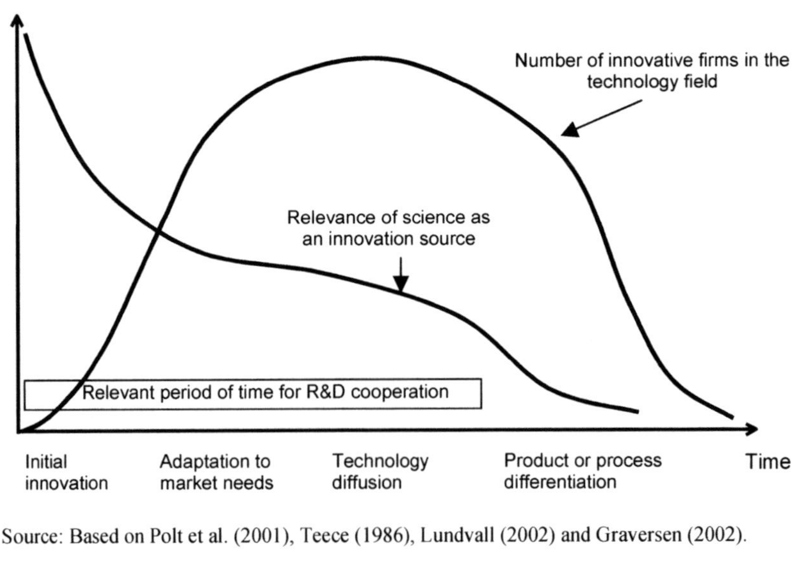

Source: Based on Polt et al. (2001), Teece (1986), Lundvall (2002) and Graversen (2002).

The knowledge included in the innovation continues to spread out through the educational system, through labour mobility and networks, and ends up being common knowledge in a scientific sense. At this point in the innovation cycle the number of firms contributing to further development of the innovation is at its highest. The knowledge of the innovation is now possessed by a relative large number of firms that are capable of implementing and using it. The innovation has now reached a point where its main users, the firms, are mainly product and process oriented. They try to gain additional benefit from improvements to the innovation, which in turn will outdo earlier versions of the product, adopting a red ocean strategy. At this stage it is still possible to make minor changes and improve the initial innovation, but at the end of the cycle the potential for further development of the product will vanish, and the only way to make any profit is to engage in product differentiation or price dumping. At this stage there is no use for research-based knowledge or R&D cooperation.

By comparing the incentives for R&D cooperation, firms will be expected to choose the R&D cooperation partners needed in a particular project. This in turn means that the type of R&D conducted in the firm, its R&D competences, and absorption capacity, as well as the need for complementary knowledge, the intensity of research and some business sector specific characteristics, will together determine the choice of R&D cooperation partners.

EMPIRICAL DATA ON R&D COOPERATION

The Danish Centre for Studies in Research and Research Policy collects and delivers data on R&D activities in Danish firms to the OECD MSTI statistics. In this present analysis the available data for 2001 is used. The database includes R&D data for 2465 firms, roughly covering all private sector firms with more than 5-10 employees, using an advanced weighting algorithm. From the sample of 2465 firms, 592 firms had R&D activities. When the figures are weighted to population level, 2641 firms had R&D activities out of a total population of 18381 firms, i.e. 14 percent.

The survey responses from the R&D active firms were used to determine whether the firms had any kind of R&D cooperation with partners either from Denmark or abroad, and whether the partners were from the public or private sector. It was found that 63 percent of the R&D active firms had R&D cooperation; see Table 1 below.

Table 1
Share of R&D active firms that cooperate on R&D, in percentage

Type of R&D cooperation partners	Share of cooperating firms
National R&D cooperation partner	54
Foreign R&D cooperation partner	28
Partners from public research institutions	28
Partners from the private sector	54
At least one partner	63

A few of the variables in the survey database have missing values for a few firms. The missing values are updated and improved with logic imputation or multiple imputation techniques. The latter method is a statistically efficient way to improve the information in data sets with missing values. Summary statistics for the variables on firm characteristics, R&D profile, R&D competence, absorption capacity, and business sector characteristics are referred to in Appendix Table A.1.

The first group of variables determines briefly the demographic profile of the firm. The R&D profile concerns the different competences the firm possesses, and therefore the R&D cooperation partners that the firm needs in order to gain the required complementary competences. As indicators for R&D capacity and absorption capacity, the analysis includes a range of variables that describe the competence stock in the firm. The variables measure capacity, continuity, and intensity of R&D.

In order to emphasize the specific business sector characteristics, four meso variables are used. These variables are defined on a two digit NACE classification level. The minimum efficiency index approximates how difficult it is for smaller firms to enter a market defined by its business sector constraints. The market concentration index tells how monopolistic the specific business sector is. The demand for capital illustrates barriers for new entrants in the business sectors. Profitability shows how competitive the specific business sector is, and it can be used as a mark up on the average economic settings of the market. Furthermore, a general business sector graduation is used. This business sector graduation is based on the NACE classification, but supplemented with information on which sector the firms' main R&D activities are performed in (if different from the main economic activity sector).

DETERMINANTS OF R&D COOPERATION

R&D firms often have several R&D cooperation partners; from the public as well as the private sector, and national as well as foreign. Distinguishing between types of R&D cooperation partners makes it possible to analyze what particular factors determine a specific choice. The probability for cooperation among the R&D firms is estimated by logistic regressions on each type of cooperation partner, and Table 2 shows the results for cooperation with at least one partner, with a national partner, a foreign partner, a private firm, and a public research institution, respectively. The general model in column one shows the factors that determine whether or not to cooperate on R&D. The results for the different type of R&D cooperation partner indicate that there is a partner specific explanation to the choice of different types, as illustrated in Figure 1.

An older firm is expected to have a more developed network than a younger firm, so older firms are expected to have a higher probability of R&D cooperation. The general result however is that age is not a significant determinant of cooperation, although it is found that older firms have a somewhat higher probability for cooperation with foreign private sector partners. These results match the findings of Smith et al. (2001). They found that older Danish firms more often participated in the 4th Framework Programme.

The size of the firm is also expected to be a positive factor encouraging cooperation, since larger firms are expected to have larger networks, more R&D personnel, and more R&D projects. However, this factor only matters for R&D cooperation with a national partner. If the partner is foreign, the size of the firm does not matter. The explanation for this may be that a significant group of small R&D intensive firms are very international in their choice of R&D partners. Similar conclusions regarding firm size can be found in Cassiman and Veugelers (2002), focusing on Belgium firms, Bayona et al. (2001) on Spanish firms, and Smith et al. (2001) on Danish firms, all of whom have shown that the size of the firms positively affects the probability of R&D cooperation. A geographical location of a firm in the capital region is furthermore believed to produce a higher probability that it will cooperate on R&D with local partners, because capital regions perform the majority of national R&D. It is found that R&D firms located in the capital region have a higher probability for R&D cooperation, but this only concerns firms that have partners from abroad and partners from the private sector. Hence, it seems that the firms in the capital region are more global in their choice of R&D partners.

R&D profile

Firms placed near public research institutions are expected to have more intensive R&D cooperation because of the expected R&D intensive clusters there. However, the data results show a negative overall result for the probability of R&D cooperation, though most coefficients for the detailed types of cooperation partners are insignificant. An explanation for these findings could be that firms cooperating with other private firms need more other complementary R&D competences than firms cooperating with public research institutions. Another explanation may be the small distances in Denmark, where everybody can be reached within a few hours. The type of R&D activity carried out in the firm seems to affect the probability of cooperation, as well as the choice of R&D partner type. At the general level the firms conducting product-oriented R&D tend to have a higher probability to cooperate on R&D. National affiliations do not change the result, although the influence on a foreign partner from the public sector is insignificant.

Table 2
Logistic model of R&D firms' choice of cooperation partners, 2001

Explanatory variables	R&D cooperation with at least one partner from								
	Total	Denmark	Abroad	Private sector			Public sector		
				Total	Denmark	Abroad	Total	Denmark	Abroad
Intercept	-7.82	-7.78	-7.18	-6.47	-5.69	-7.32	-11.95	-11.89	-10.29
Age (Years)	0.01	0.01	0.01	0.00	-0.00	**0.02**	0.01	0.01	-0.01
Size (Log number of employees)	**3.98**	**4.28**	0.68	**2.43**	**2.43**	0.97	**6.51**	**6.33**	0.63
Size squared	**-0.99**	**-1.18**	-0.02	**-0.58**	**-0.63**	-0.11	**-1.56**	**-1.49**	0.05
Size tripled	**0.08**	**0.10**	0.00	**0.05**	**0.05**	0.01	**0.12**	**0.11**	-0.00
Part of a concern	**0.59**	**0.46**	0.19	**0.51**	0.26	0.08	0.02	-0.02	-0.21
Situated in the Capital region	**0.50**	0.23	**0.40**	**0.51**	0.21	0.41	-0.21	-0.20	-0.15
R&D profile									
Located in a municipality with a knowledge centre	-0.20	-0.29	-0.44	-0.25	-0.29	-0.30	0.12	0.23	-0.06
Located next to a municipality with a knowledge centre	**-0.52**	-0.32	**-0.61**	-0.23	-0.23	-0.42	-0.34	-0.21	-0.58
Product-oriented R&D	**1.75**	**1.42**	**2.33**	**1.31**	**0.76**	**2.52**	**1.83**	**1.74**	1.16
Process-oriented R&D	**0.44**	**0.44**	**-0.60**	0.19	0.24	**-0.66**	**0.45**	**0.44**	0.68
Customer-oriented R&D	-0.11	-0.34	-0.33	-0.22	-0.01	-0.43	0.33	0.18	**1.05**
Basic research share of R&D expenditures	**0.03**	**0.02**	**0.03**	**0.02**	0.00	0.01	**0.03**	**0.03**	**0.04**
Applied research share of R&D expenditures	0.01	**0.01**	-0.00	-0.00	0.00	-0.01	**0.01**	0.01	**0.04**
Self financed share of R&D expenditures	-0.05	0.29	0.62	**0.74**	**0.74**	0.49	-0.58	-0.43	0.30

R&D competences and absorption capacity										
Own R&D department		**1.04**	**0.81**	**1.71**	**0.82**	**0.39**	**1.59**	**1.05**	**0.98**	1.37
Share of R&D employees		**1.04**	**1.26**	**1.77**	**1.79**	**1.66**	**1.51**	0.28	0.09	1.38
Female share of R&D employees		0.06	0.65	**-1.93**	0.01	**0.78**	**-2.00**	**1.43**	**1.51**	0.33
Research educated employed		**1.77**	**0.94**	**0.81**	**1.35**	0.47	**0.61**	**1.72**	**1.63**	**0.98**
Foreign researchers employed		-1.24	-0.78	**1.44**	-0.42	-0.40	**1.69**	-0.55	-0.30	**1.65**
Scientist intensity		0.55	**0.73**	0.41	0.48	**0.52**	0.42	-0.66	-0.38	**-1.28**
R&D intensity		-0.61	-0.65	0.10	-0.38	-0.47	0.41	0.10	-0.19	**1.80**
Capital intensity		**0.48**	0.32	0.15	-0.03	-0.09	0.17	**0.57**	**0.57**	-0.09
Business sector characteristics										
Minimum efficiency scale		**-0.17**	-0.07	-0.02	-0.09	-0.02	-0.01	-0.02	-0.00	0.36
Market concentration index		-0.01	-0.01	-0.00	-0.01	-0.01	-0.00	-0.02	-0.01	-0.01
Capital requirement		-0.01	-0.01	-0.13	-0.07	-0.02	-0.14	-0.01	-0.01	-0.04
Profitability		-0.05	-0.01	-0.17	-0.08	-0.02	**-0.18**	-0.02	-0.01	-0.09
Business sector	IT	0.39	0.27	-0.23	**0.51**	0.55	-0.10	-0.49	-0.46	-0.58
	Machinery	0.02	-0.13	-0.60	0.43	0.29	-0.39	**-0.95**	**-0.87**	-2.18
	Agro-food industry	-0.22	-0.17	0.21	0.35	0.03	0.19	**-0.93**	**-0.94**	0.48
	Knowledge service	0.30	-0.03	**0.62**	**0.50**	0.24	0.30	**-0.76**	**-1.12**	0.94
	Environmental industry	**1.89**	**1.02**	**1.22**	**1.38**	0.41	**1.41**	**1.87**	**1.94**	0.01
	Medical & health care	0.50	0.10	0.77	**0.87**	0.31	**0.94**	-0.42	-0.38	-1.43
Weighted number of observations		2641	2641	2641	2641	2641	2641	2641	2641	2641
-2 Log Likelihood		783	816	696	817	816	668	707	689	322
Concordant (%)		79	78	82	76	70	81	84	83	89
Share of cooperating firms (%)		63	54	28	54	45	25	28	27	8

Note: Significant parameter estimates at a 10 percent level are in bold. The reference business sector is the manufacturing industry. See further explanations of the variables in Table A.1.

Firms conducting process-oriented R&D activities also show a higher probability of engaging in R&D cooperation, particularly with partners from Denmark, in particular partners from the public sector in Denmark. Firms conducting process-oriented R&D activities have a lower probability of engaging in R&D cooperation with foreign private firms.

These results indicate that higher intensity levels of R&D activities affect the probability of engaging in R&D cooperation, and might be an indication of specialization in product innovation, where the need for complementary resources and competences increases, and therefore the need for R&D cooperation also increases. The lack of cooperation as regards the other determinants of R&D activities could be caused by the fact that process-oriented R&D activities, i.e. R&D activities developing common knowledge or customer oriented R&D activities, are either vertically integrated in the firm or gained through the market. The difference in the probability of cooperation could thus be the result of a more complex product innovation, where the market for achieving the needed resources or competences simply does not exist; c.f. Teece (2003); or that the vertical integration of partners is not economically efficient when focusing on product innovations due to opportunism, contract costs, uncertainty in the transactions etc., c.f. Williamson (1975, 1985).

Another indication of the finding that more knowledge intensity affects the probability for R&D cooperation positively is the significant coefficients regarding the share of R&D expenditures used on basic research in the firms. This shows that a higher ratio of R&D expenditures in basic research increases the probability of R&D cooperation. As expected, this effect is most significant when the cooperation partner is from the public sector, but still there is a clear general effect. This is an indicator of the importance of basic competences and absorption capacity within the firm, and supports the Innovation Life Cycle (ILC in Figure 1. Similarly, the ratio of applied research compared to overall R&D expenditures is an insignificant factor. This indicates that the more complex the R&D, the more appropriate cooperation seems to be.

The self-financed share of total firm specific R&D expenditures can be expected to have a negative influence on the probability of cooperation. This is so because a higher amount of self-financing reduces the dependence on external sources. Another reason to expect that an increased self-financing share in total R&D will negatively influence cooperation is that a broad range of external finance programs, i.e. EUREKA or the EU RTD Framework Programme, demand interaction and cooperation with private firms or public research institutions. In our analysis this result is generally insignificant, although a positive effect is found for the national private sector partner type. This is interesting because it shows that the choice of R&D cooperation, as well as partner type, is insensitive to external funding, which may indicate a complementary effect of external funding. However, firms that obtain external funding already have a high degree of internally financed R&D activities.

R&D competence and absorption capacity

Firm competences and absorption capacity make a firm more capable of identifying, collecting, and utilizing new knowledge, as well as making it increasingly attractive as a potential partner for other firms. The assumption of the significant importance of absorption capacity is verified by our analysis. The existence of a permanent R&D department within a firm increases the probability for R&D cooperation, as does the employment of research-educated employees, as was also found in Bayona et al. (2001) and Cassiman and Veugelers (2002).

A high ratio of R&D employees relative to the overall number of employees seems to affect the probability of R&D cooperation positively in general, in particular in the private sector. The effect is insignificant for public sector R&D institutions. The effect of having foreign scientists employed seems only to affect the probability when the partner type is foreign. Thus, the employment of foreign scientists indicates a strong international focus in a firm. A high share of females among total R&D employees will decrease the probability of R&D cooperation with foreign partners, but increase the probability to cooperate with a national partner.

Scientist intensity only affects the probability of R&D cooperation with partners from private firms and foreign public research institutions, respectively positive and negative. R&D intensity, on the contrary, only positively affects the probability of having R&D cooperation with foreign public research institutions. In general, R&D intensity was found to be an insignificant factor, a result opposite to that reached by Smith et al. (2001) and Graversen and Mark (2003). Finally, the capital intensity of the R&D activities also tends to affect the probability of R&D cooperation at a general level, or when the partner is from the public sector in Denmark; otherwise the coefficients are insignificant. This indicates that R&D cooperation with public sector R&D institutions demands expensive facilities and thus may constitute more capital intensive R&D projects.

Business sector characteristics

In order to control the influence of business sector characteristics on R&D cooperation, four indirect measures are used. The minimum efficiency scale seems not to have much effect. A positive effect indicates that business sectors with high entry barriers cooperate more. The effect from the market concentration index shows that a business sector with a few dominant players only rarely cooperates. One should not put much emphasis on this because all coefficients are highly insignificant. Further, the measure for required capital shows that business sectors with high requirements for capital strength cooperate less. Again, the coefficients are insignificant. Finally, the measure for profitability shows that when firms are in a market that is profitable, the probability of cooperation declines. This is particularly true when focusing on foreign private sector partners. The explanation for this is that when the market is profitable, firms do not want to share profits. Likewise they do not want to share the risk, and when a market is profitable enough this seems to be a plausible strategy.

A general division into different business sectors does not affect the probability for cooperation in the machinery industry or agro-food industry, except for a significant negative influence in the agro-food industry when the partner is from the national public sector. Contrary to an earlier study by Graversen and Mark (2003) using older data, there is only weak support for an above average IT business sector cooperation, although the weak indications concern primarily the private sector, as was also found by Graversen and Mark. The explanation for the change compared to Graversen and Mark (2003) may be found in the burst of the IT bubble in year 2000.

The probability that firms from the knowledge service sector engage in R&D cooperation is only increased if the partner is from the private sector (and is less significant if the partner is a foreign partner). Firms from the knowledge service sector engage less often in R&D cooperation with national public R&D institutions. The results in Table 2 also show that firms from the environmental industry have a higher probability of R&D cooperation. This conclusion holds true whether the partner is from Denmark or abroad. In contrast, the medical and health care industry seems only to cooperate at an above average level with foreign partners from the private sector. The reference business sector is the manufacturing industry.

CONCLUDING REMARKS

It is commonly accepted that R&D cooperation reduces the expenses of R&D activities. Thereby firms gain a better market position through cheaper, faster and easier access, and better understanding of technological progress, innovation, and development. Furthermore, cooperation in R&D will lower the expenses of specific R&D projects and thereby minimize the economic risks or help overcome the economic barriers that the firms face when they are engaged in R&D projects. The present analysis of R&D cooperation was performed on R&D active firms solely, where 63 percent had R&D cooperation with at least one partner. Those partners could be from other private firms, public research institutions; and be located either in or outside of Denmark. The basic characteristics of the firms do seem to affect their probability to engage in R&D cooperation. On the other hand the R&D profile of the firm seems to affect the probability of engaging in R&D cooperation with all types of partners.

The ability of a firm to identify, collect and commercialize new knowledge through R&D cooperation, and thereby the probability of it engaging in R&D cooperation, depends heavily on the ability of the firm to utilize the gains there from. Firms with permanent R&D competences, i.e. a permanent R&D department or research educated employees, have the competences to utilize the gains from R&D cooperation and therefore have a higher probability of engaging in R&D cooperation. This is confirmed in our analysis.

Different business sector characteristics show only minor additional influence. Though almost all coefficients are insignificant they all tend to be negative, which is by no means surprising since most firms would not share their surplus through cooperation. Different business sector dummies seem only to have little variation in their effect on the probability of R&D cooperation. Only firms from the environmental industry are more likely to engage in R&D cooperation. Further examination shows that the IT and medical sectors more often cooperate with the private sector.

Research policy implications

As it was pointed out in Section 2, R&D cooperation provides the R&D-active firm with the possibility to reduce costs, which in turn increases profits. What kind of partners such firms choose to cooperate with depends on each firm's own R&D profile and activities, as illustrated in Figure 1. Empirical evidence was presented in Section 4. One would expect that firms would always cooperate in R&D projects whenever there is an economic rationale or benefit. However, some firms still prefer to perform R&D by themselves in order to prevent others from copying their inventions. This behaviour may, from a firm's point of view, be fully rational, but it may also be possible for them to increase their benefits from R&D by improvements in the National Innovation System. Hence, there is room for policy actions if there are firms who are either not R&D active, or not yet fully R&D cooperative. In order to bridge these gaps there are a broad range of research and business policy actions that can be undertaken to create and reveal the necessary complementary competences, i.e. increase the potential number of R&D cooperation partners.[2]

The political instruments that can be used to increase R&D cooperation will initially have to fit current political priorities. Thus, the actions taken and the instruments used will vary depending on the current political agenda. A general focus on encouraging more firms to carry out R&D will naturally lead to a higher number of potential partners and thereby more cooperation.

Graversen (2002), Graversen and Mark (2003) and Polt et. al. (2001), among others, provide a list of various political instruments that can be used to increase direct incentives, which in turn will lead to increased R&D cooperation. The effect of those instruments will also depend on political priorities. An overall effort to increase the number of R&D active firms will naturally increase the number of potential cooperation partners, and as such increase the number of actual R&D cooperation partners. This could be obtained either through tax deductions for R&D cooperation expenses or more directly by R&D subsidies.[3]

---

[2] The rationale for public support to increase the potential number of R&D partners is the positive difference between social and firm specific return from R&D activities in the economy, where increased R&D is desirable for society due to the higher payoff; see Sørensen et.al. (2003) for an example.

[3] Highly intensive R&D firms with no surplus in the short run cannot use tax deductions as long as they do not pay taxes.

Another policy could be to increase the number of research competent employees in non-R&D firms, or increase the initial knowledge base at each individual firm by educating the workforce. This in turn would increase the R&D competences and absorption capacity as well, which will increase the gains from R&D cooperation and thus produce the incentives to engage in R&D cooperation. The political instruments could also include tax deductions for supplementary education and economic rewards to those who have increased their knowledge base above a certain threshold. Subsidies to hire highly educated or research competent employees will also lead to an increase in R&D competences and absorption capacity, and thus enhance R&D cooperation. Moreover, tax deduction for R&D expenses linked directly to expenses stemming from R&D cooperation will reduce the costs the individual firm faces when it cooperates with a university. This will make the access to institutional knowledge easier to obtain, which will in turn increase knowledge circulation and knowledge transfer, which further could increase the mobility between the two sectors. This will also lower the barriers linked to R&D cooperation. Another way to reduce barriers is to build and support knowledge service centres, competence centres, business campuses, and different networks and venture capital foundations. These organizations can help match private firms and public institutions and thereby reduce otherwise significant search costs. This will help firms to discover the "right" partner for R&D cooperation, if not already known.

The analysis in the present chapter indicates the need for prioritisations in both research and business policy, aimed at increasing the range of R&D cooperation among R&D active firms. By increasing the range and the intensity of R&D cooperation among firms, the gains obtained from the R&D funds invested will increase. An increase in R&D cooperation among firms will lead to positive synergies, which will increase the general social welfare of society.

The present findings support in general the earlier empirical findings on firm R&D cooperation in other studies; cf. footnote 1. Hence, the results seem to be sufficiently universal that they may be used for policy purposes in countries other than Denmark as well. For example, the results on the importance of absorption capacity can be used and implemented in other countries since it seems to have an immediate effect. Improved R&D cooperation among firms in a country will increase firm growth through an improved knowledge stock in the participating firms.

It is left for future research to test the universality of the results using the same model and data for several countries, for example using the CIS4 data in the EU. However, using CIS data makes another improvement difficult, namely increasing the robustness of a general model using panel data as input.

BIBLIOGRAPHY

D'Aspremont C. and A. Jacquemin., (1988), Cooperative and noncooperative R&D in duopoly with spillovers. *The American Economic Review,* Vol. 78, No. 5, pp. 1133-1137.

Bayona C., T. Garcia-Marco and E. Huerta, (2001), Firms' motivations for cooperative R&D: an empirical analysis of Spanish firms. *Research Policy,* Vol. 30, pp. 1289-1307.

Becker W. and J. Dietz. (2004), R&D cooperation and innovation activities of firms - evidence for the German manufacturing industry. *Research Policy,* Vol. 33, pp. 209-223.

Belderbos R., M. Carree and B. Lokshin, (2004a), Cooperative R&D and firm performance. *Research Policy,* Vol. 33, pp. 1477-1492.

Belderbos, R., M.A. Carree, B. Diederen, B. Lokshin and R. Veugelers. 2004b. Heterogeneity in R&D cooperation strategies. *International Journal of Industrial Organization,* Vol. 22, pp. 1237-1263.

Caloghirou, Y.I., S. Vonortas and S. Nicholas. 2003. Research Joint Ventures. *Journal of Economic Surveys,* Vol. 17, pp. 541-570.

Cassiman B. and R. Veugelers. 2002. R&D cooperation and Spillovers: Some Empirical Evidence from Belgium. *American Economic Review,* Vol. 92, pp. 1169-1184.

Cohen W. and D. A. Levinthal. 1990. Absorptive capacity: a new perspective on learning and innovation. *Administrative Science Quarterly,* Vol. 35, pp. 128-152.

Fontana, R., A. Geuna and M. Matt. 2003. Firm Size and Openness: The Driving Forces of University-Industry Collaboration. *SPRU Working Paper Series,* no. 103.

Fritsch M. and R. Lukas. 2001. Who cooperates on R&D? *Research Policy,* Vol. 30, pp. 297-312.

Genua A. 1998. Determinants of university participation in EU-funded R&D cooperative projects. *Research Policy,* Vol. 26, pp. 677-687.

Graversen E. 2002. Danish Firms' R&D Cooperation. (In Danish) *AFSK Report Series* 2002/5.

Graversen E. and M. Mark. 2003. Determining factors for Danish Firms' Choice of R&D Cooperation Partners. (In Danish) *Nationaløkonomisk Tidsskrift,* Vol. 141.

Graversen E. and M. Mark. 2004. Determinants of Danish Firms' Choice of R&D-cooperation Partners. CFA WP 2004/6.

Hagedoorn J. 1993. Understanding the rationale of strategic technology partnering: interorganizational modes of cooperation and sectoral differences. *Strategic Management Journal,* Vol. 15, pp. 291-309.

Hagedoorn J. and G. Duysters. 1997. Satisfying strategies in dynamic inter-firm networks. The efficacy of quasi-redundant contacts. MERIT 1997 (2/97-016).

Kastelli I., Y. Caloghirou, S. Ioannides. 2004. Cooperative R&D as a means for knowledge creation. Experience from European publicly funded partnerships. *International Journal of Technology Management,* Vol. 27, pp. 712-30.

Knudsen M. P., B. Dalum and G. Villumsen. 2001. Two Faces of Absorptive Capacity Creation: Access and Utilisation of Knowledge. *Paper presented at the Nelson and Winter Conference.* DRUID, Aalborg.

Laursen K. and A. Salter. 2004. Searching high and low: What types of firms use universities as a source of innovation? *Research Policy,* Vol. 33, pp. 1201-1215.

Lundvall B-Å. 2002. *Growth, Innovation and Social Cohesion: the Danish Model.* Edward Elgar, London.

Mora-Valentin E. M., A. Montoro-Sanchez and L. A. Guerras-Martin. 2004. Determining factors in the success of R&D cooperative agreements between firms and research organisations. *Research Policy,* Vol. 33, pp. 17-40.

Negassi S. 2004. R&D cooperation and innovation - a micro econometric study of French firms. *Research Policy,* Vol. 33, pp. 365-84.

Oerlemans L. and M. Meeus. 2001. R&D cooperation in a transaction cost perspective. *Review of Industrial Organization,* Vol. 18, pp. 77-90.

Polt W., H. Gassler, A. Schibany, C. Rammer, N. Valentinelli and D. Schartinger. 2001. Benchmarking Industry-Science Relations – The Role of Framework Conditions. *Final Report. European Commission, Firm DG, Benchmarking Coordination office.*

Robertson T. S. and H. Gatingnon. 1998. Technology development mode: a transaction cost conceptualization. *Strategic Management Journal,* Vol. 19, pp. 515-531.

Simon H. 1999. The many shapes of knowledge. *Revue d' Economie Industrielle,* No. S88, pp. 23-39.

Schmidt T. 2005. Knowledge Flows and R&D Co-operation: Firm-level Evidence from Germany. *ZEW Discussion Paper No. 05-22.*

Smith V., E. Graversen, B. Damm and J. Overgaard. 2001. International Research Cooperation, Spillovers and Competition. Empirical Evidence from the EU 4th Framework Programme. *AFSK WP Series 2001/1.*

Teece D. J. 2003. *Explicating Dynamic Capabilities: Asset Selection, Coordination and Entrepreneurship in Strategic Management Theory.* University of California, Berkeley.

Teece D. J. 1986. Profiting form technological innovation: Implications for integration, collaboration, licensing and public policy. *Research Policy.* Vol. 15 (6), pp. 285-305.

Tether B. 2002. Who co-coperates for innovation, and why: an empirical analysis. *Research Policy,* Vol. 31, pp. 947-967.

Williamson O. 1975. *Markets and Hierarchies.* New York: The Free Press.

Williamson O. 1985. *Economic Institutions of Capitalism.* New York: The Free Press.

Vinding A. L. 2002. Absorptive capacity and innovative performance: A human capital approach. In: *Interorganisational Diffusion and Transformation of Knowledge in the Process of Product Innovation.* Ålborg University.

APPENDIX

Table A.1
Descriptive statistics for R&D active firms in 2001

Explanatory variables	Average	Minimum	Maximum
Firm characteristics			
Age (Years)	16	1	145
Size (Employees)	132	1	10091
Part of a concern	60 %	0	100
Situated in the Capital region of Copenhagen	41 %	0	100
R&D profile			
Located in the same municipality as a knowledge centre[1)]	0.30	0	1
Located next to a municipality with a knowledge centre [1)]	0.28	0	1
Conducting product-oriented R&D	0.95	0	1
Conducting process-oriented R&D	0.66	0	1
Conducting customer-oriented R&D	0.79	0	1
Basic research share of operating R&D expenditures	5 %	0	100
Applied research share of operating R&D expenditures	13 %	0	100
Improvement expenditures share of operating R&D expenditures	82 %	0	100
Self finance share of total R&D expenditures	0.88	0	1
R&D competence and absorption capacity			
Own R&D department	0.41	0	1
R&D employees share of all employees	0.25	0	1
Female share of total R&D employees	0.18	0	1
Research educated employed	0.13	0	1
Foreign researchers employed	0.02	0	1
Scientist intensity (R&D man year/R&D employees)	0.63	0.04	1
R&D intensity (Operating R&D costs/turnover)	0.11	0	1.97
Capital intensity (R&D costs/R&D man year)	0.70	0.06	9.01
Business sector characteristics			
Minimum efficiency scale (log 1. quartile expenditure at business sector level)	1.92	0.63	8.63
Market concentration index	1.55	0.04	104
Capital requirement (log business sector avg. of the sum of assets)	0.46	0.01	431
Profitability (business sector avg. of ordinary return relatively to net turnover)	0.04	-10.96	13.71

Business sector	IT	19 %	0	100
	Machinery	8 %	0	100
	Agro-food industry	8 %	0	100
	Knowledge service	20 %	0	100
	Environmental industry	7 %	0	100
	Medical & health care	5 %	0	100
	Other manufacturing industry	32 %	0	100
Number of firms: 592 (Not weighted), 2641 weighted				

Note: [1)] A knowledge centre is a university, a public research institution, or something similar.

19

Champions of Growth: How Innovative Mid-Sized Firms Keep on Growing

Hermann Simon

INTRODUCTION

Growth is one of the most fundamental characteristics and continuing achievements of the Hidden Champions. We are talking about long-term, sustainable growth, not about temporary strong bursts of growth. History shows that quite a few of the spectacular growth stories celebrated by the business community, stock markets and the business press are not lasting but evaporate after three or five years. Recent examples are WorldCom, Enron or the Internet bubble of the late 1990s/early 2000s. The growth pattern of the Hidden Champions is different. They do not necessarily grow by spectacular annual rates but keep growing continuously and consistently over long periods of time. By adding one stone to the other they multiply their size over the decades. And many of them grow even though they are operating in mature markets.

Why do some companies grow while others stagnate or decline? One difference lies in the will to grow. This is definitely a foundation stone of the Hidden Champions' growth pattern. The unrelenting will to grow is deeply rooted in the convictions of the founders and the leaders of the Hidden Champions. Also, growth is an extremely important career determinant, and a growing company is much more attractive to young, ambitious people than a stagnating or shrinking firm. Growth creates ever new opportunities and challenges. By attracting energetic young people it induces a self-enforcing momentum which drives a company forward, in a virtuous circle.

Reinhold Wuerth has never ceased to emphasize the importance of growth for his company. We will describe the growth of Wuerth in more detail below. Another famous believer in the magic of growth is Berthold Leibinger, who has driven a small manufacturer of mechanical sheet metal cutting machines to become the dominant force in laser cutting technology. His company Trumpf is today the clear world market leader and has a revenue of over 1.2 billion euros. In this case the growth was propelled by the transition from mechanical to laser cutting technology. In the course of this development Trumpf has also become one of the leading firms in industrial lasers.

Or take Guenther Fielmann, who started his first optical store in the mid-70s and propelled his chain of specialty stores to over 1,000 outlets and a revenue of over 1 billion euros – all this in a stagnating market. In his main market Fielmann has now a volume market share of over 50%. So far Fielmann operates only in Europe, thus limits for further growth are not in sight. A continuous supply of energy from the leader keeps moving the company forward.

Another exemplary driver of growth is Aloys Wobben, the founder and CEO of Enercon. Enercon was founded in 1984 and is now the world's number two in wind turbines, a rapidly growing market. Its revenue is over 1 billion euros and it employs 7,000 people. Enercon is technology-driven to the extreme. The rapidly growing wind energy market offers unlimited global growth opportunities for the future.

Or take BARTEC, a world leader in explosion protection products. Founded in 1975 the company just celebrated its 30th anniversary. While not of the same size as the revenue billionaires mentioned above, it shows very consistent growth multiples of 3-4 over each ten year period. BARTEC stands for BARlian TEChnology, and the founder Reinhold A. Barlian has been the driving force behind this ongoing expansion.

FROM HIDDEN CHAMPION TO BIG CHAMPION

A Hidden Champion company that keeps growing is bound to eventually become a Big Champion. Indeed, many of the Hidden Champion companies which were mid-sized when I first looked into them in the mid 1990s have become companies with revenues of one or even several billion euros. Table 1 shows a selection of these once Hidden, now Big Champions. All champions in this table had revenues of less than 1 billion euros in 1993/4 and have more than 1 billion euros in revenue in 2003/4. While ten years ago they were mid-sized Hidden Champions, today they can be described as truly Big Champions. A firm with more than 1 billion euros or – at the current rates (2005) – 1.33 billion US-dollars or 11 billion Chinese yuan is a big company by all international standards. The two last columns in the table contain the Compound Annual Growth Rate (CAGR) and the Growth Multiple. The Growth Multiple is Revenue 2003/4 divided by Revenue 1993/4 and tells us how many times bigger the companies are in 2003/4 relative to 1993/4 - in terms of revenue. The CAGR is simply the 10th root of the Growth Multiple.

The top growth champion is SAP, whose revenue is 13.3 times higher than it was ten years ago when this company was a true Hidden Champion which I detected in the early nineties. This growth corresponds to a spectacular compound annual growth rate of 29.6%. In the early nineties SAP was only known to experts. Today it is *the* prime force and *the* clear world market leader in business standard software. Even in the case of SAP there is room for future growth. So far, mostly large companies have adopted SAP software. In the recent past, SAP has started to penetrate the segment of mid-sized companies, where the sales to an individual customer are smaller, but the number of customers is much larger. Today SAP is a giant with a turnover of 7.5 billion euros.

Table 1
Growth Champions

Rank	Company	Primary Product	Revenue 1993/94 in Mill. €	Revenue 200/04 in Mill. €	Compound Annual Growth Rate in %	Growth Multiple
1	SAP	standard business software	563	7500	30	13,3
2	Fresenius	dialysis services	895	7270	23	8,1
3	Hexal	generic drugs	160	1110	21	6,9
4	Mahle	pistons and cylinders	767	5195	21	6,8
5	Puma	sporting goods/sports lifestyle	210	1274	20	6,1
6	Porsche	sports car manufacturer	978	5580	19	5,7
7	Kamps	bakery products	305	1679	19	5,5
8	Andritz (Austria)	production systems, industrial process solutions	322	1481	16	4,6
9	Gildemeister	turning- and milling machines	255	1100	16	4,3
10	Brose	automotive parts	453	1925	16	4,2
11	Festo	pneumatics/automation	300	1220	15	4,1
12	Claas	agricultural machinery	493	1928	15	3,9
13	Ratiopharm	generic drugs	280	1076	14	3,8
14	Trumpf	industrial lasers	319	1221	14	3,8
15	Behr	automotive climate control systems and engine cooling systems	813	3005	14	3,7
16	Swarovski	crystal collectibles and accessories	476	1680	13	3,5
17	Dürr	car painting systems	614	2120	13	3,5
18	Schmitz Cargobull	trailers	306	1000	13	3,3
19	Brose	window lifters, door systems for cars	600	2000	13	3,3
20	Fielmann	eyeglasses	392	1010	10	2,6
21	Krones	bottling systems	600	1520	10	2,5
22	Stihl	chain saws	738	1530	8	2,1
23	SEW	gear units	508	1050	8	2,1
24	Lindt & Sprüngli	chocolate	584	1181	7	2,0
25	Enercon	wind energy plants	598*	1200	19**	2,0
26	Alfred Kärcher	cleaning agents and equipment	530	1032	7	1,9
27	Fuchs Petrolub	lubricants	656	1041	5	1,6
28	Eberspächer	particle filters	930	1335	4	1,4
			* figure for 2000		**2000-2004	

SAP's stellar growth is almost matched by the growth of Fresenius, the parent of Fresenius Medical Care (FMC). FMC is the world leader in dialysis, a treatment for kidney disease. There are about 1.3 million dialysis patients in the world, and FMC treats 125,000 of them. Thus, its world market share in terms of dialysis patients is close to 10%. Of course, such stellar growth comes along with a strong increase in employment. Since 1996 the number of Fresenius employees has doubled, to now 68,500. Again, the growth potential for the future is almost unlimited, as there are many more patients in the world in need of this costly high technology treatment. Fresenius sees its focus for future growth in China, as its CEO announced in the spring of 2005 (Frankfurter Allgemeine Zeitung, February 25, 2005).

While the growth of SAP and Fresenius took place in newly developing markets with technology as the growth driver, several of the companies in Table 1 are in mature markets but still show an impressive growth record. Mahle, world leader in pistons and cylinders, is a supplier to the automotive industry. Schmitz Cargobull makes truck trailers, a classical engineering product. Stihl is the world leader in chain saws, again not a market with much natural growth. How did these companies grow? Mostly by continuously expanding their market shares. They simply beat the competition and year after year take away a piece of the market from their competitors.

Almost all companies in existence have started small. From small origins giant firms can emerge. Some of our Hidden Champions from the nineties have become truly large, global corporations. They have not achieved this through extreme growth in a short period of time, but rather through continuing, steady steps forward, year after year, decade after decade. They can serve as role models for a long-term success strategy, particularly for Chinese firms, which are sometimes tempted to grow too quickly. When you want to go for a march of 1,000 miles or climb a very high mountain, it is more important to walk at an adequate, constant speed over many days than to go at the maximum speed in the beginning. Patience and endurance are required to achieve sustainable long-term growth.

In the following paragraphs we look in more depth into selected cases of Hidden Champions' growth. We consciously consider different sizes, Wuerth as an example of a Big Champion, Schmitz Cargobull and Enercon as Hidden Champions of about 1 billion Euros in revenue, as well as BARTEC, which offers an example of a much smaller company. We outline both common and specific traits in the growth patterns of these companies.

FIFTY YEARS OF GROWTH: THE EMERGENCE OF A BIG CHAMPION

Wuerth, today the world leader in assembly products, was founded as a wholesaler for screws by Adolf Wuerth immediately after World War II in 1945. It remained a very small, modest shop for years to come. When Adolf Wuerth died in 1954, the company had two employees. One was his son Reinhold Wuerth, who was 19 at that time and who took on the leadership of the little company. Growth became the mantra for Reinhold Wuerth. We already described his visions in Section 1. Reinhold Wuerth never ceased to emphasize the importance of growth for his company. He compares a company to a tree. As long as a tree keeps growing it is doing well and stays healthy. The day it stops growing it starts dying, Wuerth says. According to him growth keeps a company in good spirits and agile.

Figure 1 shows what has become of this will to grow since Reinhold Wuerth took over in 1954: fifty years of unrelenting growth.

Figure 1
Wuerth Group: Revenue Growth 1954-2004

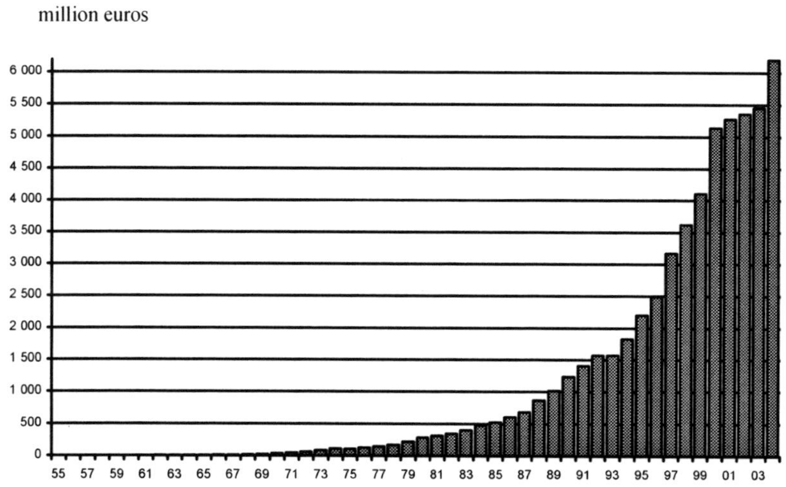

In the first two decades revenue remained at a level which is not visible at the scale we have to use to show the current size of the company. What is really spectacular is the fact that the company has grown in every single one of those 50 years. Not surprisingly there are periods with stronger growth, for instance the second half of the 1990s when revenue doubled in about four years. In the period 2000-2003 the growth rates were much lower, which is partially an effect of the extreme growth in the preceding phase. Extreme growth exhausts the spirits of a company and therefore it is not unusual that a phase of strong and rapid growth is followed by a period of smaller growth rates. But as we see in the revenue for 2004, the old spirit is coming back and higher growth rates are expected for the future.

The result of these 50 years of unrelenting growth is a global giant. Revenue has grown to over 6 billion euros (8 billion US-dollars or 66.3 billion Chinese yuans at 2005 rates). The Wuerth group employs close to 50,000 people worldwide. The company has more than 300 subsidiaries and has established its own presence in almost 100 countries. The company is very profitable relative to its peers in the industry. While obsessed by growth, Reinhold Wuerth has always emphasized the importance of profit. "Growth without profit is deadly" is one of his favourite quotes. Of course, continuing this exceptional growth pattern remains a big challenge for the successors of Reinhold Wuerth. In 2005 a young energetic CEO, Robert Friedmann, took the helm at the amazingly young age of 38. Friedmann is determined to lead the company into another long-lasting phase of continuous growth.

How did Wuerth achieve this spectacular growth? Throughout its history the company remained essentially loyal to its core competencies. These competencies comprise the ability to sell C-products to business customers and to handle the tremendous logistic challenges associated with this business. Staying true to these two core competencies, Wuerth expanded into many directions. The most important one is regional expansion, and eventually globalization. Starting as a regional business the sales territory was continuously expanded, first in Germany, later in Europe, then in the whole world. But Wuerth also expanded the target markets. Starting from an initial focus on woodworking trades (carpenters, cabinet-makers, joiners) it gradually conquered markets such as automotive, metal workers, electricians and industrial corporations. While acquisitions played a certain role they were not central to Wuerth's long-term growth. The growth was mostly internal.

GROWING THROUGH FOCUS AND EFFICIENCY

Schmitz Cargobull exemplifies a very different growth strategy which is, however, equally impressive. Schmitz was founded in 1892 by Heinrich Schmitz as a cartwright's workshop. Today Schmitz is by far the largest producers of truck trailers in Europe. Observe that this market is not yet global. It manufactures in nine highly automated plants in Germany, Spain, UK, and Latvia. The foundation for this ascendance to European market leadership emerged from a crisis and was laid in the mid 1990s when the company had to reformulate its strategy. Then Schmitz decided to exclusively focus on four product groups, which are depicted in Figure 2.

Figure 2
Schmitz Cargobull: Only Four Product Groups

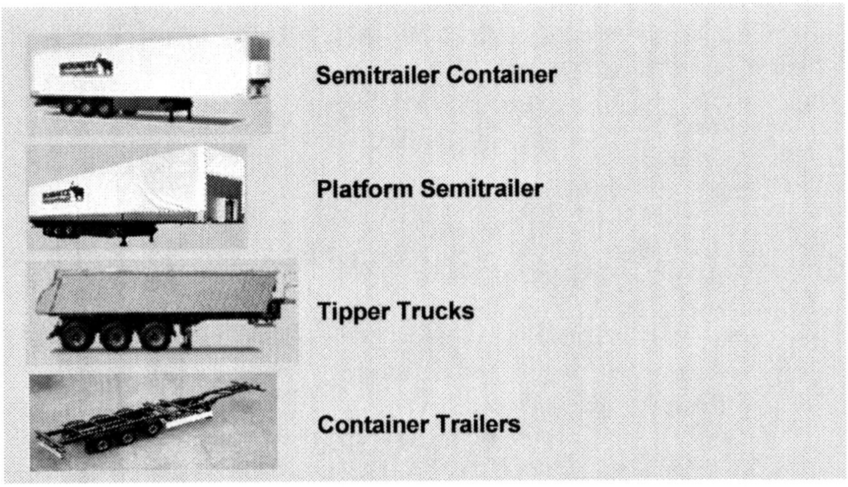

They decided to manufacture only models which sell at least 5000 units per year, a very large number in the truck trailer market. At the time this focussing meant that Schmitz had to give up 90% of its products, since only 10% of the products achieved sales number of more than 5000 units. What was the result? With such a drastic streamlining of the assortment one would probably expect a decline in revenue, while in actuality, after the implementation of its new strategy Schmitz entered into a long-lasting growth phase which catapulted it into the uncontested leadership position in Europe. How could this happen in an extremely competitive market? The reduction of complexity strengthened Schmitz' competitiveness in unbelievable ways. Here are the highlights of the strategy and the ensuing results:

- Schmitz wants to be the best partner to its customers, the trucking companies.
- It offers the lowest life cycle costs.
- It claims quality leadership.
- It claims leadership in closeness to customer (one stop shopping, European wide sales and service network, systems/modular solutions, financing, market for used trailers).
- Delivery time, a very important factor in the trailer business, has declined from 4-6 weeks to 4-5 days.

Schmitz has become a totally customer-driven supplier. And here are the results as of 2005:

- Revenue has climbed in the last ten years from 306 million to 1 billion euros, which corresponds to a compound annual growth rate of 11.8%.
- Schmitz is the most profitable company in the trailer market.
- With a compound annual growth rate of only 5.8%, the number of employees has grown much less in percentage terms than revenue and profit, thus reflecting a very strong increase in employee productivity.
- Sales per salesperson have grown by 60%, a big improvement in sales productivity.
- Schmitz' market share has been increasing year after year in the period under consideration, and is now close to 25%.
- Of the four main competitors, two have been declining in market share and the market leader of 1999 left the market in 2004. Schmitz has now a relative market share of 2.3, i.e. its more than double the size of the second in the market.
- Schmitz Cargobull has become a very strong brand. Its logo is printed in large letters on the back of all trailers. If a customer refuses to show the Schmitz brand he will not be served.
- In spite of a decline in prices by 30% in the last ten years Schmitz, could defend its margin at a very high level for the industry. With a margin of over 6% it is by far the most profitable company in the field. In Chapter ? we will discuss close competition as a performance driver. Krone is today the closest competitor of Schmitz, with a location only a few miles away from Schmitz. In the lee of Schmitz, Krone has sailed from fourth to second place in the market.

Is Schmitz satisfied with these results? Not at all. Bernd Hoffmann, who took over as CEO from Peter Schmitz in 2004, formulates even more ambitious goals:

- Capacity expansion from 30,000 to 55,000 units, essentially a duplication.
- Market share goal of 35%, up from 25% in 2005.
- Return of Sales goal of 10%, up from 6% in 2005.
- Annual growth of 10% to 1.75 billion euros in revenue, up from 1 billion euros in 2005.

The future will show whether Schmitz can retain its growth momentum. If they remain as focussed and efficiency-oriented as they have in the last ten years, there can be little doubt that Bernd Hoffmann and his team will realize their ambitions.

The growth of Schmitz Cargobull is essentially based on two pillars. Through focus and the ensuing efficiency Schmitz was able to take away market share from its competitors. The second pillar was the integration process in Europe, which allowed Schmitz to transform itself from a German to a truly European company. Behind all this was the unrelenting will to grow demonstrated first by Peter Schmitz and, more recently, by Bernd Hoffmann.

GROWN WITH THE WIND

While Wuerth and Schmitz Cargobull have achieved growth in existing markets, the Hidden Champion Enercon has been "growing with the wind". Enercon was only founded in 1984 by the engineer Aloys Wobben and is now the second largest maker of wind turbines in the world. Wind power is a fast growing sector within the renewable energy market, with Germany leading the pack. As of the beginning of 2005, 7,764 wind turbines are installed worldwide, 5,324 or 68.5% of which are in Germany. These turbines represent a worldwide capacity of 7.1 Gigawatt. Of these, 5.4 Gigawatt or 76% are installed in Germany. In 2004 the energy generation by wind power in Germany exceeded the generation by water power for the first time. In Germany Enercon has a market share of 42% and is the market leader. On a global scale the growth opportunities for wind energy and, thus for Enercon, are unlimited. Indeed, Enercon is clearly aiming at the world market with its corporate slogan "Energy for the World".

Since the year 2000 Enercon's revenue has doubled, from 598 million euros to over 1.2 billion euros. The company now employs 7,000 people, a very big number for a company which just celebrated its 20th anniversary. Enercon's wind turbines are operating in 31 countries. It has sales offices in 17 countries and supplies service and maintenance worldwide.

Enercon is a radically technology-driven company. No other company has contributed as many innovations to wind energy technologies. Its depth of manufacturing is unusually high and its quality is superb. Enercon is highly focussed and, similar to Schmitz Cargobull, it offers only four products. It also provides excellent service, and it's worthwhile to look at Enercon's Partner Konzept (EPK). Below is a quote from the Enercon.com homepage:

"The ENERCON PartnerKonzept (EPK) gives customers the guarantee of consistently high turbine availability for the first twelve years of operation with calculable operating costs. From servicing to safety work, maintenance and repairs, all eventualities are covered by a single agreement. Because of its economical service reliability, the EPK has long since become an acknowledged ENERCON quality feature. More than 90% of all E-66 wind turbines (E-70 predecessor) operators have signed an appropriate EPK agreement."

Figure 3
Enercon: Grown with the Wind – Wind Park

The spectacular growth of Enercon is essentially based on technological leadership and a steady stream of innovations. Dedicated service also seems to play a critical role. Equipped with these competencies Enercon can take full advantage of the emerging wind turbine market. Again behind all this stands an entrepreneur, Aloys Wobben, who devotes his energy and intelligence to growing with the wind.

GROWING WITHOUT EXPLOSIONS

The fourth growth case we are looking at in some depth is BARTEC. BARTEC is a global leader in safety and security technologies in the mining, oil, and gas industries. It produces explosion protection devices for machinery and engineering systems and measurement solutions for dangerous liquids. Their business slogan is: "BARTEC protects people and the environment through safe components, systems and products".

Reinhold A. Barlian had an international perspective from the beginnings of his company in 1975. In the first decade BARTEC built up a sales network in Europe following the motto "Europe is our home market". In the second decade, from 1985-95, BARTEC entered the US-market with its own subsidiary and established a worldwide network of sales offices. This process was supported by a continuous flow of innovations, with ever higher electronics sophistication. After 1995 the core competency increasingly focussed on safety, and a joint venture for the mining industry in China followed. Networks and alliances have become essential elements of the strategy of BARTEC, since safety solutions are developed together with the manufacturers of the machines. The company now has 20 international sales subsidiaries.

As so many Hidden Champions, BARTEC is a niche player and consequently not of the same size as the Big Champions discussed above. It shows, however, a superb long-term growth performance. From 0 in 1975, revenue grew to 15 million Euros in 1985, to 45 million Euros in 1995, and to 160 million Euros in 2005 (see figure 4).

Figure 4
BARTEC: Growing Without Explosions

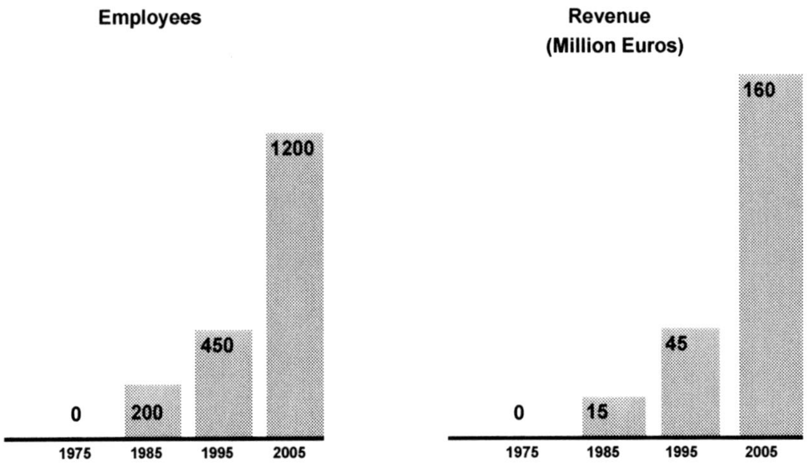

For the last two decades this has meant a growth multiple of 3-4. In 1985 BARTEC employed 200 people; today the headcount stands at 1,200, a fabulous record of job creation. In 2005 Reinhold Barlian passed the helm to Ralph Koester, the new CEO. Barlian's legacy to Koester is "to continue the BARTEC-success story under a new, stronger and future-oriented leadership". Future growth markets are mostly in Asia, with emphasis on China. Koester brings with him many years of experience in these markets and has declared as his goal "to go from No. 1 in Europe to No. 1 in the world" (Tauber-Zeitung, February 3, 2005). He sees technology leadership as the foundation for market leadership. A doubling of revenue in the next five years is expected.

BARTEC is a typical case of a start up Hidden Champion in a niche market. Unlike Enercon, which entered the new market for wind energy and profited from its strong growth, BARTEC is operating in a more slowly growing market. It grew through innovations which have, however, long cycles in this industry, as well as through steady international expansion. New regulations for industrial safety are going to foster BARTEC's growth in the future. Bottlenecks in the growth have been a lack of internationally experienced people and finances. With a growth multiple of 3-4 per decade, strains of this kind are unavoidable. Again, as in the other cases, an entrepreneur, Reinhold Barlian, played a pivotal role for the "growth without explosions".

CONCLUDING REMARKS

- Many Hidden Champions are also champions of growth. They do not stand still but successfully pursue long-term growth goals. This persistence eventually transforms some of them into Big Champions, which is the way big corporations emerge. Others grow within their niches, so that their absolute size may remain modest but their market share and their market position become stronger in the process. The will to grow is embedded in the leaders, often originating with the founders. And one challenge is to keep it alive and instill it in the successors, ideally across generations of successive leaders.
- Continuous growth with steady, unspectacular annual rates seems to produce better long-term sustainability than growth in sudden bursts.
- Growth of a company is obviously not confined to growing markets. As we have seen from the cases we've examined, there are companies like Schmitz Cargobull which mostly grow by taking market share from their competitors. This is a viable growth strategy as long as you can create and sustain clear competitive advantages.
- There is no single way to grow. A company can grow by adding new products, by entering into new regional markets, or by addressing new customer segments or industries. All these options leave room for growth and exist to a varying degree in most markets.
- The case studies show that the Hidden Champions grow along all available growth avenues. Wuerth grew mostly through regional and customer segment expansion. Schmitz Cargobull took its higher revenue mostly from competitors, even down to driving the former market leader out of the market. Enercon strongly profited from "growing with the wind", which, however, did not come automatically but required technological leadership. And BARTEC grew as a specialist in a niche market.
- It should be emphasized that growth without profit is dangerous. The growth champions in the cases examined have shown above average profitability. This is the acid test: Only profitable growth is "good" growth.

By all means try to keep your company growing. Growth makes so many things so much easier. Never forget that the growth of a company doesn't happen of its own account. The will to grow is the first and most essential ingredient. The ways to grow are manifold.

20

Dual Innovation Strategies Empirical Modelling

Marc A. Zegveld
Erik den Hartigh

INTRODUCTION

Many studies in management sciences recognise the exploitation – exploration dilemma. These studies, however, conclude either that finding a balance is essential or confirm that the dilemma does indeed exist. There are as yet no studies that have defined the exploitation – exploration dilemma in a quantitative framework (Zegveld, 2000). The creation of such a framework would provide researchers the ability to define and measure the actual development stage or current focus of a company within the exploitation – exploration dilemma. As a result, the framework could provide information about what actually drives a company within this dilemma, and offer an insight into aspects which provide the impetus for these developments. While exploitation focuses on alignment and refinement, or incremental changes, exploration focuses on search, variation or radical changes. In this study, the punctuated equilibrium paradigm, a theory intended to detect incremental and radical change, is altered in order to operationalise the exploitation – exploration dilemma.

The exploitation – exploration trade-off can be related to a broad spectrum of different perspectives on decision making. March (March, 1991) found that within the rational choice approach (Hey, 1982; Radner and Rothschild, 1975), theories of limited rationality (Cyert and March, 1963; Simon, 1955) and theories of organisational learning (Arrow, 1962; Levinthal and March, 1981; 1993; Winter, 1971) the exploitation – exploration trade-off is recognised as a fundamental issue. Also, more recent theoretical developments such as evolutionary theories (Hannan and Freeman, 1987) and systems theory (Varela, 1991) have recognised the exploitation – exploration trade-off as crucial. Recent studies which examine the dilemma (Bennet III, 1998; Garavan, 1997; Mabey and Salaman, 1995; Smith and Saint-Onge, 1996) frequently use organisation learning theories (Argyris and Schon, 1978; Fiol and Lyles, 1985; Hedberg et al, 1976) to understand the interplay between refinement, selection, and choice versus risk-taking, search and variation.

The scope of this chapter focuses on a single company and its relation with its environment, and stresses the dynamics of resources in terms of learning and adapting. The study combines generic firm processes and the stakeholder approach to analyse aspects of exploitation such as incremental innovation and aspects of exploration such as radical innovation. In this study it is assumed that the processes and content of the interactions between companies and their environment influence the company as well as the environment. This implicitly states that interaction, and thus change itself, is continuous.

The aim of this chapter is to demonstrate the interrelationship between exploration and exploitation on a company level. The chapter deploys the exploitation – exploration dilemma through the use of exploration and exploitation and by the use of existing quantitative frameworks on both exploration and exploitation on the company level. The focus is on the interrelationship of exploration and exploitation with regard to stability and change.

EXPLORATION AND EXPLOITATION: STABILITY AND CHANGE IN BUSINESS PROCESSES

The resource perspective defines resources, and not a company as an economic element, and both the punctuated equilibrium paradigm and the resource perspective stress learning, which means that a company can be defined as either a process or as a coherent group of specific processes. Based on the description of the two major schools of strategic management contained in the comprehensive literature, the existence of company processes can be aggregated towards generic firm processes. Within a company several processes can be recognised, each of which influences the other. Through aggregation five generic processes are identified: process of forming the company's perspective (1); process of forming resources (2); process of configuration of production (3); process of production (4) and the process of value creation (5).

In developing a framework to detect and measure the balance between exploitation or incremental change, and exploration or radical change, these forms of change should not exclude one another. Therefore, the quantitative model of Romanelli and Tushman (1994) needs to be adjusted. Different and thus exclusive variables for exploitation and exploration are essential to ensure that the interplay between exploitation and exploration, or the absence of such interplay, can be measured.

Figure 1
Processes of change

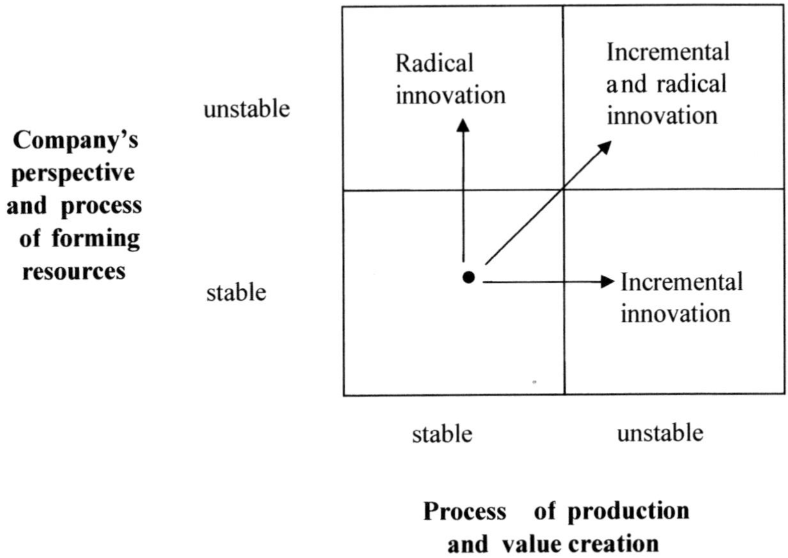

EXPLORATION

The company's perspective and the process of forming resources, especially the development of resources above the input level of the different core stakeholders, is related to the exploration of the firm. From an economic perspective the development of additional production factors above the input factors is defined as Total Factor Productivity (Solow, 1957), residual value, technology or knowledge. In economic theory a change in the build-up of residual value is a determinant of change of perspective or paradigm. By adapting the Total Factor Productivity (TFP) to companies, the residual value is defined as a resource related to the company's perspective. It is assumed that a stable build-up of residual value implies a stable company perspective.

The company's perspective implies a specific perception of the environment and as a result, through routines, stabilises the company as a whole and also stabilises the focus on the long-term dominance of the company. From the company's perspective, given the aim of long-term dominance, the company concentrates on combining and recombining resources in such a way that the existing core stakeholders' benefit and new resources are developed and operationalised. The build-up of resources is stable when, from the company's perspective, no fundamental changes occur within the firm's basic orientation towards its customers, employees, partners and shareholders. Recombining the existing resources and the development of new resources is intended to increase the productivity of resources and can be achieved by speeding up the learning curve and finding new combinations of resources that fit the company's current perspective. Resource productivity means that more resources can be developed above the input level of the resources provided by the different core stakeholders.

Within the proposed framework the findings of Penrose (1959) can be regarded as supportive of the idea that through the development of new resources, the production process can be changed from a company's perspective, which is then defined as incremental innovation. Penrose defined the company as a complex and structured combination of different resources, which combines constraints as well as cumulative development. Using the resource perspective of the company, Winter (1964) suggested that routines have a relatively durable quality over time to retain skills and other forms of knowledge. Secondly, Winter (1971) stated that companies have decision rules and retain or replace them according to the satisfaction principle, as defined by Simon (Simon, 1955) and Cyert and March (1963), which states that companies attempt to obtain satisfactory minima, rather than optimise (Hodgson, 1998). The satisfaction principle and its background of '*bounded rationality*' (Simon, 1955) fits in with the path-dependent trajectories of a company. From this perspective, stability in exploration is defined as a continuity of the high order or the perspective of a company, and hence continuity of the different core stakeholders in relation to the company. A continuity of both the company's perspective and the different core stakeholders results in a longitudinal continuous build-up of total resource productivity.

A different perspective implies a different build up of resources. Radical innovation can be measured by measuring the stability of the build-up of resources. Radical innovation implies the creation of novel 'techno-economic paradigms' (Dosi, 1984; Dosi et al, 1988; Freeman and Perez, 1989; Soete 1999 and 1987). From the company's perspective, no fundamental changes occur in the firm's basic orientation towards its core stakeholders. By combining and recombining existing resources new resources are developed which improve the resource productivity. Exploration can be related to a stable process of forming resources which is the result of 'learning by doing' or to a different process of cognition and hence to a different build-up of resources and knowledge. The development of knowledge within the company's perspective is a result of combining and recombining resources. Beyond the company's perspective, different stakeholders may also be identified besides new combinations of new resources. To detect radical innovation, a change in build-up of knowledge or the change in the type of combining and recombining resources should be observed. Radical innovation is therefore defined as: Change of the high order or a change of the perspective of a company and hence a discontinuity in the importance of the different core stakeholders in relation to the company. A discontinuity of the company's perspective results in a longitudinal discontinuous build-up of the total resource productivity.

From an economic perspective, resources are defined as 'knowledge' or 'surplus' to generate additional value, which is value above the sum of the values delivered by the initial core stakeholders. The difference between ex-post resources and ex-ante resources is defined here as 'knowledge' or 'intellectual capital'. The developed 'surplus' is the result of the company's perspective and its related developed resources. Hence a different perspective will lead to different resources and a change in the direction and the level of the 'surplus'. In the process of radical change a new perspective emerges, and the company fundamentally changes its view of its resources and related core stakeholders. A different perspective on resources leads to a situation whereby a different and new positional advantage and different competencies are developed. The result of the full deployment of this new perspective is that the share of the different core stakeholders and their resources, as well as the build-up of 'intellectual capital', changes. From an economic perspective, developing 'intellectual capital' is related to the theories concerning endogenous growth, which will be discussed in the next section.

In this chapter the quantitative analysis on exploration is based on the work of Solow (Solow, 1957), Griliches (1998) and Zegveld (Zegveld, 2000). The developed framework, adapting Total Factor Productivity to companies, will be used in order to test stability in the relationship between exploration and exploitation.

EXPLOITATION

Stability in exploitation and incremental innovation, which are two mutually exclusive aspects of exploitation, will be defined by using the concept of the value chain. The concept of the value chain, and hence value chain analysis, is chosen because this concept is a generally accepted method for analysing the process of production in relation to the process of value creation.

A generic strategy should be related to the process of production and the process of value creation and should be financially detected at the company level. Karnani (Karnani, 1984) translated two Porterian generic strategies into a quantitative model based on financial variables, and identified the following generic strategies:

- *Efficiency strategy*
  An efficiency strategy implies the aim of continuously reducing costs and thus increasing value per unit turnover. Choosing an efficiency-strategy leads to a focus on economies of scale, bringing down the experience curve and reducing overhead. Karnani (Karnani, 1984) focused on the reduction of firm specific costs, such as production costs, to detect an efficiency strategy, and found that these costs should be made independent of scale. The independence of scale can be related to a single product or a single unit turnover. This study focuses on a company as a whole and

uses time as the denominator to analyse the position of a company. Within this study turnover per unit labour cost is defined as the company's level of efficiency.

- *Added value strategy*
  An added value strategy can be detected if most of the created value can be related to an increase of added value per unit turnover. This strategy has several similarities with the Porterian strategy of differentiation, since both strategies focus on delivering premium value for the customer at a premium price. This premium value can be related to design, branding, product quality, customer service, or dealer network. A successful deployment of an added value strategy implies a trade-off between the premium price the customer is willing to pay and the increased cost development as a result of the extensive research, product design and intensive customer support used to gain additional margins. Karnani (Karnani, 1984) defined absolute demand (here translated as premium price) as a denominator for an added value strategy, which is comparable with the findings of Hall (Hall, 1980). A successful added value strategy combines uniqueness and customer value with higher margins or added value if the price premium exceeds the extra costs incurred by being unique. As a result the level of added value per unit turnover is defined as the level of differentiation.
- *Volume strategy*
  Karnani (Karnani, 1984) stated that production costs should be analysed independently of scale, and previously it was concluded that company specific costs should be analysed per unit turnover in order to analyse production costs independently of scale. A volume- strategy can be detected when most of the created value can be related to higher turnover, without changing the added value per unit turnover or company specific costs per unit turnover.

These three generic strategies constitute the basis for the stability of a company in terms of how it develops its operational value: by growth of turnover, decrease of costs, or by an increase of added value.

Given the dynamic resource framework it was chosen to analyse a company on a longitudinal basis and therefore the development along both axes, and not the position, is crucial to categorise a company within one of the defined generic strategies. Following the defined generic strategies the two excluding aspects of exploitation are defined. First, we define stability in exploitation, which implies that a substantial part of the development of the total operational value of a company can be related to a single generic strategy. Incremental innovation implies that no one single generic strategy realises a substantial part of the development of the total operational value of a company. Incremental innovation can be the result of a change of deployment from one generic strategy towards a different generic strategy or can be the result of an absence of a generic strategy within the period of analysis. A discontinuity in the deployment of a generic strategy can be the result of explicit or implicit choices by management or by core stakeholders and will affect the process of production and the process of value creation.

EMPIRICAL RESULTS

An empirical test was carried out to validate the proposed framework and used data from 85 companies listed on the Amsterdam Stock Exchange during the 1988-1997 period.

For exploitation, the normalised operational cash flow per category per variable is presented in Table 1. The categorization of companies with respect to incremental innovation related to size and industry are presented in Tables 2 and 3.

The operational cash flow measure (OCF) reflects the value arising from operational processes in which products and services are developed, produced and marketed. In this way all non-operational cash flows, e.g., as a consequence of 'financial engineering', are excluded.

The operational cash flow measure is composed of three variables (strategic value drivers): *Volume*, *Efficiency* and *Added value*. *Volume* indicates the extent to which the firm is able to internalize the size and growth of the market. *Volume* is measured as turnover. *Efficiency* indicates the costs the firm has to incur to be able to accommodate customer demands. *Efficiency* is measured as *Volume* divided by employment costs. *Added value* indicates the value that the firm's products generate for the customer compared to competitors' products. *Added value* is measured as *factor costs (= employment costs + depreciation + net profit)* divided by *Volume*. From the variables *Volume (V)*, *Efficiency (E)* and *Added value (AV)*, the operational cash flow is calculated as follows: $V*[AV-(1/E)]$.

In the Tables 1 and 4 below the impacts of innovation on operational cash flow are normalized. Normalization is implemented by defining twice the standard deviation of the cash flow as 1. This is done per company for each of the variables *Volume*, *Efficiency* and *Added value*. This normalization protocol is chosen since it generates normalized data without defining explicitly a single variable as dominant or with a value of "1". By normalizing the development of operational cash flow per company, companies can be aggregated in such a way that every company has an identical weight.

Table 1
Development of operational cash flow and categorisation

Development of normalised operational cash flow and categorisation							
	Total OCF	Differentiated by generic components					
		Volume	Significance	Efficiency	Significance	Added value	Significance
- Incremental innovation	0.733	0.637		0.115		- 0.020	
- Stable focus	0.817	0.682		0.029		0.106	
- Volume	0.745	0.840	0.025**	0.030		-0.125	
- Efficiency	0.601	0.575		0.262	0.311	-0.237	
- Added value	0.949	0.595		- 0.054		0.404	0.030**
- Total	0.768	0.656		0.079		0.034	

** Significance at 95% level per generic component

Table 2
Categorisation by size

Categorisation by size				
	Total	Size		
		AEX	Midcap	Smallcap
- Incremental innovation	49 (58%)	9 (53%)	7 (54%)	33 (60%)
- Stable focus	36 (42%)	8 (47%)	6 (46%)	22 (40%)
- Volume	13 (15%)	2 (12%)	2 (15%)	9 (16%)
- Efficiency	6 (7%)	1 (6%)	1 (8%)	4 (7%)
- Added value	17 (20%)	5 (29%)	3 (23%)	9 (16%)
- Total	85 (100%)	17 (100%)	13 (100%)	55 (100%)

Table 3
Categorisation by industry

Categorisation by industry				
	Total	Industry		
		Physical production	Distribution and communication	Services
- Incremental innovation	49 (58%)	25 (64%)	13 (52%)	11 (52%)
- Stable focus	36 (42%)	14 (36%)	12 (48%)	10 (48%)
- Volume	13 (15%)	6 (15%)	2 (8%)	5 (24%)
- Efficiency	6 (7%)	3 (8%)	2 (8%)	1 (5%)
- Added value	17 (20%)	5 (13%)	8 (32%)	4 (19%)
- Total	85 (100%)	39 (100%)	25 (100%)	21 (100%)

It can be concluded that a volume and an added value oriented strategy have a significantly different structure of operational cash flow components than the other generic strategies. Secondly, a certain relationship between size and industry type and the deployment of one of the generic strategies can be detected; large companies active in relative older industries have a more stable focus than small companies within young industries.

The exploitation – exploration dilemma means that incremental and radical innovation are intertwined. Given the dilemma a tension between maximising operational cash flow and maximising the build-up of residual value is apparent. It is assumed that radical innovation and hence a change of perspective create new opportunities to gain long term dominance and as a result, will also positively influence the company's ability to increase the development of its operational cash flow. Besides a distinction between companies with a positive and negative radical innovation, a second distinction is made between the period before and after the 'break' or radical innovation. A calculation of the average annual growth of the operational cash flow is calculated before and after the break. The impact of radical innovation on the development of operational cash flow provides evidence (Table 4) that radical innovation has a significant influence on the development of OCF; positive radical innovation accelerates the development of the operational cash flow and a negative radical innovation decreases the development or even diminishes the operational cash flow.

Table 4
Radical innovation and its impact on operational cash flow

Radical innovation and operational cash flow				
	Delta OCF per year			
	Average	Before radical innovation	After radical innovation	Significance
-Radical innovation	0.505			
-Build-up (Beta>0)	0.593	0.438	0.943	p = 0.034**
-Tear down (Beta<0)	0.260	0.491	-0.397	p = 0.012**
-Stable build-up	0.815			
-Total	0.735			

** Significance at 95% level

The categorisation of companies with respect to radical innovation, related to size and industry is presented in Tables 5 and 6. These include a subset of 4 companies (out of 22) based on a more strict definition of radical innovation.

Table 5
Categorisation by size

Categorisation by size				
	Total	Differentiated by size		
		AEX	Midcap	Smallcap
- Radical innovation	22 (26%)	4 (24%)	1 (8%)	17 (31%)
-Strict Definition	4 (5%)	0 (0%)	0 (0%)	4 (7%)
- Stable build-up	63 (74%)	13 (76%)	12 (92%)	38 (69%)
-Total	85 (100%)	17 (100%)	13 (100%)	55 (100%)

Table 6
Categorisation by industry

Categorisation by industry				
	Total	Differentiated by industry		
		Physical production	Distribution and communication	Services
- Radical innovation	22 (26%)	7 (18%)	7 (28%)	8 (38%)
- Strict Definition	4 (5%)	0 (0%)	2 (8%)	2 (10%)
- Stable build-up	63 (74%)	32 (82%)	18 (72%)	13 (62%)
- Total	85 (100%)	39 (100%)	25 (100%)	21 (100%)

In Table 7 the categorisation of incremental innovation is combined with the findings about radical innovation.

Table 7
Combined results

Combined results: exploitation and exploration		
	Total	Incremental innovation
- Radical innovation	22 (26%)	15 (31%)
- Strict definition	4 (5%)	4 (8%)
- Stable build-up	63 (74%)	34 (69%)
- Total	85 (100%)	49 (100%)

	Volume	Efficiency	Added value
- Radical innovation	4 (31%)	0 (0%)	3 (18%)
- Strict definition	0 (0%)	0 (0%)	0 (0%)
- Stable build-up	9 (69%)	6 (100%)	14 (82%)
- Total	13 (100%)	6 (100%)	17 (100%)

The normalized cash impact of companies with a stable focus on exploitation is 0,817, whereas the normalized cash impact of companies with an unstable focus on exploitation is 0,733 (Table 1). On exploitation 49 companies (58%) changed their focus on the process of production and value creation and are categorised as incremental innovation companies (see column 'total' in Tables 2 and 3). The normalized cash impact of companies with a stable focus on exploration is 0,815, whereas the normalized cash impact of companies with a unstable focus on exploration is 0,505. As regards exploration 22 companies (26%) changed their focus on the process of forming resources and are categorised as unstable on exploration (see column 'total' in Tables 5 and 6). Based on the findings it can be concluded that the deployment of exploration is more stable than the deployment of exploitation, and that stability pays off.

On both exploration as well as exploitation the cash-impact analysis was repeated analysing the cash-impact before and after the shift or change per company. From the perspective of exploitation the impact remains similar; however from the perspective of exploration the cash impact (0,505) altered from 0,438 before the shift towards 0,943 after the shift.

Exploration is dominantly stable, however companies which are able to change their exploration focus gain the highest cash-impact.

CONCLUSIONS

Given the scope of the exploitation – exploration dilemma, companies have to balance exploitation and exploration, and hence the framework should not exclude developments in a given time frame, but must allow for changes in both dimensions at the same time. An empirical test was carried out to validate the proposed framework and used real data from 85 companies listed on the Amsterdam Stock Exchange. Based on the boundaries and limitations defined, the data sample was limited; the selected companies were listed for at least ten years or more.

Stability and change occur both with regard to exploration and exploitation. It can be concluded that stability on exploration is more durable than on exploitation. In both cases stability pays off, however a shift in exploration generates a significantly higher impact on the development of the cash flow of companies.

Based on the findings, 58% (49 out of 85 companies) achieved incremental innovation and thus changed their focus on the organisation of their value chain in relation to the process of value creation. At the same time 42% (36 out of 85 companies) can be categorised as having a stable focus. This implies that in at least half of the years analysed, the value of the variable related to one particular generic strategy improved, and that at least half of the development of the operational cash flow can be traced to the development of that specific generic strategy. Based on the findings it may be concluded that consistency pays off: companies with a focus have a much higher development of operational cash flow than companies that changed their perspective. In this sense it is concluded that from the perspective of exploitation a deployment of a single generic strategy is more useful than incremental innovation. Secondly, it is also be concluded that the pay-off from a added value strategy is higher than the development of the operational cash flow of efficiency or volume-driven companies.

Radical innovation is related to the change in the perspective of a company and the process of forming resources. By using endogenous growth theory (Solow, 1957) it is expected that the development of residual value, or intellectual capital, which cannot be related to one of the core stakeholders, will be stable for most companies. A change of perspective implies a different orientation towards existing or different core stakeholders and generic firm processes and results in a different build-up of residual value.

Radical innovation occurs when the direction of the residual build-up in a given period of time differs from the overall development and is contrary to the development in the previous period. Following this definition 22 companies (26%) changed their perspective in twelve years. Using the proposed model it is possible to determine whether radical innovation occurred in each company and to determine the typology of radical innovation or the typology of stability of high order. Based on the analysis it can be concluded that radical innovation is rare and that the perspective of most companies is stable.

Companies should manage both exploitation as well as exploration. Optimizing both aspects, we found that changes in the process of forming resources has a higher impact on the cash flow developments of companies than a change in the deployment of one of the generic strategies.

BIBLIOGRAPHY

Argyris, C., and D. A. Schon, (1978), *Organizational Learning*, Addison Wesley, New York.

Arrow, K., (1962), The economic implications of learning by doing, *Review of Economic Studies*, Volume 29, Issue 3, pp. 155-173.

Bennett III, R.H., (1998), The importance of tacit knowledge in strategic deliberations and decisions, *Management Decision*, Volume 36, Issue 9, pp. 589-597.

Cyert, R.M., and J. March, (1963), *A behavioral theory of the firm*, Prentice Hall, New York.

Dosi, G., (1984), *Technical change and industrial transformation*, St. Martin's Press, New York.

Dosi, G., (1988), Sources, procedures and microeconomic effects of innovation, *Journal of Economic Literature*, Volume 26, September, pp. 1120-1171.

Fiol, C.M., and M.A. Lyles, (1985), Organizational Learning, *Academy of Management Review*, Volume 10, Issue 4, pp. 803-813.

Freeman, C.,, and C. Perez, (1989), Structural crisis of adjustment, business cycles and investment behaviour, in: G. Dosi, C. Freeman, R. Nelson, G. Silverberg and L. Soete (eds.), (1988), *Technical change and economic theory*, Columbia University Press, New York.

Garavan, T., (1997), The learning organization: a review and evaluation, *The Learning Organization*, Volume 4, Issue 1, pp. 18-29.

Hall, W.K., (1980), Survival strategies in a hostile environment, *Harvard Business Review*, September-October, 58 (5), pp. 75-85.

Hannan, M.T., and J. Freeman, (1987), The ecology of organizational foundings: american labor unions, 1836-1985, *American Journal of Sociology*, Volume 92, pp. 910-943.

Hedberg, B.L.T., Nystrom, P.C. and Starbuck, W.H., (1976), Camping on seesaws: prescriptions for a self-designing organisation, *Administrative Science Quarterly*, Volume 21, Issue 1, pp. 41-65.

Hey, J.D., (1982), Search for rules for search, *Journal of Economic Behavior and Organization*, Volume 3, pp. 65-81.

Hodgson, G.M., (1998), Evolutionary and competence-based theories of the firm, *Journal of Economic Studies*, Volume 25, Issue 1, pp. 25-56.

Levinthal, D.A., and J. March, (1981), A model of adaptive organizational search, *Journal of Economic Behavior and Organization*, Volume 2, pp. 307-333.

Levinthal, D.A., and J. March, (1993), The myopia of learning, *Strategic Management Journal*, Volume 14, pp. 95-112.

Karnani, A., (1984), Generic competitive strategies - an analytical approach, *Strategic Management Journal*, Volume 5, pp. 367-380.

Mabey, C., and G. Salaman, (1995), *Strategic human resource management*, Blackwell Business: London.

March, J.G., (1991), Exploration and exploitation in organizational learning, *Organization Science,* Volume 1, Issue 1, pp. 71-81.

Penrose, E., (1959), *The theory of the growth of the firm*, Wiley: New York.

Radner, R., and M. Rothschild, (1975), On the allocation of effort, *Journal of Economic Theory*, Volume 10, pp. 358-376.

Rosenberg, N., (1976), *Perspectives on technology*, Cambridge University Press, Cambridge.

Romanelli, E., and M.L. Tushman (1994), Organizational transformation as punctuated equilibrium: an empirical test, *Academy of Management Journal*, Volume 37, Issue 5, pp. 1141-1166.

Simon, H.A., (1955), A behavioral model of rational choice, *Quarterly Journal of Economics*, Volume 69, pp. 99-118.

Smith, P.A.C., and H. Saint-Onge, (1996), The evolutionary organization: avoiding a titanic fate, *The Learning Organization*, Volume 3, Issue 4, pp. 4-21.

Soete, L.L.G., (1987), *Technologie en economie*, Rijksuniversiteit Limburg, Maastricht.

Soete, L.L.G., and B.J. ter Weel, (1999), *Innovation, knowledge creation and technology*, MERIT, Maastricht.

Solow, R.M., (1957), Technical change and the aggregate production function, *Review of Economics and Statistics*, Volume 39, pp. 312-332.

Varela, F.J., (1991), Organism, cognitive science and the emergence of selfless selves, *Revue Européenne des Sciences Sociales*, Volume 29, Issue 89, pp. 173-198.

Winter, S.G., (1964), Economic 'natural selection' and the theory of the firm, *Yale Economic Essays*, Volume 4, Issue 1, pp. 225-272.

Winter, S.G., (1971), Satisfying, selection and the innovation remnant, *Quarterly Journal of Economics*, Volume 85, pp. 237-261.

Zegveld, M.A., (2000), *Competing with dual innovation strategies*, werk-Veld, The Hague.

21

The Verdoorn Law, Firm Strategy and, Firm Performance

Erik den Hartigh
Fred Langerak
Marc Zegveld

INTRODUCTION

In many markets and firms, positive feedback effects play an important role. Positive feedback effects mean that there is a tendency for that which is ahead to get further ahead, and for that which loses advantage to lose further advantage (Arthur, 1996). Such positive feedback effects result from four mechanisms: (1) social interaction effects; (2) network effects; (3) scale effects; and (4) learning effects. Of these mechanisms social interaction effects and network effects are market-bound, while scale effects and learning effects are firm-bound (Den Hartigh and Langerak, 2001). Prior research has highlighted the influence of market-bound positive feedback mechanisms on market structure and market outcomes, but has largely ignored the impact of firm-bound positive feedback mechanisms on firm performance.

In this study we aim to fill a part of this gap in extant knowledge by: (1) investigating the extent to which Dutch firms listed on Euronext have been able to exploit scale and learning effects between 1983 and 2002; (2) examining the extent to which the realization of scale and learning effects has affected firms' performance during this period; and (3) investigating the competitive strategies used by firms to increase their productivity and performance. This will enable us to assess the success of these strategies and to advise firms as to how they should change their strategy in order to become more successful by making better use of productivity gains.

SCALE EFFECTS AND LEARNING EFFECTS

Scale effects and learning effects are the basic mechanisms which bring about firm-bound positive feedback effects. Scale effects occur when there is a positive *static* relationship between the size of a firm's output (the number of goods and services produced) and its productivity measured in financial terms (Scherer and Ross, 1990). Thus, scale effects are cost advantages that result from a growth in the scale of output.

Learning effects occur when there is a positive *dynamic* relation between the growth in output and the growth in productivity measured in financial terms (Amit, 1986). Thus learning effects are reflected in the downward shift of the firm's average total cost function. Learning effects may be induced, exogenous, or autonomous.

- *Induced learning* results from conscious managerial action aimed at improving the productivity of input factors (Li and Rajagopalan, 1998).
- *Exogenous learning* results from effects that are external to the firm and that cannot be controlled by management. Still, management can position the firm in the right networks to enable it to optimally internalize these effects and therefore benefit from them.
- *Autonomous learning* involves the automatic improvement in productivity of input factors as accumulated production increases. It is often referred to as *learning-by-doing*, or as the *experience curve* (Alberts, 1989). As such, learning is a product of experience, which means that the level of productivity is a function of cumulative output, and thus the faster the output expands the faster productivity grows. In other words, autonomous learning causes productivity to rise in response to, or as a by-product of, the increase in total output.

VERDOORN'S LAW, FIRM STRATEGY AND FIRM PERFORMANCE

The realization of scale and learning effects can be measured through a firm-level application of Verdoorn's law (Den Hartigh, 2005). Verdoorn's law states that in the long run there is a linear relationship between the growth in output and the growth in productivity (Verdoorn, 1949; Kaldor, 1966). This relationship can be explained by the theory of cumulative causation. In this theory it is primarily the growth in effective demand that stimulates technological growth, through a larger potential for division of labor and through learning-by-doing. The resulting productivity increase stimulates higher outputs through the extension of existing markets and the opening up of new markets (Kaldor, 1966). This suggests that productivity gains and growth in output are mutually reinforcing mechanisms. Following this line of reasoning, Verdoorn's law has been empirically observed across nations, regions and industries using country-level, region-level and industry-level data (e.g., Verdoorn, 1949; Fingleton and McCombie, 1998; Harris and Lau, 1998).

To date Verdoorn's law has not been assessed using firm-level data. The argument of cumulative causation, however, also applies at the firm level. The firm-level reasoning is that a firm's growth in output over time causes productivity improvements through the realization of scale effects and autonomous learning effects. At the same time firms can realize productivity improvements as the result of induced and exogenous learning effects. Firms can use these productivity gains to increase their output by stimulating additional market-demand for their products. This can be done by using the cost advantage to win market share either by: (1) offering products at lower prices than competitors for equivalent benefits (a *volume-efficiency strategy*); or (2) by providing customers with unique benefits at higher prices than competitors (a *volume-differentiation strategy*). Through these strategies productivity gains and growth in output become mutually reinforcing, causing a positive feedback loop (see Figure 1).

Figure 1
Verdoorn's law at the firm level

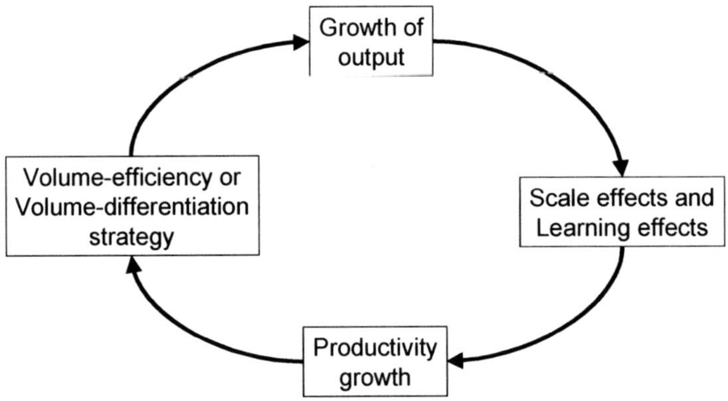

Analogous to the relationship between the growth in output and the growth in productivity, there is also a relationship between the growth in productivity and the growth in firm performance. This is logical because output, i.e., quantities sold, prices and costs, are known determinants of firm performance. Through the realization of scale and learning effects the *volume-efficient* firm achieves a higher output with a better unit margin and with prices below those of competitors. The *volume-differentiated* firm achieves a higher output with a better unit margin with prices higher than those of competitors. Thus both strategies improve firm performance.

ESTIMATING VERDOORN'S LAW AND THE PRODUCTIVITY-PERFORMANCE RELATIONSHIP

Verdoorn's law can, as previously mentioned, be used to measure the realization of scale and learning effects. The law is formulated as follows:

$$\Delta P = a + b^{*}\Delta Q \qquad (1)$$

In this formula ΔP is the change in productivity (i.e., output divided by input) between year t and year $t+1$ and ΔQ is the change in output between year t and year $t+1$. This relationship can be estimated provided that longitudinal firm-level data are available. If the value of the Verdoorn coefficient (b) is larger than 0, this means that scale effects and/or autonomous learning effects have been realized as a consequence of the growth in output over time. The value of the intercept (a) indicates whether there has simultaneously been growth in productivity as a consequence of induced and/or exogenous learning effects. Thus Verdoorn's law can be used to measure scale effects, in the sense that output in year $t+1$ is larger than output in year t, causing the productivity in year $t+1$ to be higher than in year t, and learning effects in the sense that productivity grows as the cumulative output increases.

The relation between productivity growth and the growth in firm performance can be formulated as follows:

$$\Delta FP = c + d^{*}\Delta P \qquad (2)$$

In this formula ΔFP is the change in firm performance between year t and year $t+1$ and ΔP is the change in productivity between year t and year $t+1$. A value of the coefficient (d) larger than 0 means that firms have realized better performance as a consequence of the growth in productivity over time. The value of the intercept (c) indicates whether there has been a growth in firm performance independent of productivity growth.

METHODOLOGY

We assessed Verdoorn's law and the productivity-performance relationship for 13 industries based on pooled data from 118 firms listed on Euronext between 1983 and 2002. The data was derived from annual reports and corrected for inflation. The yearly growth in output (ΔQ) was measured as the growth in operational cash flow. The yearly growth in productivity (ΔP) was measured as the ratio between the operational cash flow (output) and the number of employees (input). The yearly growth in firm performance (ΔFP) was measured as the growth in net profits. We focused on operational cash flow as output measure for two reasons. First, cash flow reflects the value arising from operational processes in which products and services are developed, produced and marketed. This way all non-operational cash flows, e.g., as a consequence of 'financial engineering', that would distort output and productivity figures are excluded. Second, operational cash flow enables us to analyze which strategy firms used to realize scale and learning effects.

We analyzed the strategic value drivers that firms used to generate operational cash flow: volume, efficiency and differentiation (see Box 1). Volume (V) indicates the extent to which a firm is able to internalize the size and growth in the market. Volume is measured as turnover. Efficiency (E) indicates the costs the firm has to incur to be able to accommodate customer demand. Efficiency is measured as volume divided by employment costs. Differentiation (D) indicates the added value that the firm's products generate for the customer in comparison to competitors' products. Differentiation (D) is measured as added value divided by volume. Together, efficiency and differentiation make up the value-performance indicator (i.e., a measure of the operational value created per unit turnover). This is calculated as $D-(1/E)$. The firm's operational cash flow is calculated by multiplying the value-performance indicator by turnover: $V*[D-(1/E)]$.

Three strategic value drivers used to generate operational cash flow. Based on Porter's (1980) theory of competitive strategy, Van Asseldonk (1998) derived three strategic value drivers with which the strategic direction followed by a company over time can be measured and characterized.

Volume: Some firms choose volume as the dominant driver for value creation. Such firms often focus on strong autonomous growth, often operate in growing markets and strive to grow faster than the market. This strategy is reflected in statements such as: "Our strategy is to grow 25% per year." With volume as the dominant driver firms manage operational efficiency and product differentiation within the context of their aspired volume growth.

Efficiency: Other firms select efficiency as their dominant driver of value creation. Their strategic objective is to improve productivity by certain percentages every year. Quite often these firms compete in stagnating markets and they compete with other firms on the basis of operational excellence by investing in automation and mechanization. These firms try to maintain or increase their market share by offering lower prices than competitors. These firms create their value in large part by improving their cost structure.

Differentiation: Lastly, there are firms that select differentiation as their dominant driver of value creation. These firms emphasize product development and branding. They introduce new products with a higher added value, either by improving the quality of the product itself or by investing in brand image and brand leadership. Such firms build a broad assortment of products to serve various segments in the markets in which they operate.

It is possible to combine multiple value drivers. By producing high volumes ("Volume strategy") of standardized products ("Efficiency strategy"), a firm takes advantage of economies of scale and experience curve effects. By producing high volumes ("Volume strategy") of differentiated products ("Differentiation strategy"), a firm provides superior customer value. Combinations like efficiency with differentiation are difficult to implement due to the potential for conflict between cost minimization and the additional cost of value-added differentiation.

RESULTS

The results of our assessment of Verdoorn's law are summarized in Table 1 below.

Table 1
Estimation of Verdoorn's law for Dutch industries, based on firm-level data

Industry	No. of firms	Coefficient (b)	T-value of coeffi-cient	Intercept (a)	T-value of constant term	R square
Basic industry	14	0.672 (***)	17.268	0.008	1.132	0.571
Food industry	8	0.322 (***)	6.056	0.003	0.476	0.221
Media	6	0.177 (**)	2.719	0.024 (*)	2.287	0.074
Engineering industry	14	0.261 (***)	7.468	-0.002	-0.334	0.199
Construction industry	9	0.234 (***)	5.967	0.004	0.621	0.199
Wholesale	14	0.346 (***)	7.951	0.002	0.189	0.236
Transport	8	0.566 (***)	8.552	-0.001	-0.079	0.371
Telecommunications	1	0.652 (**)	3.354	0.008	0.291	0.429
Financial services	15	0.606 (***)	16.981	-0.021 (*)	-1.984	0.582
IT services	5	0.100 (**)	2.916	0.000	0.004	0.095
Other business services	7	0.449 (***)	8.313	-0.041 (**)	-3.217	0.375
Retail	10	0.297 (***)	4.024	-0.006	-0.557	0.114
Electronics industry	7	0.511 (***)	11.600	-0.018	-1.123	0.619
Averages		**0.399**		**-0.003**		**0.314**

Significance at the 5% level is indicated by (*), at the 1% level by (**) and at the 0.1% level by (***)

In this table, the results of the calculations of equation 1, Verdoorn law, are shown for the 13 industries investigated. These calculations are based on pooled firm-level data. For every industry the number of firms supplying the data included in the assessment is indicated. The value of the coefficient *(b)* is shown in the table, which indicates the extent to which firms in this industry are able to realize scale effects and autonomous learning effects. The T-value of the coefficient indicates the validity of the assessment. Further, the value of the intercept *(a)* is shown, which indicates the extent to which firms in this industry are able to realize induced and exogenous learning effects. Here also, the T-value is provided to indicate the significance of this assessment. Finally, we provide the R square of the entire equation, indicating the explanatory power of the regression model.

The analysis shows that the average Verdoorn coefficient was 0.399 ($p<0.01$) during the period 1983-2002. This means that every 1% increase in output caused a growth in productivity of 0.399%. Firms have therefore realized scale effects and autonomous learning effects because their output has increased over time. It is noteworthy that the Verdoorn coefficient varies greatly across industries. The coefficients range from 0.100 (IT services) to 0.672 (basic industry). Interestingly, our analysis also shows that firms were simultaneously confronted with a negative value of the intercept of -0.003. This means that over this period firms were confronted with an average decline in productivity of 0.3%. This decline is brought about by the inability of firms to realize induced and exogenous learning effects. Again, this percentage varies greatly across industries. The intercepts range from -0.041 (professional services) to 0.024 (media).

The results of the assessment of the productivity-performance relationship are summarized in Table 2. In this Table, the results of the calculation of Equation 2, the productivity-performance relationship, are shown for the 13 industries investigated. These calculations are based on pooled firm-level data. For every industry the number of firms providing data included in the assessment is indicated. In the table, the value of the coefficient *(d)* is shown, which indicates the extent to which firms in this industry have increased their profits as a result of productivity growth. The T-value of the coefficient indicates the validity of the estimate. Further, the value of the intercept *(c)* is shown, which indicates the extent to which firms in this industry have increased their profit independent of productivity growth. Here also, the T-value is provided to indicate the significance of this calculation. Finally, we provide the R square of the entire equation, indicating the explanatory power of the regression model.

Table 2
Assessment of the productivity-performance relationship for Dutch industries

Industry	No. of firms	Coefficient (d)	T-value of the coeffi-cient	Intercept (c)	T-value of constant term	R square
Basic industry	14	4.086 (***)	12.811	-0.073	-1.776	0.470
Food industry	8	1.535 (***)	7.953	0.079 (***)	5.316	0.331
Media	6	1.738 (***)	6.952	0.071 (*)	2.469	0.340
Engineering industry	14	2.900 (***)	9.575	0.084 (**)	3.145	0.298
Construction industry	9	1.221 (**)	3.160	0.125 (***)	4.778	0.069
Wholesale	14	3.484 (***)	13.725	0.072 (*)	2.511	0.501
Transport	8	4.760 (***)	9.814	0.102	1.699	0.474
Telecommunications	1	0.475	0.682	0.164	1.824	0.027
Financial services	15	1.638 (***)	20.704	0.089 (***)	5.150	0.679
IT services	5	2.819 (***)	4.811	0.205 (***)	4.686	0.238
Other business services	7	2.496 (***)	4.743	0.105 (**)	2.783	0.191
Retail	10	1.983 (***)	7.392	0.053	1.825	0.309
Electronics industry	7	3.812 (***)	9.941	0.019	0.339	0.607
Averages		**2.534**		**0.084**		**0.349**

Significance at the 5% level is indicated by (*), at the 1% level by (**) and at the 0.1% level by (***)

The analysis shows that during the period 1983-2002 there was a positive relationship (*b=2.534; p<0.01*) between productivity growth and growth in firm performance. This means that every 1% increase in productivity caused a growth in net profits of 2.534%. The values of the coefficients vary between 1.535 (food industry) and 4.086 (basic industry). The values of the intercepts range from 0.071 (media) to 0.205 (IT services). This means that firms in these industries increased their net profits, independent of their rise in productivity, by 7.1% and 20.5% respectively.

The results of the strategic value driver analysis are presented in Tables 3 and 4.

In Table 3 we show the average values, measured across all firms, of the volume and the value-performance indicators for the beginning of the measurement period (1983), and at the end of the measurement period (2002). We performed a t-test over all firms to determine the significance of the difference between the 1983 and 2002 values. The results of this analysis reveal that the volume indicator, i.e., turnover corrected for inflation, significantly increased from €1941 to €5492 million between 1983 and 2002, but that the value-performance indicator, i.e., operational value per unit turnover, did not change.

Table 3
Independent-sample t-test for volume and value-performance

	Mean 1983	Mean 2002	t-value
- Volume (in million Euro)	1941	5492	-2.425
- Value-performance	0.106	0.110	-0.183

Note: T-values with an absolute value above 1.96 are significant.

However, as explained earlier, the value-performance indicator consists of an efficiency component and a differentiation part, which might cancel each other out. To disentangle these effects we compared the indicators for efficiency (turnover/employment costs) and differentiation strategy. In Table 4 we show the average values, measured across all firms, of the efficiency and differentiation indicators for the beginning of the measurement period (1983) and the end of the measurement period (2002). We performed a t-test over all firms to determine the significance of the difference between the 1983 and 2002 values. The results reveal that the indicators did not change significantly between 1983 and 2002.

Table 4
Paired-sample t-test for efficiency and differentiation

	Mean 1983	Mean 2002	t-value
- Efficiency	5.810	5.592	0.360
- Differentiation	0.354	0.353	0.039

Note: T-values with an absolute value above 1.96 are significant.

Together these findings show that firms chose, on average, a volume-driven strategy to generate operational cash flow, and not a strategy directed at enlarging value performance (through efficiency and/or differentiation).

EXAMPLE

In order to illustrate the method used, we provide the data of a Dutch based multinational firm that is exemplar of the sample as a whole. In Figure 2 below, we present a graph of the Verdoorn law for this firm, covering the period of 1986-2002. On the horizontal axis one can observe ΔQ, the change in output between year t and year $t+1$, and on the vertical axis one can observe ΔP, the change in productivity between year t and year $t+1$. From the regression line drawn in the Figure, it can be seen that the Verdoorn coefficient *(b)* over the period 1986-2002 is 0.65, indicating that any 1% increase in output has caused a growth in productivity by 0.65%. This means that the firm has realized considerable scale effects and autonomous learning effects over the period analysed. The value of the intercept *(a)* is -0.09, indicating that the firm has seen an average annual decline in productivity of 9%, because it has been unable to realize induced and exogenous learning effects.

Figure 2
The Verdoorn law 1986-2002

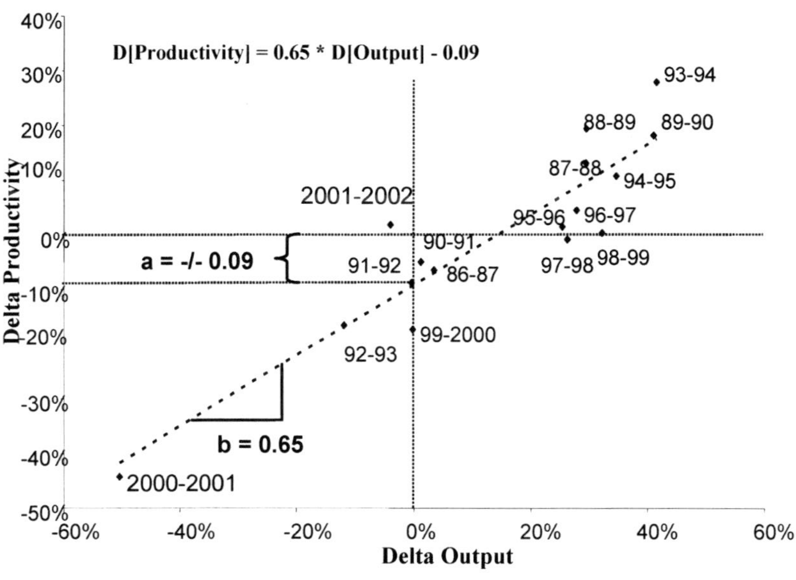

In Figure 3 below, we present a graph of the productivity-performance relation for the same period. On the horizontal axis one can observe ΔP, the change in productivity between year t and year $t+1$, and on the vertical axis ΔFP is shown, representing the change in firm performance between year t and year $t+1$. From the regression line drawn in the Figure it can be seen that the value of the coefficient *(d)* is 1.71, indicating that any 1% increase in productivity has caused a growth in net profits by 1.71%. This means that the firm has realized considerable better performance as a consequence of the growth in productivity over the period analyzed. The value of the intercept *(c)* is 0.15, indicating that the firm has realized an average annual growth in net profits of 15%, independent of productivity growth.

Figure 3
The productivity-performance relationship, 1986-2002

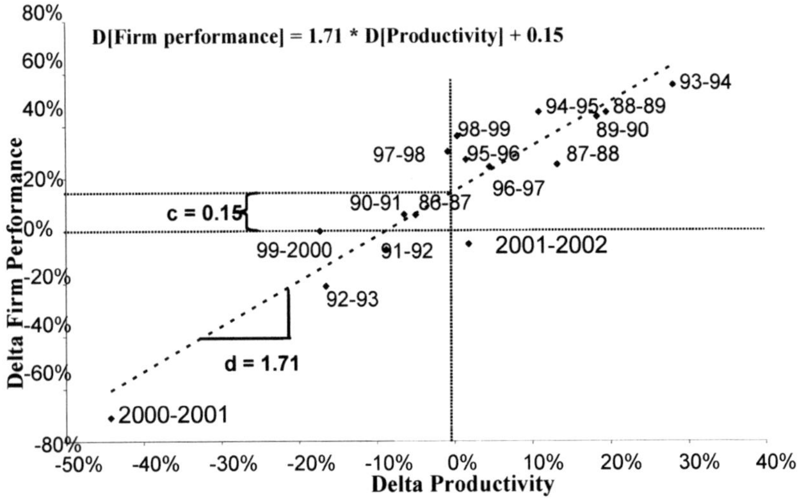

Figure 4
Strategy graph: Volume versus value-performance

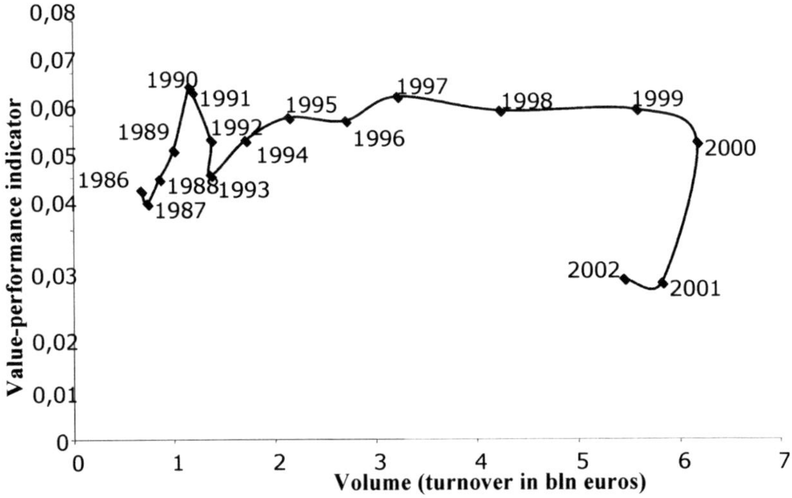

In Figure 4 we present a graph of the firm's strategy over the period 1986-2002, as measured by the value driver "Volume" and the value-performance indicator (operational value created per unit of turnover). It is easy to see that, on the whole, over the period analyzed the firm has chosen volume growth as an important driver for value creation.

In Figure 5 below we present a graph of the firm's strategy over the same period, but now the value-performance indicator has been split into its components, "Efficiency" and "Differentiation". This graph presents us with a fuzzy image of the strategy over the period analyzed. It can be seen that in different periods efficiency and differentiation have alternated as important value drivers, but that neither of them has been dominant over the years. Indeed, in this graph the firm in 2002 finds itself back not far from where it started in 1986.

Figure 5
Strategy graph: Efficiency versus Differentiation

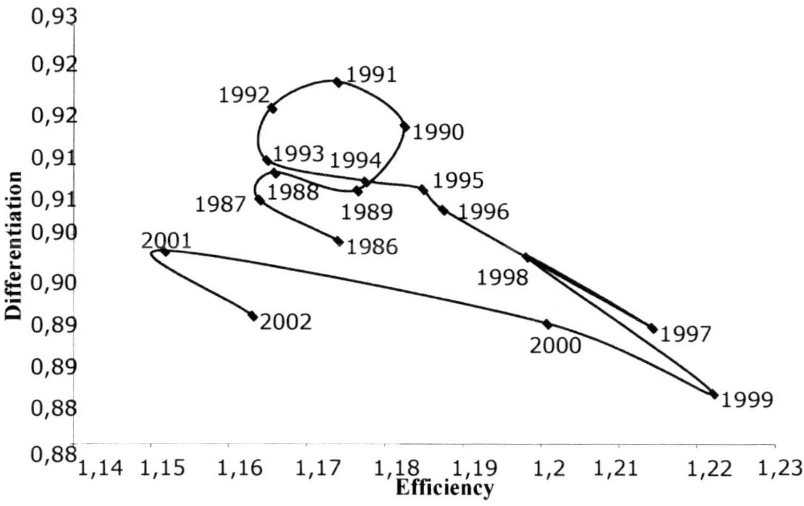

From the strategy graphs presented above it can be concluded that, taken over the entire period analyzed, volume has been the primary strategy driver of this firm.

DISCUSSION

The results of both our overall analysis and of the sample firm analysis show that firms have incurred productivity losses through negative induced and exogenous learning effects. At the same time they have realized productivity growth, and hence performance growth, through the realization of scale and autonomous learning effects. These effects were realized because firms used a

"Volume" strategy aimed at internalizing market growth to generate operational cash flow. This choice of strategy makes firms vulnerable to changes in market growth. As soon as demand declines they are confronted with lower or even negative productivity growth, because they are no longer able to realize scale effects and autonomous learning effects. This has a detrimental impact on firms' performance, as witnessed in the economic downturn since 2000.

In order to reduce their dependence upon market growth as well as to improve productivity and performance, firms have two alternatives. The first alternative is to improve their induced and exogenous learning capabilities, whereby managers attempt to implant the learning efforts directly by: (a) increasing investments in R&D (which steadily declined between 1983 and 2002) to an internationally competitive level; (b) mounting efforts in recruitment and training of employees to ensure that their knowledge and skills keep pace with the firm's technological requirements; (c) gaining access to and participating intensively in (international) knowledge exchange networks; and (d) forming alliances with universities to ensure the inflow of highly trained employees.

The other alternative is to modify their volume strategy to one that reinforces the relationship between productivity growth and output growth. They can do this by adopting "Volume-efficiency" or "Volume-differentiation" strategies. Based on the three strategic value drivers presented earlier, two self-reinforcing strategies can be identified, which are based on combinations of the volume and efficiency drivers and of the volume and differentiation drivers.

Volume-efficiency strategy: This strategy requires a no-frills product that is produced at a relatively low cost and made available to a very large customer base. The cost advantage is used to win market share by pricing the product lower than competing products. The associated distribution strategy is designed to obtain the most extensive distribution, and the promotional strategy involves trying to make a virtue out of low cost products. By winning market share and producing higher volumes of products firms take even more advantage of economies of scale and experience curve effects. Maintaining an efficiency strategy requires a continuous search for cost reductions throughout the organization and preferential access to raw materials, components, labor and other important input factors. Without these advantages, competitors can easily mimic a firm's volume-efficiency strategy.

Volume-differentiation strategy: This strategy uses productivity gains from scale and autonomous learning effects to create a product that is perceived as unique. The unique features should provide superior value to the customer. It is through this creation of superior value that firms win market share. By winning market share and producing higher volumes of products, firms can take even more advantage of economies of scale and learning effects. Because customers see the product as unrivalled and unequalled, the price elasticity of demand tends to be reduced and customers tend to be more loyal. This enables firms to adopt a premium pricing strategy. In order to maintain this strategy firms should continuously search for new sources of differentiation in all aspects of their business.

The objectives of firms following these strategies is to use the realized cost advantages through scale and learning effects to stimulate additional market-demand, either by charging customers lower prices and/or by providing additional benefits at higher prices. This additional demand results in higher market share, higher output growth, higher productivity growth, and hence higher performance growth. Implementation of either strategy will set in motion a positive feedback loop, and therefore strengthen the positive link between productivity growth and performance growth.

GENERALIZATIONS AND FURTHER RESEARCH

This research is limited by the fact that only Dutch-based firms that were listed on Euronext have been investigated. This means that there is a geographical limitation as well as a limitation in firm size, inasmuch as only larger firms are listed on the stock market. Still, these limitations do not preclude us from making some careful generalizations. First, while the firms in our sample are Dutch-based, most of them are multinationals, performing their operational activities around the globe. They can hardly be characterized as "Dutch firms", but are better characterized as "Western multinational firms". Therefore we could, with some care, state that the results obtained give an indication of the relation between firm strategy and firm productivity and firm performance that could be valid for a broader sample of Western multinational firms.

As regards our specific findings on the Verdoorn law, there is another reason for us to make some careful claims about the validity of our findings for other regions and for smaller firms. The values we found in our research are in accordance with the values found in previous cross-industry and cross-regional research, e.g., with the values found by Verdoorn himself (in a cross-country and cross-industry study, 1949), by Kaldor (a cross-industry study for the UK, 1966), by Fingleton and McCombie (a study across the European regions, 1998), by Harris and Lau (a study across UK regions, 1998), by León-Ledesma (a study across Spanish regions, 2000), by Bianchi (2002) (a cross-industry study for Italy), by Aspergis and Zikos (a cross-industry study for Greece, 2003) and by Naastepad (a study of the Dutch economy as a whole, 2006).

Still, these studies were executed with industry, regional or country level data, and not, as ours was, with firm level data. The values they found may differ from the values found at the firm level. Therefore, in order to further validate the results obtained from our analysis and to enable us to make valid generalizations, more research should be carried out at the firm level, including firms based in other countries or regions, and including smaller firms as well.

BIBLIOGRAPHY

Alberts, W. W., (1989), "The experience curve doctrine reconsidered", *Journal of Marketing*, Vol.53, July, pp.36-49.

Amit, R., (1986), "Cost Leadership Strategy and Experience Curves", *Strategic Management Journal,* Vol.7, no.3, pp.281-292.

Arthur, W. B., (1996), "Increasing Returns and the New World of Business", *Harvard Business Review*, Vol.74, July-August, pp.100-109.

Aspergis, N. and Zikos S., (2003), "The law of Verdoorn: evidence from the Greek disaggregated manufacturing time series data", *The Economic and Social Review*, vol.34, no.1, pp.87-104.

Bianchi, C. (2002) "A reappraisal of the Verdoorn law for the Italian economy, 1951-97", [in:] McCombie, J., M. Pugno and B. Soro (eds.), (2002), *Productivity Growth and Economic Performance: Essay's on Verdoorn's Law*, Palgrave MacMillan, Houndmills, Basingstoke.

Den Hartigh, E. (2005), *Increasing returns and firm performance: an empirical study,* ERIM PhD Series Research in Management 67, Erasmus Institute of Management, Rotterdam.

Den Hartigh, E. and Langerak F., (2001), "Managing increasing returnsą, *European Management Journal*, Vol.19, no.4, pp.370-378.

Fingleton, B. and McCombie J.S.L., (1998), „*Increasing Returns and Economic Growth: Some Evidence from Manufacturing and from the European Union Regions*", *Oxford Economic Papers*, Vol.50, pp.89-105.

Harris, R.I.D. and E. Lau (1998), "*Verdoorn's Law and Increasing Returns to Scale in the UK Regions 1968-91: Some New Estimates Based on the Cointegration Approach", Oxford Economic Papers*, Vol.50, pp.201-219.

Kaldor N. (1966), "Causes of the Slow Rate of Economic Growth in the United Kingdom", [in:] F. Targetti and A.P. Thirlwall (Eds), (1989), *The Essential Kaldor,* Cambridge University Press, Cambridge, pp.282-310.

León-Ledesma, M.A. (2000), "Economic growth and Verdoorn's law in the Spanish regions, 1962-91", *International review of applied economics,* Vol.14, no.1, pp.55-69.

Li, G. and S. Rajagopalan (1998), "Process improvement, quality, and learning effects", *Management Science,* Vol.44, no.11, pp.1517-1532.

Naastepad, C.W.M. (2006), "Technology, demand and distribution: a cumulative growth model with an application to the Dutch productivity growth slowdown",. *Cambridge Journal of Economics,* Vol.30, pp.403-434.

Porter, M.E. (1980) *Competitive strategy: techniques for analyzing industries and competitors.* The Free Press, MacMillan, New York.

Scherer, F.M. and D. Ross (1990), *Industrial Market Structure and Economic Performance*, 3rd ed., Houghton Mifflin Company, Boston.

Van Asseldonk, A.G.M. (1998), *Mass-individualisation: business strategies applying networked order to create economic value in heterogeneous and unpredictable markets*, TVA management bv, Veldhoven.

Verdoorn, P.J. (1949), *Fattori che Regolano lo Sviluppo della Produttività del Lavoro*, L'Industria, 1: 3-11. An English translation is available in: McCombie, J., M. Pugno and B. Soro (eds), (2002), *Productivity Growth and Economic Performance: Essay's on Verdoorn's Law,* Palgrave MacMillan, Houndmills, Basingstoke.

22

Innovation Strategies of Polish Companies and the Environment in which they Operate

Edyta Gwarda-Gruszczyńska

INTRODUCTION

The competitiveness of companies themselves, as well as whole economies, is becoming more and more dependent on innovation. The degree of innovativeness of the Polish economy and Polish companies is much lower than in the other EU countries. There are a number of facts which confirm this, including the following:

- Over 1.5 million Poles have left the country to work abroad (most of them are well educated and highly qualified, and sadly this trend is increasing);
- we can observe a low level of export of advanced technology products (generating 1 mln Euro/year in lost opportunities) and very few patents;
- insufficient growth of the innovative share of GDP and a tendency to decrease expenditures for R&D generate another 1.8 mln Euro/year in lost opportunities;
- a large amount of foreign direct investments with ready to use technologies and innovations;
- an insufficient level of EU funding consumption.

All these factors have caused the EU to regard Poland as "a country which is losing ground".

Because Polish companies operate under such difficult and changeable conditions, it seems that the most important activity that should be undertaken is to put strong emphasis on strategic approaches to the process of creating, planning and implementing innovations on both the institutional and company levels. What is most important in this respect is to treat innovation as part of both governmental and company strategy.

This chapter focuses on the strategic aspects of building competitive advantage for Polish companies using innovative ideas and strategies and on constructing a climate which would spur the creation of innovations and new technologies in Poland. The author, in four sections, tries to show the institutional and industrial environment for innovations and new technologies in Poland and presents a SWOT analysis and strategy undertaken by the Polish government to increase the level of Polish companies' innovativeness. Further, the actual innovative and technological status of Polish companies is presented, and recommendations are made suggesting possibilities to improve companies' strategies by following a process of innovation creation and implementation of changes in mentality.

THE INSTITUTIONAL AND INDUSTRIAL ENVIRONMENT FOR INNOVATIONS AND NEW TECHNOLOGIES IN POLAND

The Polish economy is not competitive and its innovation growth is too slow (see Table 1). Some basic barriers which are obstacles to success need to be overcome in order to accelerate economic growth. These barriers are mainly connected with R&D funding, cooperation between business, academia and research institutes, and the poor innovative conditions in Polish companies.

Table 1
Comparative view of the high technology potential of Poland (2005)

Country	Export of high-tech products	R&D share in GDP
USA	36%	3%
EU-15	18%	2%
Poland	**2%**	**0.58%**

Source: http://bz.premier.gov.pl

In Poland 1 billion Euro is paid out annually for R&D to public institutions such as: Research and Development Centers (JBRs) – approximately 120 institutions; Polish Academy of Sciences (PAN), over 100 entities; and higher education institutions. It can be seen that the amount of funds allocated to R&D is not a problem, but rather the fact that they are inefficiently spent. For instance JBRs use up approximately 450 million Euro annually - more than the universities (circa 400 million Euro). The structure of distributing money is also imperfect. Most of it is spent for primary research (See Table 2).

Table 2
Structure of Polish R&D funding

Type of R&D expenditures	Poland	USA
Expenditures on primary research (theoretical work not targeted at achieving particular practical applications)	38,5 %	18%
Expenditures on developmental research (enhancement of existing products, processes and services)	36%	61%
Expenditures on applied research (research initiated to achieve new knowledge which will have practical applications)	26%	21%

Source: Polish Agency for Entrepreneurship Development (PARP).

Up to 65% of R&D money in Poland comes from public sources. In countries such as Japan, the USA, Sweden, Germany and Finland private companies finance a much larger portion of R&D – up to 60-70% of R&D expenditures. The European Union also demands this kind of proportion in financing R&D. This means that Poland needs to find a way to increase the financing for new technologies and innovations from private business, especially venture capital.

Despite the fact that JBRs and PAN receive the most funding they rarely collaborate with the business community. Their results can hardly ever be used in the production process. What's worse, those institutions operate without cooperation with universities (Ipsos-Demoskop, 2001).

Another reason for the relatively low innovation level is lack of support for companies. Companies tend not to apply to proper governmental sources for the financing of research projects. Legal regulations also omit this element of support for entrepreneurs. In the documents regulating scientific financing, the words "company" or "entrepreneur" do not even appear. At the same time Polish companies, especially small and medium-sized ones, do not invest in R&D as it is too expensive and too risky an endeavor. In addition it's also difficult for companies to find R&D institutions that could collaborate with them.

Another important element of the poor situation of innovation development in Poland is the lack of financing for innovations and new technologies at the early stages of development (Tamowicz, 2005). There is practically no venture capital financing in Poland (only 5 funds) or business angels investors (approximately several dozen persons). The available data shows that venture capital prefers to invest in Polish companies that are at the expansion stage of their development (See Table 3). This creates a problem and makes it necessary to change governmental strategy to make it more innovation and new technology oriented.

Table 3
Value of venture capital investment in Poland in 1999-2004 (mln Euro)

Stage of company's development	1999	2000	2001	2002	2003	2004
Seed	1.8	2.9	2.2	0	15.1	0
Start up	16.2	38.8	22.7	9.8	2.0	0
Expansion	131.2	158.2	73.9	53.7	40.4	44.1
Replacement	24.2	1.9	47.8	34.6	42.7	59.5
Buyout	10.6	0	3.8	19.5	32.9	30.8
Total	**184.1**	**201.8**	**150.4**	**117.6**	**133.1**	**134.4**

Source: European Venture Capital Association (EVCA), 2005.

Funding, ranging from several to tens of thousands of Euro can be obtained from the Polish Agency for Entrepreneurship Development (PARP) or Microfund, but for companies that need a few hundred or several thousands of Euro to launch their business, there are no sources of financing.

As the data presented above shows, the access of Polish companies to external sources of innovation is limited and a model of innovation support in Poland needs to be developed and implemented by the Polish government. Politicians and administrative workers have to be aware that this is a priority issue and that the welfare of the country depends on their actions. These actions should be targeted at:

- separating out two streams of financing from public money allocated to R&D: one stream where cash flow would be monitored by Government; and a second stream where cash flow would be monitored by the Ministry of the Economy.
- introducing changes in the system of financing for universities, which should be spending more money on students studying at those faculties which are strategic for technological and innovative progress in Poland.
- transforming JBR and PAN institutions into centers that are conducting real R&D work, termination of most of the redundant JBRs and institutes, and merging institutes and JBRs with universities.

Though the system of financing for R&D activities in Poland is far from being perfect, positive changes can be seen in the process of developing regional innovation strategies, which contribute to the fact that the innovative potential of small and medium-sized companies is being recognized. A big role in that process is played by a new area, which emerged in the realm of supporting competitiveness and innovation in regional economies. And though the opinion that small and medium-sized firms (SMEs) are the locomotives of regional innovation is not unanimous – according to the survey "Main Science & Technology Indicators in Poland in 2000-2004" 60% of the respondents point to them as key enablers in that field - it is beyond doubt that it is precisely SMEs which will be first to respond to the newly created conditions for innovation development (GUS, 2005). As a result they will be receivers of future support and they will be generating growth.

In order for small and medium-sized companies to be able to fulfill those expectations, heavy emphasis should be placed on initiatives that aim to help facilitate the development of institutions which support business development: academic and technology incubators, technology parks, delivery and guaranteed funds, as well as consulting, training and educational systems targeted at companies' needs. Despite some optimistic claims regarding this issue, in reality such institutions encounter tremendous difficulties when it comes to expanding their scope of activities.

Polish companies also encounter difficulties in the industrial sector environment. An important element of the Polish economy is direct foreign investments, i.e. having a share of foreign capital in the company (Kotowicz-Jawor, 1998) (Table 4).

Table 4
Foreign Direct Investment in Poland in 2000-2004

	2000	2001	2002	2003	2004
Foreign Direct Investment (mln USD)	9 341	5 713	4 131	4 123	6 159

Source: Polska: Bilans płatniczy na bazie transakcji, Narodowy Bank Polski (NBP), 2005.

The inflow of those investments is not always a positive stimulator of technological development in companies involved in one of the innovation areas. It happens very often in Poland that as a consequence of the appearance of foreign investors, in-house research departments are liquidated. Investors replace them with their own proven innovations, and use their own advanced R&D centers (Orłowski, 2004). Another huge obstacle to innovation development in Polish companies is the barrier regarding national market demand for their products. Companies, both private and state-owned, are not able to cope with competitive high-quality imports, which limit demand for their products. At the same time, companies with foreign capital are increasing imports, as they prefer to import raw materials rather than buy them on Polish market (even though they are often cheaper).

THE INNOVATIVE AND TECHNOLOGICAL STATUS OF POLISH COMPANIES

Companies which want to achieve their goals in a very competitive environment should implement different types of innovations: product, technological, market and organizational innovations. The two first categories are called technical innovations and they are the ones often addressed by the general idea of innovation (Haffer and Haffer, 2004). But one should keep in mind that in this new competitive situation the success of a company depends equally on all effectively implemented innovations, whether they are technical, market (marketing) or organizational (Bettis and Hitt, 1995). Research done in small and medium-sized companies in advanced countries shows that "the fastest growing companies are not the ones technically innovative, but they grow as a result of market innovations and international alliances. High-tech ventures do not show quick growth. Most jobs are created in traditional companies, which are conducting small investments and thus growing slowly" (OECD, 1996). This means that companies can achieve excellent results, effectively managing the processes of creating and implementing marketing and organizational innovations, even if they are not able to efficiently introduce significant technological and product innovations. Nevertheless, well-managed companies look for every efficient method of gaining competitive advantage in the marketplace, by implementing all types of innovations; product and technological as well as market and organizational.

The innovation potential of Polish companies in comparison to other countries in the European Union is quite low. Innovations that are being introduced in Polish firms are mainly organizational and product oriented; we can see little market or technological innovation.

Being innovative, which is usually included in a company's strategy, is often defined by Polish companies as their ability to come up with new products. Therefore the innovation potential of a given company, as well as its attitude towards innovation, are often indicated by factors such as: strategic planning as a tool in the company; having a special budget for research and development purposes; cooperation with a scientific center, R&D institute, or a university; number of patent applications put forward by the company; implementation of new technologies; and finally the knowledge, experience and innovation management skills of the company's management. These indicators allow for the classification of a company as ready for innovation implementation (See Table 5).

Table 5
Innovative status of Polish companies

	Small **Up to 5 workers**	**Medium and Large**
Strategic planning	8%	20%
Use of bought patents, licenses and know-how	5%	15%
Own patents created in the company or sold	3%	10%
Special funds for research and development	5%	13%
Collaboration with R&D institutes, universities	Low a)	20%

Notes: a) however, small companies are looking for collaboration more frequently than larger companies.
Source: Adapted from Żołnierski, 2005.

When analyzing innovation culture and the innovation ratios of Polish businesses, it should be emphasized that despite the fact that the level of innovation is low, at the same time it is well balanced (IPSOS-DEMOSKOP, 2001). Żołnierski, in his report for the Polish Agency for Entrepreneurship

Development (PARP), shows that among companies employing up to 5 workers 8% use strategic planning and treat innovation as a part of their operational strategy; 5% make use of patents, licenses and know-how they bought; and 3% own patents created in their company or sell their own licenses and know-how (Żolnierski, 2005). Five percent of small firms allocate special funds to research and development. Most of them collaborate with scientific centers, R&D institutes, or universities. The previous results demonstrate the low level of innovation in small Polish companies. Implemented technologies are mainly basic information and telecommunication technologies (internet, basic software).

What is disturbing in this picture is the fact that management in these firms often lacks the necessary knowledge and awareness of how to manage innovations, while at the same time they view their companies as innovative, competitive and modern. But the reality is far from satisfying when we look at the actual expenditures by Polish SMEs on modern technologies and innovation (Żołnierski, 2005).

In the group of medium-sized and large companies, the share which uses long-term planning for more than a 5-year period is 20% on average. However, it is also positively dependent on how large a given company is – among companies which employ over 250 workers this percentage soars to 40%. There is no tangible relation between the frequency of using long-term planning and the size of the company. The lowest percentage of medium-sized and large companies using long-term planning is found in the construction industry (8%), and the highest percentages in services (27%), commerce (24%) and the general sector (34%).

An average of 15% of medium-sized and large companies use patents and licenses which they bought. There is a positive relation between the usage of patents and licenses and employment levels in a company. A large number of the medium and large companies which use patents and licenses are firms with a turnover of over 1 250 000 Euro, with headquarters in big cities (21%), and are the recipients of foreign capital investment (32%).

The fact of purchasing and using patents and licenses signals that a company is open to innovation. Polish firms which decide to buy ready-made licenses are open to innovations – purchase of a license constitutes an element of innovation implementation strategy and is being applied along with in-house research and development. Therefore, the purchase of a license is not treated as an alternative to in-house research which could enable the implementation of innovation in the company, but is used as complementary means in a situation where innovation has to be introduced quickly and in-house research is unprofitable.

Less than half of medium-sized and large companies utilizing patents also implement strategic planning. A third of medium-sized and large companies utilizing patents also own their own patents and licenses. Approximately 40% of them conduct in-house research and development, and one-third cooperates with a scientific center, R&D institute or university. This result shows that the purchase of ready-made licenses is part of a wider strategy to welcome innovations in a company.

An average of 10% of medium-sized and large companies have their own, in-house developed patents or sell their own licenses or know-how (companies employing over 250 staff, turnover exceeding 250 000 Euro, located in big cities, engaged in foreign trade, with foreign capital). Large companies are main players here. Approximately 13% of medium-sized and large companies allocate specific funds to research and development. Most of them are industrial and agricultural companies, with smallest numbers coming from commerce and services. Among those companies the biggest group is state-owned enterprises and commercial partnerships with a turnover exceeding 1250000 Euro and engaged in foreign trade. Medium-sized and large companies most often allocate from 5000 to 12000 Euro. Some (17%) allocate 250 000 Euro a year.

Twenty percent of large and medium-sized companies collaborate with a scientific center, R&D institute, or a university. This cooperation most often concerns modernization and the development of new technologies (21% of cooperators), tests on products' quality, as well as consulting and advisory services (18%). Among those companies involved in cooperation with a scientific center, R&D institute or university, 12% stated that they cooperate with a particular R&D center but without specifying the scope of the cooperation.

Despite the fact that the innovation potential of Polish companies in comparison to other countries in the European Union is classified as quite low, one can find some success stories. In such an environment some Polish companies are able to build and improve their competitive position not only on the Polish but also on the international market. Most of these companies use strategic planning, explore the innovative potential of their employees, and take opportunities to form alliances with institutions that can influence their level of innovation. Positive examples include firms which found a market niche thanks to the proper implementation of an innovation strategy, and which have achieved a strong position on the market due to the effective implementation of innovations. Examples of such companies include: ComArch, Vigo-System, Advanced Digital Broadcast (ADB), Solaris Laser, Ascor, and others.[1]

POLISH STRATEGY TO INCREASE THE LEVEL OF COMPANIES' INNOVATIVENESS AND COMPETITIVENESS

The data presented above confirm the thesis that the barriers to innovativeness in Poland are connected with R&D funding, companies' cooperation with academia or research institutes, and the poor innovative condition of Polish companies. The larger picture and the scale of the problem can be seen by examining the SWOT analysis prepared by the Ministry of Regional Development (2006). According to the document "Operating Program – Innovative Economy 2007-2013" Polish companies and their partners in the business environment have to work a lot to change the situation and turn the economy into a quickly developing one (See Table 6).

---

[1] http://www.comarch.pl; http://www.adb.pl; https//:www.solarislaser.com.pl; https//:www.vigo.com.pl; https//:www.ascor.com.pl

Table 6

SWOT analysis of the Polish environment for innovativeness of Polish companies

Strengths	Weaknesses
– Change in export structure to more manufactured products – Changes in Direct Foreign Investment into greenfield investment direction – Participation in EU and growing activity and credibility of Poland – Openness of economy and possibility for free movement of capital and goods – Dynamic growth of sales of companies operating in Poland and EU-25 – Existence of a group of companies with high potential to compete in the global market – Large supply of well-educated workers – Large and not declining share of researchers in total employed in R&D in research institutes – Increasing quantity of doctorates and people with PhD in R&D industry – Relatively new and regularly developed informatics infrastructure of science	– Low level of manufactured and high-tech goods in export – Low level of new and modernized goods in sold production – Low working capacity – High level of tangible assets in use and high undercapitalization of research machinery – Low level of companies' investment – Insufficient development of service industry – Low level of productivity of companies operating in Poland comparing to EU-15 companies – Insufficient knowledge of "Poland" and Polish company trademarks, low level of internationalization of companies – Low preparedness of companies to cooperate with other companies and academia and research institutes – Business institutional system has not adjusted to companies' and R&D needs – Low level and ability of companies to innovate – Reduced access to capital for innovative activities, especially for SMEs. Reduced access to financing for the development of innovative activities in the early stages – Lack of pro-innovative and pro-entrepreneurial education at Polish Universities – Insufficient affinity of companies' industry with new technologies (information and communication technologies) – Low level of inventiveness and low level of knowledge and use of intellectual property rights in the economy – Small quantity of registered patents – Low level of expenditures for R&D, development and applied research in

	the public and private sectors – Unfavorable age structure of research and academic workers – Unfavorable structure of R&D industry – Limited experience of Polish researchers in participation in large international research projects – Insufficient interest of researchers to provide development and applied research.
Opportunities	**Threats**
– Identification of activities aiming at realizing the Lisbon Strategy – Concentration of EU funding in investments aiming at economic growth and the creation of new employment positions – Access to world achievements in entrepreneurship and innovativeness – Changes in industrial structure of the economy – Creation of a coherent business environment system meeting the companies' needs – Increasing working capacity – Increasing modernization and efficiency of industries through transfer of innovative solutions – Enforcing the research networks, consortia and technological platforms – Poland's participation in European Research Space building – Development of National Innovation System – Implementation of Regional Innovation Strategies stimulating the idea of building knowledge regions – Restructuring of research institutions and consolidation of best scientists to develop strategic research programs – Coherent policy of innovativeness – Services development based on well-educated workers.	– High budget deficit compared to GDP – Workers' qualifications do not fit the economy's needs – High costs of economic activity of companies – Unstable and unclear legal environment – Deregulation of economic processes – Administration not friendly to entrepreneurs – Difficulties with some industries' restructuring – Public assistance directed to mature industries – Lack of reform of the public finance system – Inelastic labor market and high costs of employment, poor environment to open new businesses – Lack of innovative attitudes among entrepreneurs – Flow of the best educated and experienced work force to other countries – brain drain – Low level of security in economic turnover – Lack of consequence in policy of increasing expenditures for R&D – Difficult to reorient the research of Universities to become more company and economy oriented

Source: Ministry of National Development, Program Operacyjny Innowacyjna Gospodarka, 2007-2013. Narodowe Strategiczne Ramy Odniesienia 2007-2013, Warszawa 2006.

Taking into account all the opportunities, threats, strengths and weaknesses, a strategy to make the Polish economy more innovation and technology oriented was established at the end of 2006. The main aim of this strategy is to develop the Polish economy and create innovative companies. More detailed aims are as follows:

- Increase in companies' innovativeness
- Growth of Polish scientific competitiveness
- Increase of science's role in economic development
- Increase of innovative Polish products' share in the international market
- Creation of stable and better employment positions.

These strategic aims mean big changes for both companies and science, especially because in 2007-2013 Poland will receive billions of Euros in EU financial support to spend for raising innovativeness and creating new technologies. For Polish companies and scientists this will be a big challenge. More control of EU financing support will be given to the regional administration, which will also be responsible for the distribution of funding. Altogether all this means that the next steps to make good use of EU support will require changing the mentality of entrepreneurs and sharing information about existing opportunities. A sound education in and deep understanding of the processes of innovation creation and implementation will be necessary to build companies' competitive advantage.

In the next section I will, on the basis a review of the literature, try to show the directions and areas in which entrepreneurs should change their mentality regarding activities connected with innovation and the creation and implementation of new technologies.

POSSIBILITIES TO IMPROVE COMPANIES' STRATEGIES BY FOLLOWING THE PROCESSES OF INNOVATION CREATION, IMPLEMENTATION, AND CHANGES IN MENTALITY

First of all Polish managers, mostly SMEs' managers, will have to learn strategic thinking. According to the new approach to competitiveness companies maintain or enhance their competitive position on the market mainly thanks to successful innovations. Therefore innovations become a standard for modern companies which are successful on a very competitive market. As Peter Drucker wrote "successful entrepreneurs try to create value and make some kind of contribution – regardless of what their driver is – money, power, curiosity, or aching for fame and recognition. But effective entrepreneurs aim high. They are interested not only in improving or modifying already existing things. They attempt to create new, different values, new and different ways of satisfying needs, transform material into resources or combine existing resources into a new, more effective configuration" (Drucker, 1992, pp. 43-44). In the economic practices of modern companies this means spreading of the "entrepreneurial spirit, which is visible in implementation of innovations in all aspects of the company".

Each kind of innovation has a strategic value for a company. Practical implementation of innovation strategy requires that decisions and actions are taken systematically in a company, and all the decisions and actions should comply with the general strategy of the company.

The process of creating and implementing innovations starts with the establishment of a general strategy for a company. This general strategy, as a coherent concept of how the company should operate in the competitive environment, sets long-term goals for the organization and defines methods of achieving them. Such strategy constitutes a conceptual behavior pattern of the company, and is a result of a SWOT analysis of opportunities, threats arising from the existing environment, and the strengths and weaknesses of the company. The general aims, course of actions, and corresponding budgets associated with such a strategy are built to achieve competitive advantage and create new ways for the company to grow or remain sustainable (Obłój, 2002).

Innovation strategy should be subordinate to the general strategy, while being an important part of it and a tool for its implementation. The decision to choose a particular form of innovation results from an analysis of trends in the environment and the internal capabilities of the company (Christensen, 1997). This decision triggers and directs a process of creating the desired form of innovation and its market commercialization (Jolly, 1997). The research discussed in the previous sections show that Polish companies in many cases ignore this. Strategic thinking about innovations as a part of long term development is usually missing.

The characteristics of an innovative venture determine the area of the search for ideas for a new solution, i.e. attributes which guarantee its success (Christensen, 1997). Generating and screening ideas in this area is the most creative and meaningful phase of the innovation development process. If the generation of ideas is to be effective it should integrate a variety of knowledge and capability areas of the company. This process should result in many new ideas for finding solutions. At the same time these ideas should be applicable and correspond with the goals of adopted strategic schemes.

During the phase of developing the concept of a new solution, newly chosen ideas should be integrated in such a way that their market, technical and organizational aspects constitute a coherent whole. During the integration process, regardless of how excellent single ideas are, they should be being modified and improved until they become a coherent conception of a new solution, which will allow for realization of the strategic goal that has been set out. These stages in the innovative process seem to be very important for the strategy of implementation and for a company's success. Polish scientists and R&D workers are not well paid, which is one of the reasons for the "brain drain". Hopefully the Polish and EU financial assistance will follow the directions mentioned in the general strategy for Polish economic development and will enable managers to find better ways to motivate specialists and innovative personnel to work harder in this important area.

The final acceptance of the chosen concept of an innovative solution requires a multifaceted economic and a financial analysis. These are concerned mainly with evaluating market opportunities and cost and risk assessment of implementation of the concept. Therefore, this stage of economic and financial analysis contributes to a further modification of the concept of the new solution, providing a feasibility structure and indicators of success. This stage also helps decide which variant of the new solution should be pursued, i.e. the one showing the greatest potential for success. Accuracy of the assessment depends on the quality of data collected by analysts and managers during the analysis (Jolly, 1997).

The decision to choose the least risky concept of a new solution activates its development phase, during which the innovation adopts a real form, e.g. a prototype or preliminary versions of new solutions are created. Further, the innovation developed must be tested on a limited scale before full-scale implementation. Such testing should yield answers to the following questions: do all the assumed attributes of the new solution work in practice? And what should be further changed in the solution to achieve the desired result?

The last stage is the implementation of the new solution at the desired scale – i.e. its commercialization. It is this phase which consummates the whole innovation process, and at the same time the strategic aim which the implemented innovation serves.

There are good opportunities to improve the above indicated stages in the innovation process. The strategic directions set out by the Polish government include supporting activities in this field. Polish companies will be able to obtain funding for testing, building prototypes and implementation of new solutions. But still the distribution of knowledge about intellectual property rights, patents, licenses, etc. seems to be a very important task for training institutions and regional administration. Many Polish companies should also keep in mind that effectiveness of the innovation process depends not only on the creation of strategies and completing successive stages of creating and implementing the innovation, but, as Trucker's research on U.S. innovative companies shows, it depends also on obeying certain rules (Trucker 2005).

In order for the innovation process to be efficient and bring about the desired results, the following principles must be obeyed:

- treat innovation as a discipline. People should be taught how to evaluate ideas and decide which ones comply with a business's goals, as well as how they can win and sell their ideas;
- look at innovation in a broader sense. It shouldn't be limited to one department or one group of innovators. It should include new products, services, processes, strategies, business models, distribution channels, and markets (Obłój, 2002);
- the innovation process should include an organized, ongoing and systematic search for new possibilities. Innovative companies are promoting a profound understanding of social, demographic and technological changes as a way of constant search for new opportunities;
- involve the entire team in the innovation processs;
- target innovations at the customer (Wise and Baumgartner, 2005).

It can be concluded that the competitiveness of companies which implement innovations and consequently an innovation process is strongly linked to internal activities and internal relations within the company. This does not mean however that the externals can be neglected. The relation of a company with the outside world can have significant influence on the development of innovation and commercial success of the company. An effective process for creating and implementing innovations should take those relations into account.

An innovative company should plan, organize and control relations with two kinds of external environments, which are the industry sector and the institutional environment. Within the industry sector environment, particularly important for innovation processes are relations with:

- customers and end users
- suppliers and collaborators
- competitors

Within the institutional environment cooperation with the following institutions is essential:

- scientific centers and research and development centers
- business support institutions
- government institutions

In the Polish government strategy the cooperation between companies and science is strongly emphasized, but all companies in the 21st century have to consider cooperation with suppliers, clients, competitors and other stakeholders as a chance to be more innovative and an opportunity to create a competitive advantage. Polish companies are not good at this, so taking into consideration worldwide tendencies they should try to build cooperative relations based on trust and transparency.

CONCLUSIONS

From the standpoint of creating and implementing innovations, Polish companies do not compare well with companies in advanced countries. Such a situation stems from two sources. The first one is the external environment both in terms of industrial sector and institutional realm. In Poland it is still believed that Poles are no good at creating innovation. Such convictions are reinforced by foreign concerns which implement their own behavioral patterns, products and technologies. Innovative activities of Polish companies are additionally hampered by an inadequate national system for supporting innovations and the lack of proper sources of financing, like venture capital sources or business angels.

The second reason, strictly connected with the internal limitations of companies, is the low level of abilities and awareness, as well as the financial and strategic long term shortcomings of Polish entrepreneurs and their supervising managers. Examples of successful companies which skillfully build and implement innovation strategies are quite rare among Polish companies.

It seems that activities aimed at improving the situation and spurring the innovativeness and competitiveness of Polish companies should be directed at creating a system that could support these companies. Taking into account the real technology gap between Polish (especially SMEs) and European or global companies should be a starting point for making changes. Creating the ability to effectively use technology resources would seem to be an important part of implementing new technological ideas. Activities supporting the development of the innovative and competitive potential of Polish companies should follow two directions.

These activities should be, and according to the new government strategy for 2007-2013 will be, directed toward increasing awareness of both the need and ability to implement innovations in Polish companies, through:

- publicity and spread of information about the Polish best practices in this field,
- training for managers and owners in intangible resources' management; training in motivating for innovation should show different tools for different stages of the innovation process,
- training for entrepreneurs that will show them how important it is to engage employees in the innovative process,
- training concerning the building of cooperation networks, especially between business and academia but also with other institutions connected with the business environment.

These activities should also deliver tools for companies to make the innovative process easier. These tools should include financial aid such as: subsidies for innovations, and creation of a system of lower taxes for companies that create and implement new products, technologies or other innovations. Intensive support for newly created ventures or networks of research institutes, academia and business also plays a very important role in the process of building an environment for innovation in Poland.

BIBLIOGRAPHY

Bettis, R. A. and M. A. Hitt, (1995), "The New Competitive Landscape", *Strategic Management Journal*, 1995, vol. 16 (S1), pp. 7-19.

Christensen, C. M., (1997), *The Innovators Dilemma. When New Technologies Cause Great Firms to Fail*, HBS Press, Boston- Massachusetts, pp. 161-182 and 143-160.

Drucker, P. F., (1992), *Innowacja i przedsiębiorczość. Praktyka i zasady*, PWE, Warszawa, pp. 43-44.

EVCA (European Private Equity & Venture Capital Association), (2005), *EVCA Yearbook – Annual Survey of Pan European Private Equity & Venture Capital Activity*, EVCA, Zaventem.

GUS Central Statistical Office, *Main Science & Technology Indicators In Poland in 2000-2003*, Warszawa.

Haffer M. and R. Haffer R., (2004), „Zarządzanie procesami innowacyjnymi przedsiębiorstw", [in:] M. Romanowska, M. Trocki, *Podejście procesowe w zarządzaniu*, Wyd. SGH, Warszawa, pp. 219-228.

https//:www.ascor.com.pl

http:// www.adb.pl

http://www.comarch.pl
http://bz.premier.gov.pl
https//:www.solarislaser.com.pl
https//:www.vigo.com.pl
IPSOS-DEMOSKOP, (2001), *Wskaźniki innowacyjności przedsiębiorstw w Polsce*, Warszawa - badania przeprowadzone na zlecenie Polskiej Agencji Rozwoju Przedsiębiorczości.
IPSOS-DEMOSKOP, lipiec 2001, http://www.portalinnowacji.pl
Jolly, V. K., (1997), *Commercializing New Technologies. Getting from Mind to Market,* HBR Press, Boston – Massachusetts.
Kotowicz-Jawor J., (1998), „Rozluźnić gorset akumulacji", „*Nowe Życie Gospodarcze*, Vol. 23, pp. 5-6.
Ministerstwo Rozwoju Regionalnego, (2006), *Program Operacyjny Innowacyjna Gospodarka 2007-2013, Narodowe Strategiczne Ramy Odniesienia 2007-2013*, Warszawa.
Polska: Bilans płatniczy na bazie transakcji, (2005), Narodowy Bank Polski, Warszawa.
Obłój K., (2002), *Tworzywo skutecznych strategii*, PWE, Warszawa.
OECD, (1996), *SMEs: Employment, Innovation and Growth- The Washington Workshop*, OECD, pp. 57-58.
Orłowski W. M., (2004), „Czy Polska jest innowacyjna?", *Business Week*, Vol. 17, pp. 21-24.
Ramisiewicz, M., M. Kruszewska, (2006), „Innowacyjność", *Manager Magazine*, 12/06, pp. 30-50.
Tamowicz, P., (2005), Raport: *Zjawisko luki kapitałowej (equity gap) w gospodarce polskiej*, opracowane na zlecenie Ministerstwa Gospodarki i Pracy, Gdańsk.
"The Innovative Enterprise- Turning Ideas into Profits", (2002), *Harvard Business Review* - Special Issue, August, Vol. 80, No 8, pp. 117-124.
Trucker F, Five Essential Principles of Innovation, http://www.thinksmart.com/inmembership/bookshelf.
Wise R., P. Baumgartner, (2005), „Bliżej klienta. Warunek uzyskania zysków w przemyśle", *Harvard Business Review Polska*, September, pp. 108-125.
Zlepka M., (2005), *Efekty regionalnych strategii innowacji w Polsce*, Wyd. Krajowy Punkt Kontaktowy Programów Badawczych UE, Warszawa.
Żołnierski A., (2005), *Potencjał innowacyjny polskich małej i średniej wielkości przedsiębiorstw – raport z badań, PARP*, Warszawa.

About the Contributors

Gen Amano – Manager of the Finland Project, International Promotion Office, Economic Affairs Bureau, The City of Sendai, Japan.
Contact: gen_amano@city.sendai.jp

Massimo G. Colombo – Professor of Economics of Technical Change, Politecnico di Milano, Italy.
Contact: massimo.colombo@polimi.it

Vincent Dautel – Researcher, CEPS/INSTEAD, Differdange, L-4501 Luxembourg
Contact: vincent.dautel@ceps.lu

Tomasz Domański – Professor of International Marketing and Distribution, Faculty of International and Political Studies, University of Łódź, Poland.
Contact: tomdom@uni.lodz.pl

Susana Elena - PRIME Network of Excellence Researcher, Assistant Professor of Business Administration, Pablo de Olavide University, Sevilla, Spain
Contact: seleper@upo.es

Marina van Geenhuizen - Professor of Innovation and Innovation Policy in the Urban Economy, Faculty of Technology, Policy and Management, and OTB Research Institute, Delft University of Technology, The Netherlands.
Contact: m.s.vangeenhuizen@tudelft.nl

Ebbe Krogh Graversen – Research Director, The Danish Institute for Studies in Research and Research Policy, University of Aarhus, Denmark
Contact: eg@cfa.au.dk

Luca Grilli – Assistant Professor of Industrial Economics and Organization, Politecnico di Milano, Italy.
Contact: luca.grilli@polimi.it

Edyta Gwarda-Gruszczyńska – Assistant Professor of Strategic Management, Faculty of Management, University of Łódź, Vice-director, Innovation Center University of Łódź, Poland
Contact: egwarda@uni.lodz.nl

Erik den Hartigh – Assistant Professor of Technology Strategy and Entrepreneurship, Faculty of Technology, Policy and Management, Delft University of Technology, The Netherlands.
Contact: e.denhartigh@tudelft.nl

Marta Hereźniak – PhD Candidate, University of Łódź, Poland
Contact: marta_h@poczta.onet.pl

Hugo Horta – Researcher, Center for Innovation, Technology and Policy Research IN+, Instituto Superior Técnico, Technical University of Lisbon, Portugal.
Contact: hugo.horta@dem.ist.utl.pt

Nurul Indarti – Vice–director, Center for Research and Development Management, PhD Candidate, Faculty of Economics, Gadja Mada University, Yogyakarta, Indonesia.
Contact: nurulindarti@ugm.ac.id

Linda M. Kamp – Assistant Professor of Technology Assessment, Faculty Technology, Policy & Management, Delft University of Technology, The Netherlands.
Contact: l.m.kamp@tudelft.nl

Fred Langerak – Professor of Management of Product Development in the Department of Technology Management at Eindhoven University of Technology in The Netherlands.
Contact: f.langerak@tue.nl

Han van Loon - Leistungs Consult GmbH, Walchwil, Switzerland
Contact: welcome@lc-stars.com

Michael Mark – Researcher, The Danish Institute for Studies in Research and Research Policy, University of Aarhus, Denmark.
Contact: mm@afsk.au.dk

Tadeusz Markowski – Professor of City and Regional Development, Faculty of Management, University of Łódź, Poland.
Contact: Tadeusz_Markowski@wzmail.uni.lodz.pl

Jose Luiz Moutinho – Researcher, IN+ Center for Innovation, Technology and Policy Research, Technical University of Lisbon, Portugal.
Contact: jmoutinho@dem.ist.utl.pt

David C. Mowery - Professor of New Enterprise Development, Walter A. Haas School of Business, University of California, Berkeley, USA
Contact: mowery@haas.berkely.edu

Hermann Simon - CEO of Simon, Kucher & Partners, Strategy and Marketing Consultants (Bonn, Boston, Frankfurt, London, Milan, Munich, Paris, Tokyo, Warsaw and Zurich).
Contact: hermann.simon@simon-kucher.com

Maciej Urbaniak – Associate Professor of Quality Management, Faculty of Management, University of Łódź, Poland.
Contact: murb@uni.lodz.pl

Dariusz Trzmielak – Assistant Professor of Market Analysis, Faculty of Management, University of Łódź, Director, Innovation Center University of Łódź, Poland.
Contact: darekt@uni.lodz.pl

Francisco M. Veloso – Assistant Professor of Engineering and Public Policy, Carnegie Mellon University, Faculdade Ciências Económicas e Empresariais, Universidade Católica Portuguesa, Portugal.
Contact: fveloso@cmu.edu

Marc A. Zegveld – Associate Professor of Strategic Management, Department of Technology, Strategy and Entrepreneurship, Faculty of Technology, Policy and Management, Delft University of Technology, Delft, The Netherlands.
Contact: m.a.zegveld@tudelft.nl